Exceeding the Goal

Exceeding the Goal

Adventures in Strategy, Information Technology, Computer Software, Technical Services, and Goldratt's Theory of Constraints

John Arthur Ricketts

INDUSTRIAL PRESS, INC.

Industrial Press, Inc.
32 Haviland Street, Suite 3
South Norwalk, Connecticut 06854
Phone: 203-956-5593
Toll-Free in USA: 888-528-7852
Fax: 203-354-9391
Email: info@industrialpress.com

Author: John Arthur Ricketts
Title: Exceeding the Goal: Adventures in Strategy, Information Technology, Computer Software, Technical Services, and Goldratt's Theory of Constraints
Library of Congress Control Number is on file with the Library of Congress.

ISBN (print): 978-0-8311-3656-7
ISBN (ePDF): 978-0-8311-9570-0
ISBN (ePUB): 978-0-8311-9571-7
ISBN (eMOBI): 978-0-8311-9572-4

Editorial Director/Publisher: Judy Bass
Copy Editor: Janice Gold
Compositor: Patricia Wallenburg, TypeWriting
Proofreader: Michael McGee
Indexer: WordCo Indexing Services Inc.

industrialpress.com
ebooks.industrialpress.com

10 9 8 7 6 5 4 3 2 1

To Art & Milt.

CONTENTS

Preface xxiii

PART 1
OVERVIEW

PROLOGUE
The Enterprise Constraint 3

CHAPTER 1
Introduction 11

Picking Your Battles 14
Strategy 14
Information Technology, Computer Software, and Technical Services 15
Theory of Constraints 17
Executive Priorities 18
System of Systems 18
Adventures 19
Conclusion 20

CHAPTER 2
Executive Priorities 21

Chief Executive Officer 23
Chief Operating Officer 23
Chief Financial Officer 24
Chief Marketing Officer 24
Chief Human Resources Officer 24
Chief Information Officer 25
Other CXOs 26
Chief Science Officer 26

Chief Technology Officer 26
Chief Data Officer 26
Chief Innovation Officer 26
Comparing Other CXOs 27
Central Information versus Shadow Information 27
Digital Reinvention Archetypes 28
Disruptive Innovation 28
Ambiguous Boundaries 29
Living Laboratories 29
Technology Shifts 30
Business Platforms 30
Enterprise Agility 31
Conclusion 31

CHAPTER 3
Strategy **33**
Technology Research 33
Business Opportunity 34
Technology Partnership 36
Proof of Concept 37
Lessons Learned 39
Where We Want to Be 39
Horizons 40
Blue Ocean 42
Profit Patterns 42
OODA Loop 43
Alignment 44
Strategy Traps 44
Strategy Principles 46
Conclusion 47

CHAPTER 4
Information Technology, Computer Software, and Technical Services **49**
Does Information Matter? 50

The Role of Information 53
Central Information versus Shadow Information 53
System Types. 53
Technical Services 54
Information Trends 56
Dematerialization and Virtualization 56
Commoditization and Value Migration 57
Moore's Law, Brooks' Law, and Jevons Paradox 57
Enterprise Computing versus Personal Computing 59
Cloud Computing versus Traditional Computing. 60
Machine Learning, Artificial Intelligence, and Cognitive Computing 61
Internet of Things. 62
Role Reversal 62
Fantasy Information 63
Technical Debt 64
Obsolescence 64
Legacy Systems. 65
Leapfrogging 66
Information Principles 66
Conclusion 67

CHAPTER 5
Constraint Management **69**
Weakest Link Analogy. 70
Flowing Water Analogy. 70
The Goal 74
Systems 75
Flows 76
Limits. 77
Constraints 78
Buffers 79
Constraint Management Solutions 80
Production 81
Distribution 84

Projects 85
Sense and Respond 89
Focusing Steps 89
Conflict Resolution 90
Buy-in 91
Strategy 92
Technology 93
Decision-making 94
Principles 95
Information Constraints 97
Conclusion 98

PART 2
TECHNOLOGY

CHAPTER 6
Hardware **101**
Layers 101
Performance 105
Availability 105
Compatibility 106
Security 106
Storage 107
Smartphones and Cloud Computing 107
Virtual Hardware 108
Robots 108
Internet of Things 108
Time 109
Remote Control 109
Research and Development 111
Manufacturing 111
Distribution 112
Constraint Management 112
Profile: Engineering Manager 113

Profile: Production Manager 113
Profile: Sales Manager 114
Profile: Buyer 114
Profile: Distribution Manager 115
Profile: Service Manager 116
System of Systems 116
Conclusion 118

CHAPTER 7
Software **119**
Strange Familiarity 119
Code 120
Software Industry 121
Software Origins 122
Software as a Service 122
Microservices 122
Functional Requirements 123
Developmental Requirements 124
Operational Requirements 125
Never Used versus Unexpected Use 125
Software Engineering 126
Metrics 127
Size 127
Complexity 128
Defects 129
Estimating 131
Benchmarks 131
Life Cycle 135
Testing 135
Work Breakdown Structure 136
Languages 137
Object-Oriented Programming 137
Internationalization 138
Promotion 138

Operations 139
Legacy Systems...139
Application Understanding and Impact Analysis 140
Technical Debt 143
Constraint Management...148
Profile: Research Manager 149
Profile: Development Manager...149
Profile: Operations Manager 150
Profile: Support Manager 151
Profile: User 151
System of Systems 152
Conclusion 154

CHAPTER 8
Data **157**
Data, Information, and Knowledge...157
Misinformation, Disinformation, Malinformation.....................158
Data at Rest, Data in Motion, Data in Process 158
Data Types...159
Data Structures...159
Formulas and Expressions 160
Data Administration 164
Extract, Transform, Load 164
Transformation 165
De-duplication...165
Compression...165
Encryption 166
Streaming 166
Logging...166
Data Errors 167
Information Biases...170
Analytics...172
Security...175
Privacy...175

Retention, Backup, Recovery, Archiving, Disposal 176
Technical Debt 177
Open Data 178
DataOps 178
Constraint Management 181
Staff Functions 182
Profile: Chief Data Officer 183
Profile: Chief Privacy Officer 184
Profile: Chief Science Officer 184
Profile: Business Process Owner 184
Profile: Data Analyst 185
Profile: Design Manager 185
Profile: Database Administrator 186
Profile: Development Manager 186
Profile: Operations Manager 187
Profile: Data Producers 187
System of Systems 188
Conclusion 189

CHAPTER 9
Knowledge **191**
Knowledge Work 192
Machine Learning 192
Artificial Intelligence 193
Cognitive Computing 193
Fairness 194
Data Cleansing 196
Digital Robots 198
Hype Cycle 202
Cold Start 203
Pseudo-AI 203
Technological Singularity 203
Constraint Management 204
Conclusion 205

CHAPTER 10

Networks **207**

Voice Networks 208
Data Networks 208
Video Networks 209
Specialty Networks. 209
Security 210
Disasters 210
Content Delivery Networks 214
Software Defined Networks 214
Network Policies 214
Braess' Paradox. 216
Constraint Management 216
Conclusion 217

CHAPTER 11

Architecture **219**

Why Technical Architecture Matters. 219
Technical Architecture 220
Hardware Architecture 221
IT Architecture. 221
Data/Information/Knowledge Architecture. 224
Application Architecture. 225
Enterprise Architecture. 231
Frameworks 231
Constraint Management 232
Conclusion 233

CHAPTER 12

Skills **235**

Education 236
Interviews 240
Imposter Syndrome 242
Modifying versus Rewriting Code. 243
Specialists versus Full Stack Developers 243

Introverts, Extroverts, and Ambiverts 245
Conservatives, Liberals, and Centrists 247
Collaboration versus Concentration 250
Remote Work 250
Wrap-around Days 251
Non-technical Work 253
Solo versus Team Work 253
Productivity versus Anti-productivity 254
Stack Ranking 254
Legacy Skills 255
Age, Jobs, and Careers 256
Skill Sets 256
Constraint Management 256
Conclusion 259

CHAPTER 13

Methodology **261**

Why Methodology Matters 261
Domains 264
Information Projects 264
Iron Triangle 266
Choosing a Methodology 267
Planned Methodologies 267
Critical Path 269
Critical Chain 271
Agile Methodologies 275
Scrum 277
Kanban 278
DevOps 279
Hybrid Methodologies 280
Dark Agile 283
Technical Services 284
Constraint Management 285
Conclusion 286

CHAPTER 14

Projects **289**

Success versus Failure 289
Work Rules 290
Status of Planned Projects.....292
Status of Agile Projects 294
Spin 295
Scope Creep versus Scope Surge 298
Expediting.....299
Replanning 301
Nontechnical Executives.....305
Constraint Management.....308
Conclusion 309

CHAPTER 15

Processes **311**

Projects versus Processes 311
Manufacturing Processes 312
Service Processes 314
Business Processes.....317
Information Processes.....319
Constraint Management.....325
Conclusion 326

CHAPTER 16

Portfolio **327**

Life Cycle Management.....327
Portfolio Management.....328
Prioritization 334
Feasibility 334
Justification.....337
Initiation 337
Termination 338
Governance.....339

Multi-project Management 340
Project Management Office 345
System of Systems 346
Constraint Management 347
Conclusion 348

CHAPTER 17
Services **349**
Professional, Scientific, and Technical Services 351
Service Organization 351
Service Methodology 355
Service Disruptions 356
Service Engagements 359
Constraint Management 361
Conclusion 362

PART 3
SYNERGY

CHAPTER 18
Constraint Management Redux **367**
Constraint Management for Information 367
Information for Constraint Management 368
Focusing Steps 370
Global Optimization 371
Conflict Resolution 372
Decisive Competitive Edge 374
Strategy and Tactics 375
Continuous Improvement 377
Constraint Management 378
Lean 378
Six Sigma 379
Trio 379
Quality 380

Value 380
Limits 381
Duality of Constraints 384
Ultimate Limit 388
Paradigms 391
Backsliding 392
Conclusion 392

CHAPTER 19
Strategy Redux **393**
Executive Perspectives 393
Traditional Strategy 394
Strategic Initiatives 399
Digital Disruption 399
Role of Technology 404
Technology Questions from Constraint Management 405
Technology Questions from Strategy Consultants 406
Technology Questions from the Information Field 406
Strategy Traps 411
Enterprise Scenarios and Strategies 412
Technical Scenarios and Strategies 414
Dynamic Strategy 418
Conclusion 421

CHAPTER 20
Conclusion **423**
Information Constraints 423
Constraint Management in the Information Field 425
Capacity Constrained Resources and Bottlenecks 429
Technology Rediscovery 430
Induced Demand 431
Receding Goals 431
System of Systems 434
Sports Team 436

Strategy, Information, and Constraints 441
Some Assembly Required 442
Exceeding the Goal 444

EPILOGUE
Administration, Academia, and Constraint Management Lite 447

PART 4
APPENDIX

APPENDIX A
Strategy Principles 459
Alignment Principle 459
Blue Ocean Strategy Principle 459
Dynamic Strategy Principle 459
Execution Principle 460
Horizons Principle 460
Innovation Principle 460
OODA Loop Principle 461
Profit Patterns Principle 461
Strategy Traps Principle 461

APPENDIX B
Information Principles 463
Agile Principle 463
Commoditization Principle 463
Dematerialization Principle 463
DevOps Principle 463
Legacy Principle 464
Longevity Principle 464
Metrics Principle 464
Polarization Principle 464
Role Reversal Principle 464
Technical Debt Principle 465

Variety Principle. 465
Virtualization Principle. 465
Waterfall Principle 465

APPENDIX C
Constraint Principles **467**
Accounting Principle. 467
Aggregation Principle 467
Attention Principle 467
Buffer Principle 468
Capacity Principle 468
Chain Principle. 469
Change Principle 469
Competitive Edge Principle 470
Conflict Principle. 470
Constraint Principle 471
Decision Principle 471
Demand Principle 472
Elevation Principle. 472
Exploitation Principle 472
Flow Principle. 473
Focus Principle. 473
Goal Principle. 473
Holistic Principle 474
Improvement Principle 474
Leverage Principle 474
Location Principle 475
Measurement Principle. 475
Multitasking Principle. 475
Noise Principle 476
Optimization Principle 476
Pull Principle 476
Relay Race Principle 477
Replenishment Principle. 477

Sales Principle. 477
Segmentation Principle 478
Simplicity Principle 478
Strategy and Tactics Principle. 478
Subordination Principle 479
Supply Chain Principle 479
Technology Principle. 479
Thinking Principle 480
Time Principle 480
Utilization Principle. 481
Win-Win Principle 481

APPENDIX D
Strategic Decisions **483**
Fundamentals. 483
Decision-making 483
Cause and Effect. 484
Information Technology 484
Computer Software 486
Technical Services 487
Strategic Decision Scenarios. 488
Moving to a Strategic Constraint 488
Relaxing a Capacity Constrained Resource 489
Reducing Technical Debt 490
Portfolio Management. 491
Conclusion 496
References **497**
Index **507**

PREFACE

As an executive and consultant, I have devised Technical Strategies that align with Enterprise Strategies for more than a few businesses and government entities. Most of my work has been in Information Technology, Computer Software, and Technical Services, which I'll refer to collectively as the Information field.

In this context, I have found Theory of Constraints (TOC) to be beneficial. What I like most is its focus on leverage points. Otherwise, a manager's instinct is to strive to control everything. That's not just ineffective; it is a practical impossibility. So, the question always comes back to where to commit finite resources to achieve the enterprise's mission.

The Goal is a widely read business novel, having sold millions of copies and gone through multiple revisions [Goldratt, 2014]. It's a great place to start, but a terrible place to stop, because most readers can't put their newfound knowledge to use. In my informal survey of readers, the majority admire its concepts, but only a few can apply them unless they purchase software with TOC already implemented. Even then, changing business models and culture to embrace TOC is a formidable task. The author himself estimated that only about three percent of readers implement any TOC.

Therefore, I was intrigued, yet a bit cautious, when asked many years ago to adapt TOC to Services. Because it was invented for Manufacturing and Distribution, applying it to Services was a challenge, to say the least. Nevertheless, my colleagues and I set out to adapt TOC to Professional, Scientific, and Technical Services. *Reaching the Goal* explains how we did it [Ricketts, 2008]. Those methods are still in use today.

Because that book came from outside the avid TOC community, I couldn't predict how it would be received. But Eli Goldratt, TOC's founder, called it "beautiful work, clearly written, and one of the best books on TOC." Then he suggested that we write a series of books together. Unfortunately, he passed away before we could collaborate. This is not a book in that series, but I hope it nonetheless satisfies his scientist's appeal to "stand on my shoulders, not in my shadow."

When I first considered writing this book, my intention was to adapt TOC for Software, much like my previous book adapted TOC for Services. But my intervening years as a business process owner in Technical Services, chief technology officer in Computer Software, and technical strategist in Corporate Strategy made me realize this:

- TOC applications created for Manufacturing and Distribution are not as directly suited to managing Software, but the principles underlying TOC do apply.
- Managing Software is just one part of the much larger management problem in the Information field.
- The Information management problem can't be solved only at the operations level. It should be tackled at the strategic level, too.

This book is intended to bring a broader understanding of Strategy and Information to the TOC community, while at the same time introducing TOC principles to the Strategy and Information communities. By doing this, I hope to close (1) gaps between Enterprise Strategy and Technical Strategy, (2) gaps between the Information field and the organizations it supports, and (3) gaps between reading about TOC and doing it.

Exceeding the Goal is this book's title because reaching a goal may be sufficient for operations, but it's insufficient for strategy when competition is intense. Exceeding the goal is the path to extraordinary results.

This is not a book about standard TOC. Volumes have been written about that already. This book deliberately pushes boundaries and maybe some buttons, too. Merely calling it Constraint Management instead of TOC is bound to agitate some pundits. I hope, however, that readers enjoy thinking unconventionally. After all, that's how TOC got started.

Continuous improvement is the essence of TOC. The Information field has evolved faster and further than any other technology, thereby enabling achievements in other fields that were inconceivable a few decades ago. Thus, well-thought-out strategy, technologies and TOC are complementary ingredients for exceeding the goal.

My own adventures in Manufacturing, Research, Consulting, Software, and Strategy are the basis for this book. As told here, the Adventure sections are true stories about situations where there was significant business and personal risk that things would go pear-shaped. Many of my adventures were successful. Others, not so much. But valuable lessons can be learned from both kinds, and the spectacular failures serve as cautionary tales.

These adventures happened in a variety of contexts, including large enterprises, small start-ups, governmental units, and academia. Those in the Services realm are more often business-to-business than business-to-consumer.

Some technical aspects of my adventures are dated. That's what happens over a lengthy career in a rapidly evolving field. However, the lessons learned are as relevant today as when they were new.

Why write an episodic memoir? Over a career with more than a few twists and turns, I've learned a few things, and I hope that sharing my experiences and observations—good and bad—will help others avoid rediscovery of the harshest lessons. It's also a chance to help readers use TOC in ways that might feel out of bounds. Furthermore, my adventures not only illustrate various principles, they may help my family and friends appreciate what I was doing when I was away on all those business trips.

Some books of this ilk imply that they will solve all your problems. I make no such claim. Indeed, the breadth of this book is meant to recognize that large enterprises have complex problems with no easy solutions. But there are solutions. And oftentimes the best solution turns a complex problem into a simpler one.

Before we get started, here are the usual disclaimers. I speak for myself, not for others. All mistakes are mine. Your mileage may vary.

Finally, thanks go to my family, friends, and colleagues. Throughout my adventures, they have been an inspiration for how to exceed the goal.

Exceeding the Goal

PART 1

OVERVIEW

PROLOGUE

THE ENTERPRISE CONSTRAINT

Constraints hold organizations in check. If there were no constraints, productivity would be easy, and organizations could grow without bounds. But in most enterprises, survival and growth are perpetual struggles. This is especially true in the Information field, where the half-life of some technologies can be counted on one hand and most enterprises succumb within a few decades, if not sooner.

Whether managers recognize it or not, their role is to manage constraints. It's far better to recognize constraints than to soldier on unaware. Otherwise, hidden constraints dominate managers rather than the other way around.

Many books have been written about Constraint Management, but few about Constraint Management of Information Technology, Computer Software, and Technical Services. This book addresses that gap because every modern organization is both lifted and limited by technology.

As will be seen in the later overview of Constraint Management, the enterprise constraint is key to overall organizational performance because it's the dominant constraint. Every other system in an organization may have a local constraint, but by accident or design, those local constraints should be subordinate to the enterprise constraint.

At the start of my career, I stumbled into an enterprise constraint without appreciating its significance or knowing how to find my way out of the thicket. For me, it began here.

ADVENTURE: **Not in Kansas Anymore**

My first job after college was an adventure in Manufacturing. The job title was production coordinator, and it amounted to (1) keeping track of work in process as it made its way through the Heat Treat department, and (2) expediting jobs destined for favored customers or urgent stock replenishment. All steel parts went through Heat Treat at least once. All metal parts went through multiple times to degrease, deburr, soften, harden, strengthen, or straighten them. Heat Treat was therefore the primary convergence point.

Heat Treat in summer was like a medieval painting of hell, with soaring flames, roiling quench pits, sooty equipment, and tortured souls. In winter, it was like hell frozen over because snow and sleet would gust in through broken windows and settle among the furnaces and equipment, creating a scene of ice and fire. Though mine was technically a white-collar job, it often was sweltering or frigid down in the Heat Treat department where I spent most of my workday.

Heat Treat was grueling, gritty work, so it employed ex-cons, weekend alcoholics, and guys tough enough to eat a thick-sliced ham sandwich without teeth. The foreman was known as "The Rat" due to his reputation for untrustworthiness. I don't think anyone in the office tower expected me to last down in deeply blue-collar territory.

Constraint Blindness

I did not realize it at the time—and neither did the management ranks above me apparently—but I was coordinating the output of the entire factory. Let me explain. In that plant, as in many other such plants, Heat Treat was the enterprise constraint. That meant no matter how much any other department produced, everything the factory produced overall was ultimately governed by Heat Treat. With the benefit of hindsight, frequent expediting—my job—was a symptom of constraint blindness.

No one saw the constraint because we weren't looking for it. Consequently, considerable management attention went into attempting to optimize the entire plant by driving every department to have maximum utilization. Time and motion studies had been done for every job in the factory, and the executive mandate was not to waste time anywhere. This policy was common back then, and still is in some plants today—but it can have disastrous consequences.

The Way Things Work

When I started the job, work-in-process inventory was stacked everywhere. The inventory manager explained it this way: "Despite efforts to minimize inventory by producing only for firm orders or stock replenishment, the walls of the building are what really limits the inventory."

Release of jobs into the shop was supposed to be governed by start date, which was calculated as due date minus expected production time, unless first availability was later due to congestion in the factory. Therefore, if a job was expected to take six weeks to complete, it was released into the shop with that much lead time before the due date, or the customer was notified of a due date later than what they requested.

Once in the shop, jobs with the least time remaining to their due date were supposed to get priority, but workers would sometimes do the easy tasks first, regardless of due date. Furthermore, if some departments didn't have enough work to stay busy, some jobs were started early to create utilization. The resulting unpredictability meant that planned lead times included plenty of wait time, with relatively little actual work time.

Amateurs Fumbling Along

When the steelworkers went on strike during their contract renegotiation, we office workers were sent out to run the plant as best we could as a skeleton workforce. I discovered that running a punch press every day was relentlessly boring, but driving a forklift to get and put thousands of pounds of steel parts on shelves well above my head was starkly terrifying. I had no mishap, but my concern was not irrational because a couple other office workers driving electric transport vehicles managed to collide head-on even though they were the only two people working on that floor. Consequently, no one was happier than me to see the steelworkers come back to work.

Nevertheless, this involuntary assignment yielded invaluable insight. My fellow office workers also had no prior experience doing factory jobs, yet we managed to assemble twice as many finished goods per day as the standard productivity benchmark. So much for the factory bottleneck being final assembly, as managers and the union believed.

Bullwhip Effect

MRP

Eventually my job expanded into making the new Material Requirements Planning system work because this would presumably eliminate stockouts, late delivery, and expediting. Getting bills of material and existing orders into the MRP system was the first step. To my surprise, the MRP system assumed that the plant had unlimited capacity and complete flexibility, so it would accept new order due dates that were utterly unrealistic.

Almost immediately, we began suffering daily reschedules from a favored customer using its own MRP system. Oftentimes, those reschedules moved orders in or out by several months without regard for our actual production lead time or whether an order had already been started. Apparently, their MRP system had no concept of what was realistic either, because Finite Capacity Planning was yet to be implemented.

Today we know that disruptions downstream in a supply chain are amplified as they ripple backward. This is called the Bullwhip Effect. But back in the day, we just waited for Friday and rescheduled our production as best we could to that effectively random change request. Nevertheless, another wave of changes arrived every day during the following week because our customer's other suppliers couldn't handle the radical, random changes either. Something was terribly wrong. We were counting on MRP to save the day when, in fact, MRP was the root cause.

Shifting Priorities

I wish I could say that we were able to synchronize our MRP with our favored customer's MRP, but an economic recession intervened. Our order backlog of many months dropped steadily until there was no backlog whatsoever. As orders came in, they were dispatched immediately into the factory, regardless of due date, so expediting was less about accelerating laggards and more about finding useful work to do.

The enterprise constraint had moved externally because the factory could no longer sell everything it had the capacity to produce. That, of course, triggered considerable executive angst.

Measure Twice, Cut Once

Soon the white-shoe consultants arrived. Their mission was to advise the executives on how to make the plant more productive and more profitable.

Here's what the consultants saw. The plant produced two complementary product lines: every sale of product line A required a corresponding product from line B. Indeed, products A and B could not be used separately under any circumstance. However, competitors produced their own versions of both product lines. Because off-the-shelf products were standardized, and custom products were low-tech, customers could buy from any manufacturer or its distributors. However, the firm prided itself on higher quality than its competitors.

Product A was produced in an assembly line, while product B was produced in a job shop. What's the difference? On the assembly line, thousands of parts were assembled into products that rolled off the line continuously. In the job shop, however, batches of products were delivered intermittently to finished goods inventory or the loading dock. But both product lines' parts converged on the Heat Treat department repeatedly.

To my astonishment, executives took the consultants' advice and made a baffling strategic decision: They dropped product line B! Why? The management accounting system, with its cost allocation scheme, told them that they were losing money on product B.

What happened next? The loss on product B went away as expected, but sales of product A declined, which was not anticipated. Instead of buying A and B from different manufacturers, customers could just order both from one competitor or its distributor. Apparently, convenience beats quality for some customers—or they weren't seeing a difference in quality. Furthermore, the perception that product B was unprofitable was based on a cost accounting illusion. Dropping B meant that its allocation of costs on shared resources then had to be covered entirely by A, which caused its profit margin to plunge. D'oh!

Postscript

Shortly thereafter, I resigned to finish my degree in Information Systems before the new strategy played out fully. The last time I checked, this firm was still in business, though it had a reputation as a revolving door for executives. I've often wondered where it would be today if its executives and I had understood Constraint Management well enough to know that cost accounting had lured them into a reality distortion field and the consultants were guiding them toward unintended consequences.

MRP software has evolved to embrace some Constraint Management principles, so I probably wouldn't recognize that factory today. However, when people keep their own spreadsheets to make a process work, as still happens today with MRP, that's a sign that the system is flawed. As for me, this Manufacturing adventure led me into the Information field because I could see the potential and thought the obstacles we'd encountered with MRP could be overcome.

Lessons Learned

I learned several lessons from this factory adventure that are covered today in Constraint Management principles:

- **Constraint Principle.** The enterprise constraint exists even when managers don't see it.

- **Utilization Principle.** High utilization everywhere makes it harder to see the constraint.
- **Supply Chain Principle.** MRP systems affect entire supply chains, not just individual firms.
- **Location Principle.** The constraint can move between internal and external.
- **Measurement Principle.** Mismeasurement can mislead executives into fateful strategic decisions.

In this adventure, the operations constraint was the Heat Treat department. However, the strategic initiative that valued cost reduction more than revenue from complementary products was the greater problem.

On a personal note, this adventure taught me to be careful both when making and taking advice. Later in my career, I was an executive consultant making strategy recommendations myself, and was forever wary of unintended consequences.

Around that time, Eli Goldratt invented Constraint Management solutions for factories. We didn't cross paths until years later when those solutions had matured, and my career had taken me from Manufacturing through Academia to the Information field.

Conclusion

One of the reasons Constraint Management immediately appealed to me was I had been a member of factory management years earlier. It also appealed to me because Constraint Management principles seemed as though they ought to apply not just to Manufacturing and Distribution, but to many other industries. More about that later.

The following chapters cover fundamentals of Information constraints. In cases where Information supports the core business, those constraints are usually local. But when Information is the core business, an Information constraint can be the enterprise constraint.

CHAPTER 1

INTRODUCTION

Some readers skip a Prologue and jump straight into the Introduction. If you did, I encourage you to circle back and read the Prologue first, because it describes the beginning of my journey into Constraint Management. All lessons in this book have their roots in that adventure.

If you don't know the difference between flammable and inflammable, they say you shouldn't play with fire, or anything that might catch fire, or anything that might start a fire. With that admonition in mind, let me tell you a story about the difference between experience and expertise in the Information field.

ADVENTURE: **Surprise Expert**

After becoming an executive consultant, I traveled to a government site to meet a client for the first time. A team of our consultants had already been on site for a week, though I did not know any of them. I went because they requested help with the Technical Assessment and Technical Strategy of their project, which had already kicked off.

Wake-up Call

The vice president I worked for had casually asked me to see if I could lend a hand because the team was tackling a new technical problem and I had experience with assessments and strategy. He didn't seem at all concerned about the project, so I thought I'd be in and out in a few days once I got the team organized. Then I would come back six weeks later, stay for a few days to wrap up the project, and present our findings and recommendations.

As I dozed in the hotel shuttle from the airport, I overheard two fellows in the seats behind me talking about a project they were doing. At first, I didn't think much about it because there were undoubtedly many projects going on in the city. But they got my attention when they said they were relieved that an expert was coming in the following day to guide their Technical Assessment and Strategy project. Through my mental haze, I realized they were talking about me! I figured that the VP had oversold my expertise, or the team had been generous in their interpretation of what he said. Both were possible.

Cold Start

When I introduced myself to the full team the next morning, they were enthusiastic, despite the technical issue being one that no one in the world, including me, had previous experience with. It was totally greenfield territory, so we would be doing a cold start. Our consulting partner and the team that responded to the request for proposal had sold the client on our firm's expertise over the half dozen other firms that had bid on the work, which wasn't entirely inappropriate because technical consulting was our core business and we had extensive experience in related areas. We had just never done a project to resolve this specific technical issue before—and neither had any competitor—which the client understood and accepted, fortunately.

Dark humor in the Information field goes like this: Deep expertise is so hard to find that companies are always anxious to hire people with five years' experience on technology that's only one year old. Accordingly, when a new problem comes along, anybody with even tangential experience can claim to be the resident expert.

Thus, my expectation of a quick in-and-out visit was gone. As we met the client, I could see that we were going to have to roll up our sleeves and figure things out as we went. There was no playbook for what we were doing. And when in unfamiliar territory while consulting, it's best to look like a duck: calm on the surface while paddling furiously beneath.

Confidence Complete

Six busy weeks later, we finished the Technical Assessment and Strategy. Then we presented our final report to the government officials, as well as to a financial analyst. He was more than casually interested because his role was to figure out how to fund the project, which we estimated would take more than a year and millions of dollars, given that the scope of work covered their entire computer hardware and software portfolio.

After the room cleared and it was just the two of us, he confided that he was less concerned with the estimate than with ensuring that it wasn't an under-estimate, because going back to the legislature for more funding later would be politically untenable. So, I drew an asymmetric probability density function on a flip chart, and showed him that our estimate was at the 0.95 upper confidence limit because it included a reasonable margin for contingency in anticipation that none of the firms that would bid on the implementation project had experience with this issue.

To prevent conflict of interest, we could not bid on the remediation work ourselves because we had prepared the strategy and estimate. But the chief financial officer accepted our recommendations, the legislature funded the project, and it was completed on time and within budget. In consulting, it doesn't get much better than that.

Lessons Learned

What I couldn't foresee was that this project was the first step on a strategic initiative that would require my full attention for the next several years, eventually expanding to a worldwide program where we did many assessments, strategies, and technical projects. We went on to develop a comprehensive methodology, build a tool set, and train thousands of consultants. Industry analysts rated our capabilities as being at the highest level.

Mark Twain said good judgment comes from experience, and experience comes from bad judgment. Innocent decisions had led our client into a severe technical problem. Bad judgment led us to deploy a team with insufficient preparation. However, the ensuing experience in solving the technical issue enabled us to use good judgment with hundreds more clients. We gained the requisite expertise, but my personal motto became "think before you leap." And I stopped riding hotel shuttles.

Picking Your Battles

Strategy, Information, and Constraints aren't topics you often find together in one book. Volumes have been written about each of those subjects, so addressing all of them in one book is ambitious. But it's their relationship, not the finer details, that is really of interest here.

This book therefore assumes that most readers are already familiar with at least one of those topics and are curious about how they enable or impede each other. For readers who need an introduction or a refresher, there are summaries.

If some sections are familiar, keep in mind that this book is written for a wide range of readers whose experience may be in a different field than your own. Skim or skip over familiar sections, because you should eventually run across unfamiliar sections, no matter what your background.

The title, *Exceeding the Goal*, refers to the potential benefits gained from combining Strategy, Information, and Constraints instead of treating them as separate subjects. In today's world, if you aren't exceeding your goal, your competitors may be closer than they appear, and your customers may be looking elsewhere.

The reality of working in large organizations or consulting with them on Strategy and Information is that you win some and you lose some, so it pays to pick your battles. Constraint Management, a practical nickname for Theory of Constraints, is the best way I know to do that.

Strategy

In simplest terms, strategy is a plan to deploy resources and direct their activities so that a goal will be achieved within a specified time frame using finite resources. But in practical terms, strategy depends on execution more than plans because things rarely go according to plan.

An Enterprise Strategy is supported by strategies for major functions, such as human resources, finance, research, engineering, production, marketing, sales, distribution, and information. Those functional strategies must be aligned with the Enterprise Strategy for it to be achievable. This book focuses on Technical Strategy and how to align it with Enterprise Strategy by considering the goal and the means to achieve it.

Although many adventures cited here are from the business world, this book also applies to government entities. You will see adventures concerning city, state, and national governments. Although governments' overall goals are different from those of businesses, their information, goals, constraints, and strategies are similar, if not the same.

The Strategy chapter summarizes:

- The Horizons Model
- Blue Ocean Strategy
- Profit Patterns
- OODA Loop
- Alignment
- Strategy Traps
- Strategy Principles

Information Technology, Computer Software, and Technical Services

In the age of personal computers, smartphones, smart televisions, smart automobiles, video games, and home automation, we are all information consumers. In addition, this book covers what it takes to deliver Enterprise Information, which can be vastly different from Consumer Information due to stronger requirements for reliability, availability, scalability, security, and privacy, among other things. That is, "enterprise" in this context refers not to an aircraft carrier or a starship, but to a business or government organization. The larger the enterprise, the more complex its Information function is. However, an enterprise of any size can be buoyed or sunk by its Information. As employees bring consumer products into the workplace, effectively integrating them into the enterprise is just one more challenge.

This book uses "the Information field" as the short name for everything associated with Information Technology, Computer Software, and Technical Services. Providers are spread across multiple industries in the North American Industry Classification System:

- **Technology.** Computer and Electronic Manufacturing (hardware)
- **Publishing.** Computer Software (packaged software, such as video games)
- **Services.** Professional, Scientific, and Technical Services (custom software)
- **Telecommunications.** Telephone, Internet, Television
- **Infrastructure.** Data Processing and Hosting
- **Content.** Information Services (search, publish, broadcast)

Buyers, clients, users, and developers are found in every industry, of course. Many organizations operate their own computers, networks, and information systems. Some even develop their own software. Thus, if anything you do relies on Information, it is in scope for this discussion of Strategy and Constraints.

We will discuss specifics only in as much detail as necessary for the topic at hand for a couple of reasons. First, the half-life of specific technologies is short, so it's an ever-changing landscape that would quickly feel outdated. Second, and more importantly, Strategy is enabled by Information more than shaped by it. That is, Information restricts what you can do (the constraint) and shapes how you do it (the strategy) more than what you should do (the goal).

Information can be the enterprise constraint when manual processes won't scale. They're just too slow, too expensive, and too inconsistent. However, even if Information isn't the enterprise constraint, it's usually a capacity constrained resource because it sometimes impedes the enterprise constraint.

The Information Technology, Computer Software, and Technical Services chapter previews these areas where technical constraints are managed:

- Hardware
- Software
- Data
- Knowledge
- Networks

- Architecture
- Skills
- Methods
- Projects
- Processes
- Portfolio

Theory of Constraints

Theory of Constraints, also known as Constraint Management, is a continuous improvement method for management that originated in Manufacturing and Distribution, and later extended into Services. By identifying the one element of a system that restricts what it can produce, and orchestrating system operations around it, Constraint Management optimizes the system.

In the absence of Constraint Management, systems too often operate in a somewhat chaotic manner. For instance, in a factory, it may be difficult to predict when a production job will be completed because the shop is clogged with inventory generated in the misguided pursuit of high utilization. Constraint Management solutions bring predictability, reduce inventory, and unlock capacity.

Enterprises in the Information field sometimes operate like a factory, and when they do, standard Constraint Management solutions can be used. Indeed, if your business is the manufacturing of computer hardware, your business is a factory. Those situations are, however, the exception. The Information field, overall, often does not operate like a factory. Software, for instance, is not built in any way resembling a factory, nor does it behave like a manufactured product. And Enterprise Information both delivers and utilizes more services than Consumer Information. Nevertheless, Constraint Management principles can be applied even when Information operations are not like a factory.

The Constraint Management chapter previews these topics:

- Applications for Production, Distribution, and Projects
- Focusing Steps
- Conflict Resolution
- Buy-in
- Technology

Executive Priorities

Because this book applies to issues from the shop floor to the executive suite, the next chapter summarizes global executive studies. Executive roles studied are chief executive officer, chief financial officer, chief marketing officer, chief human resources officer, and chief information officer. Collectively, they are called CXOs or "the C-suite."

The priorities expressed by those executives are addressed by the later chapters on enterprise and technical strategy. Executive concerns therefore provide a target to shoot for.

Over two-thirds of CXOs believe their business success is shaped by, if not entirely defined by, Information. Chief marketing officers already spend more on Information matters than chief information officers do. In fact, over a thousand companies have chief data officers whose responsibilities have little or no overlap with chief information officers. Those responsibilities include strategy, sourcing, governance, partnerships, and skills for data.

System of Systems

Constraint Management is best known for optimizing what a Manufacturing or Distribution system can produce. However, when organizations are viewed from an executive perspective, what's seen is not just one system, but many interdependent systems. There are systems for engineering, procurement, production, delivery, marketing, sales, finance, legal, human resources, information, and more.

A system goal can be reached by optimizing within that system, but to reach the enterprise goal, the enterprise must optimize its system of systems. A change to one affects the others, and the effects aren't always proportional or consistent. Indeed, the systems may be pursuing conflicting local goals, and thereby hindering mutual progress. Thus, a holistic solution is harder because the systems are integrated, but the payoff is potentially much greater.

This system-of-systems conundrum is explored later, but it's a reason this book is entitled *Exceeding the Goal*. To reach a system goal, a well-chosen and carefully implemented operations solution will do nicely. But solving the system-of-systems conundrum at the strategy level opens the door to exceeding the entire enterprise's goal.

Why describe an organization as a system of systems rather than a hierarchy of super-systems and subsystems? Systems of systems do not necessarily form a hierarchy. The human body is a system of systems without a hierarchy. Our circulatory, respiratory, digestive, lymphatic, nervous, excretory, endocrine, immune, skeletal, muscle, and reproductive systems work together. Sickness or failure in one endangers them all.

Likewise, in organizations, a change to one system affects others. For instance, strategy, marketing, sales, and production must be coordinated. Moreover, the same element can be part of more than one system. For instance, one resource manager may supply workers to multiple projects, one project manager may manage more than one project, and one executive can be responsible for more than one business unit.

Adventures

The dictionary defines "adventure" as a hazardous and exciting activity, especially the exploration of unknown territory. In the business world, an adventure can also be described as efforts to bring about change in the face of deep and sustained resistance—with significant business and personal risk of failure.

Rather than a business novel, which can only cover a few basic concepts, or a tutorial, which can cover many concepts in depth if readers have the patience for it, real adventures in business and government realms are a recurring theme in this book. Given our tribal history as human beings, true stories are often more compelling than fiction or tutorials. To distinguish stories from exposition, my adventures, like the one earlier in this chapter, are clearly marked.

I've been blessed with a career spanning three fields: Manufacturing, Academia, and Information. And I have been fortunate to work mostly with people and projects that have succeeded. That has not always been the case, however, so the adventures include examples of things gone right and things gone wrong.

As I recount these adventures, it may seem that I have had many jobs, and I have. That's partly due to having a lengthy career spanning 10 employers and three fields, but it's also because I have usually had multiple roles concurrently. The last 20 years were executive roles.

Although every story has its own chronology, the adventures overall are not told chronologically. They are instead organized topically. If that creates a sense of skipping through time, I hope it's compensated by adventures directly relevant to the topic at hand. And if some of the adventures seem like they're from another era—they are. But those adventures are included specifically because their lessons are timeless. Therefore, the usual admonition about history repeating itself seems appropriate.

Should you suspect that an adventure is about your own company, it is probably coincidence. I've spoken about these adventures many times, but this is the first time I have written about them. Whenever someone wonders if an adventure happened at their company, that's almost never the case. These situations are not unique. They happen in a lot of places, but the most egregious failures are hidden, if not forgotten, so those stories can feel like revelations.

Conclusion

Constraint Management has been around since the 1980s, which makes it the best-known performance improvement method you have probably never used. It's taught in universities and professional seminars and business books, so for some readers it feels familiar. On the other hand, many people have not yet been exposed to its ideas, and for them it feels brand new. Of course, the pace in the Information field is such that practically every day brings news of some kind. Strategy has been around longer than either Constraint Management or Information, but even it is evolving, in part because Information enables new strategies.

Therefore, keep an open mind. Nobody introduced to Constraint Management thinks it will work, but it does. Conversely, everybody has enough experience with Personal Computing to assume they know how Enterprise Computing works, but they often don't. And more people execute Strategy than develop Strategy, so there may be some fresh principles there, too.

CHAPTER 2

EXECUTIVE PRIORITIES

What better way to begin a discussion of Strategy, Information, and Constraints than by understanding executive priorities? In global C-suite studies, researchers interviewed over 12,500 executives in 70 countries and 21 industries [IBV, 2018]. The studies covered chief executive officers, chief operating officers, chief financial officers, chief marketing officers, chief human resources officers, and chief information officers. I'll refer to C-level executives collectively as CXOs. Before summarizing what they said, consider this C-level adventure from multiple consulting engagements.

ADVENTURE: No Clean Slate

When new chief information officers step into that role, unless they are in start-ups, they generally inherit technology put in place by predecessors and sometimes by other CXOs and line-of-business executives. If the current portfolio of technologies is not aligned with the current needs of the organization, those Chief Information Officers (CIOs) get hit by conflicting demands: to keep the old technology running while buying or building new technology, then migrating from old to new. It's no wonder that average CIO tenure can be counted on one hand.

CIO Inheritance

When consulting with CIOs, knowing the current technology status and the Technical Strategy that led there is vital for understanding CIO goals and options. Consider the following CIOs' situations:

- CIO #1 inherited a portfolio built with the *homegrown approach*, so the bulk of the central budget was for maintenance rather than building new applications.
- CIO #2 inherited a portfolio assembled with the *best-of-breed approach*, which meant that top-rated technology had been acquired from multiple vendors, then integrated and operated by the central staff, which was a different maintenance problem.
- CIO #3 inherited a portfolio acquired with the *pre-integrated approach*, which meant that technology was licensed from as few vendors as possible because vendors did the integration and some customization, then the central group operated the technology.
- CIO #4 inherited a migration of portions of the technology portfolio to *Cloud Computing*, which meant the technology was pre-integrated and operated by the vendor, but the central role was more vendor management and end-user support.

Lessons Learned

Each CIO's technology options were limited by the hand they were dealt:

- CIO #1 was burdened with technology supporting yesterday's business.
- CIO #2 was stuck with an assortment of disparate technologies.
- CIO #3 was reliant on the lead vendor's Technical Strategy and customization services.
- CIO #4 faced a substantial re-skilling challenge, plus managing multiple cloud environments.

Thus, none of these CIOs could pursue a different Technical Strategy without considering their current predicament and launching a strategic change initiative.

Chief Executive Officer

Chief Executive Officers (CEOs) say advances in technology are triggering upheaval. For instance, industry value chains are morphing into cross-industry ecosystems as customer expectations are escalating. Oftentimes, speed trumps polish: a minimally viable product today is worth two polished products next year. Of course, technology can accelerate innovation for incumbents as well as new entrants, but business model innovations are just as hard as technical innovations. Innovation entails risk and requires patience.

While innovations are coming faster, the payoff in business value is coming slower. Thus, CEOs find themselves as chief strategists in an environment where the pace of change is accelerating while the margin for error is shrinking. System-level failures require system-level thinking. A substantial number of CEOs describe their organizations as data rich, but insight poor.

Chief Operating Officer

Chief Operating Officers (COOs) are becoming less concerned about physical assets, and more concerned about digital assets, which include all the information associated with physical products. COOs are also striving for real-time operation because it pleases customers as much as it creates operating efficiencies. However, real-time operation reveals the hidden costs of global supply chains, such as long lead times. Agile supply chains and demand-driven management are antidotes to cost variability. Sensors and smart machines are the heart of digital operations. But advanced analytics and modeling are needed to respond to markets by synchronizing operations.

Industry convergence is a top COO concern because it shifts the focus from stand-alone products and services to combinations of physical and digital. Encroachment is the flip side of convergence, so together they generate opportunities as well as threats. For instance, banks and retailers are not as separate as they once were. Other top concerns are "anywhere" workplaces, which promote autonomy, and cybersecurity, which threatens digital assets. Although globalization continues, some COOs are rethinking lengthy supply chains, which may not be sufficiently agile.

Chief Financial Officer

Chief Financial Officers (CFOs) say their biggest challenge is integrating information across the enterprise. That makes it hard to prosper during economic downturns and competitive disruptions. Moreover, it's hard to evaluate the business impact of innovative technologies. Those CFOs who have solved the information integration problem are able to pursue analytics that drive better strategic decisions. And those analytics go beyond just financial data to also include operational and external data.

Headlights are better than taillights, particularly when CFOs take the long view of trends and turning points. Automated production of key metrics is essential for managing enterprise risk, but that requires deep business analytics skills.

Chief Marketing Officer

Chief Marketing Officers (CMOs) say industry convergence shifts the focus from products and services by single entities to cross-sector customer experiences from multiple entities. This creates new growth opportunities, but it also opens the door to new entrants. Thus, new revenue models, such as licensing and subscriptions, are challenging conventional product and service sales. Moreover, pioneering CMOs are striving to be the first into new ecosystems with the goal that their ecosystem will achieve dominance. CMOs say deeper, richer, personal customer experiences engage customers because mass markets are disappearing. Thus, data-driven insights are shaping marketing campaigns, but the explosion of data means external data and expertise are supplementing internal data and expertise. The marketing funnel has become a series of loops.

CMOs face several challenges, some of them technology-related. They have more influence on promotion than products, place, and price. They need a new skills mix, especially with the advent of web and mobile apps. Finally, CMOs are relying more on external partnerships.

Chief Human Resources Officer

Chief Human Resources Officers (CHROs) are already developing workforces with different capabilities, and partnering with other organizations for specific skills. CHROs are also investing in automation, though that too requires new skills, such as system integration and real-time operations. And new skills require new ways of learning. Indeed, Cognitive Computing is at the intersection between automation and learning. Capitalizing on collective intelligence is hard, but many organizations fail to use the knowledge-sharing resources they already possess.

Enterprises undergoing continuous change need workforces that adapt to change, if not drive it. Of course, CHROs seek to build and retain creative, diverse, nimble workforces, and that requires reimagining the employee experience. The anywhere workplace literally breaks down walls. As expertise is built, employee retention is necessary to prevent loss of expertise to poaching. However, factors attracting employees differ from those retaining employees.

Hiring isn't CHROs only option: freelancers increasingly provide services on demand. Of course, allocating human resources effectively depends on demand for products and services, which can be unpredictable, thereby making a flexible workforce more valuable.

Chief Information Officer

Chief Information Officers (CIOs) have these priorities: enhancing intelligence and insight, digitizing the front office, and strengthening Information skills. Doing all this at scale is hard when digital content doubles every two years, customers expect anytime access, and the half-life of critical skills is as little as two years. Moreover, analog services, such as filing tax returns, are increasingly digitized. Consequently, CIOs cite security twice as often as the next most worrisome risk: regulatory compliance violations. The scope, complexity, and pace of technology challenge even the most advanced organizations, which is why CIOs are expecting Cognitive Computing to learn rather than be programmed.

Overall, CIOs have a vision, but many struggle to implement it. Among the various CXO types, CIOs feel the most pressure to transform their organization. Being a technology visionary still is their primary role, but leading digital reinvention and building new platforms are rising roles. Instilling an agile culture has the advantage of driving innovation at scale. Scale matters because there are already over a billion transistors per person and digital content will grow thirty-fold over the next decade.

Other CXOs

Not included in the C-suite study are some CXO types relevant to this book. The chief science officer, chief technology officer, chief data officer, and chief innovation officer play key roles in the creation of Information Technology, as well as its integration with other technologies and alignment with business and government.

Chief Science Officer

Chief Science Officers (CSOs) lead scientific research organizations. Thus, CSOs evaluate and set research priorities, then coordinate the administrative structure that supports scientists. CSOs are usually scientists themselves with advanced degrees.

Chief Technology Officer

Chief Technology Officers (CTOs) lead development of technical products and services. Thus, CTOs evaluate and set development priorities, then coordinate the administrative structure that supports developers. CTOs often have degrees in engineering or business development.

Chief Data Officer

Chief Data Officers (CDOs) are responsible for strategy, sourcing, governance, partnerships, and skills pertaining to data. They must also ensure compliance with data regulations. This reflects increased attention to data as a commodity rather than the technology used to gather, store, and retrieve the data. Thus, CDO is more of a business role than a technical role, so CDOs are likely to have degrees in data analysis, math, or statistics.

Chief Innovation Officer

About 30% of Fortune 500 companies have a Chief Innovation Officer (CINO), but their roles vary widely because innovation objectives vary widely. Roles include researcher, engineer, investor, advocate, motivator, and organizer [Lovric, 2019].

Comparing Other CXOs

These CXO roles are not always distinct, but when they are, here are some differences:

- CSOs lead research, while CTOs lead development. In other words, CTOs build products and services based on knowledge that CSOs discover.
- If they are in the same firm, CIOs focus on business issues while CTOs focus on technical issues. Thus, these CIOs and CTOs align Enterprise Strategy with Technical Strategy.
- If they are in different firms, CIOs acquire and use current technologies, while CTOs oversee delivery of future technologies. Thus, these CIOs buy what CTOs build.
- CDOs are responsible for data no matter where it comes from or goes to. Thus, CDO responsibilities require coordination with all other CXO roles.

Where the CIO reports in the executive hierarchy also makes a difference:

- If reporting to the CFO, CIOs tend toward cost take-out and efficient operations.
- If reporting to the CEO, CIOs tend toward revenue growth, competition, and innovation.

Central Information versus Shadow Information

Information spending is migrating from the core to the edge of some enterprises. CIOs still spend on central technology, of course, but other CXOs and line-of-business executives spend as much as half the total spent on Information for their own organizations. Sometimes they each have their own local CIOs, but the central CIO organization does not develop this Shadow Information, does not support it, and may not even be aware of it, which makes compliance with regulations harder.

When the buyers and consumers of Information are in separate parts of the organization, line-of-business expectations about price, performance, and delivery may remain unmet, thus motivating Shadow Information. On the other hand, when technology is spread across the enterprise, it can be harder to procure economically, share data, and enforce standards.

Digital Reinvention Archetypes

Some CXOs have made more progress toward digital reinvention:

- Reinventors (27%) outperform their peers on revenue and profit, their Technical Strategy is aligned with their business strategy, they manage disruption, and they are effective at identifying and serving unmet customer needs.
- Practitioners (37%) have capabilities that don't match their ambitions, so they are playing catch-up.
- Aspirationals (36%) are still seeking the right vision, strategy, and execution.

The following sections cover opportunities and threats that every CXO faces: Disruptive Innovation, Ambiguous Boundaries, Living Laboratories, Technology Shifts, Business Platforms, and Enterprise Agility.

Disruptive Innovation

Disruptive innovation happens when a competitor with a different business model enters an existing industry. New entrants often target the bottom of a market, and then move upmarket until incumbents are ousted. This was once a rare occurrence, but it's becoming more common. However, the predicted deluge of digital invaders hasn't happened because incumbents acquire some invaders and reduction in venture capital starves others.

The solution to disruptive innovation is to prepare for digital invaders, even if the industry is capital intensive, because digital invaders can be stealthy and nimble. Start-ups often invest in intangibles to support their business model, such as skills and software, rather than physical assets. However, some new entrants are digital giants with far more resources.

CXOs say that preparing for digital invaders relies as much or more on technology factors as market factors. In addition, CXOs are:

- Driving more digital customer interaction
- Focusing more on individual customers and less on market segments
- Expanding their business partner network
- Seeking more external innovations

- Decentralizing decision-making to take advantage of interpersonal networking

Ambiguous Boundaries

Who counts as competitors nowadays isn't the simple question it once was. Some of the biggest threats to established businesses come from new entrants who aren't recognized as competitors until they already have a substantial foothold.

Furthermore, as industries converge, new entrants may have a solid foundation in their original industry. Where competitors once created a better product or service, digital disintermediation can now deflate established business models by targeting a specific segment of the value chain.

The solution to ambiguous boundaries is to create a panoramic perspective across the competitive landscape and over time. In other words, CXOs look sideways while looking ahead. Thus, CXOs are concerned about:

- Industry convergence
- Anywhere workplace
- Cyber risk
- Consumer purchasing power
- Sustainability

Living Laboratories

The majority of CXOs are considering different business models, but they approach major changes cautiously because picking the right business model isn't easy. Testing new business models is particularly difficult for three reasons. First, there is nothing to compare a new business model to. Second, the current organization often pushes back on change. Finally, timing matters—a lot.

Too bold. Too timid. Too near. Too far. Too early. Too late. Those are all legitimate concerns. The solution to business model decisions is a living laboratory. That's an environment where ideas are developed and tested, where the best ones are nurtured, and the winners are launched when poised for success. That said, most CXOs are exploring the inner region rather than outer edges of where new business models could take them.

CXOs say that technology is leading them toward the reassessment of business models. In decreasing order, these parts of the business are most affected:

- Product/service portfolio
- Operating model
- Partnerships
- Delivery channels
- Revenue model
- Customer types

Technology Shifts

Technology enables disruptive innovations, but it also enables panoramic perspectives and living laboratories. Thus, it's an essential element of strategic offense and defense. The volume and diversity of data available today are unprecedented. Nonetheless, CXOs lament an insights gap because so much of that data is not integrated in ways that support effective analysis.

While cost effectiveness has always been a benchmark for technology, agility and insight are the new frontier. The top technologies that CXOs anticipate are:

- Cloud Computing and services to reduce lead times and share resources
- Mobile solutions to access real-time data and improve customer service
- Internet of Things to utilize assets better and convert products to services
- Cognitive Computing to understand natural language and learn instead of being programmed
- Advanced manufacturing technologies for efficiency and mass customization

Business Platforms

Business platforms orchestrate assets instead of owning them, which creates a network effect with customers and more collaboration with competitors. For example, a digital identity verification service used across an industry benefits both the builders and operators of the platform by eliminating redundancy and sharing information.

Business platforms thus have a larger multiplier than previous business models: Technology Creator, Service Provider, and Asset Builder. Business platforms generate lots of heterogenous data that can be used for co-creation and product design with customers.

Historically, companies have allocated their capital to the same activities year after year. Now, however, platform builders and operators are reallocating capital.

Enterprise Agility

Industry leaders, the Digital Reinventors, are shifting away from top-down control toward autonomy and empowerment. Experimentation is encouraged. Failures are expected. This, however, creates a skills gap because employees are expected to work in cross-functional teams rather than traditional silos.

Conclusion

Although CXOs face common opportunities and threats, in Constraint Management terms CXOs have different priorities because they are facing different constraints:

- For CEOs, the constraint is somewhere in the formation or execution of strategy.
- For COOs, the constraint is increasingly in digital assets.
- For CMOs, the constraint is external, in the market for products and services, or internal, in sales.
- For CFOs, the constraint is increasingly in data rather than dollars.
- For CHROs, the constraint is internal, among employees, or external, in the job market.
- For CIOs, CSOs, and CTOs, the constraint can be just about anywhere in a sea of technology.
- For CINOs, the constraint is whatever blocks innovation, which may be business or technical factors.

The remaining chapters in this section provide overviews of Strategy, Information, and Constraints. Then in the following section we will see how principles from those areas handle Information constraints and CXO priorities.

CHAPTER 3

STRATEGY

Peter F. Drucker, famed management consultant, said, "Strategy is a commodity, execution is an art." The wisdom of that saying is often underappreciated. Various schools of thought on strategy have much to say about planning, but less about execution [Cooper, 2010].

The longest stretch of my career has been in the Information field. I've developed Technical Strategies for Information sellers and buyers. I have also created product, line-of-business, and corporate strategies. However, Strategy has evolved considerably during my time. Adventures in this chapter illustrate some of the changes in Strategy for providers. Strategies for buyers will be covered in later chapters.

Technology Research

Over the years, I have worked both in Research and in Development as separate business functions. As a research manager, I steered concepts into prototypes. As development manager, I received prototypes and advanced them into products. Later, as Chief Technology Officer, I ensured that the research pipeline aligned with product strategy and dealt with acquisitions that took product strategy in a different direction.

When a technology research project reached or surpassed minimum viable product status, we would show a prototype to customers. If they saw potential, we would do an initial field implementation. But there was no guarantee of success, either in terms of the technology working as hoped or in terms of its actual business benefits.

Nevertheless, this approach is a useful bridge out of the research lab, and customers love being invited to see technology on the cutting edge. Furthermore, relatively small scale means that it's practical to create and assess many prototypes at once.

These experiments take about a year once started, but could take several years to find a customer, and some concepts never do find a customer, which makes it hard to justify proceeding to develop them into products. I am always delighted, however, to see innovative technology get traction with customers because that makes it much easier to proceed to product development.

ADVENTURE: **False Start**

Two similar research projects show how unpredictable the outcomes can be. Both were software solutions tested in the same industry at about the same time.

Prototype A was deployed and worked as expected, and the customer not only embraced the technology, they had terrific ideas about how to use the information it produced to manage more widely than we anticipated. No one was able to quantify future business value, but it didn't matter because the customer immediately saw enough to justify implementation.

Prototype B was also deployed and worked better than expected, with estimated revenue growth of several percent. But the customer became concerned that automation would displace workers, so it did not progress further. Whether the growing revenue would create new jobs was a question thus left unexplored.

Prototype A proceeded out of Research and into Development. B went back to the lab while we searched for another interested customer.

Business Opportunity

Rather than leading with technology, another approach is to begin with a business opportunity and investigate where technology might apply. When strategists saw an especially promising customer opportunity that aligned with our own business strategy, they would create a quasi-independent business unit to pursue it.

My job was to create business plans for such units. Sometimes existing technology could be repurposed and extended, but often these opportunities required innovative technology products and associated services to fulfill specific requirements.

These endeavors lasted around five years, but some were terminated without spawning an ongoing business. Others graduated into going concerns in as little as one year. Each successful endeavor eventually became self-funding, but because these were multi-year big bets requiring corporate funding from launch to graduation or cessation, we could not fund many at once.

ADVENTURE: **Stumbling Out of the Gate**

Two similar business opportunities show how unpredictable the outcomes can be. Both opportunities were tested in multiple industries at about the same time.

Opportunity A was a new line of services launched by reassembling a team that had previously launched one successfully. Consequently, they knew their roles and tasks. The timing of the new offerings was good, because clients were ready to buy services to address a pain point that was getting worse every year. This was one of those business units that became a going concern, without further corporate funding, in about a year.

Opportunity B was a new line of products and services launched by a research team and a team of industry experts. Consequently, they had to sort out their roles and tasks. The timing of the new offerings was no better than average, because clients had to envision possible uses. As the business unit matured, so did the technology, and so did the clients. The business unit graduated to self-funding after about five years, but the technology took about 10 years to reach headline status.

Both of these opportunities were successful. However, one took a lot less time to reach its goal, in part because it was driven by an experienced team, but also because the other's technology took longer to mature. Thus, having a great team helps, but it's not the only thing that limits a business opportunity.

Technology Partnership

Yet another strategy, if you're in the Information field, is to become a technology partner with start-up companies. My role was technical strategist. Like a traditional venture capital (VC) firm, we had funds to invest in promising start-ups, but we also had technology to offer. Unlike a traditional VC firm that intends to cash out when a start-up goes public, our objective was to remain the start-up's technology partner, growing right along with them indefinitely.

As we received proposals, we performed a multi-step evaluation process that looked at each start-up's business model, business plan, and technology plan, among other things. Their business model was supposed to tell us how they intended to make money. Their business plan was supposed to tell us how they would implement and execute that model. And their technology plan would tell us how we would enable that plan.

As we screened proposals, we learned that most start-ups at the time were just a concept, and not a particularly well-formed concept. The majority could not articulate a convincing business model and plan. Therefore, our concerns were less about technology and more about the nuts and bolts of starting a business. Moreover, most founders had no idea what evaluation criteria we would apply because they didn't do any research into our organization. They were mostly looking for funding, when what they needed more was leadership. Therefore, we put executives into the larger start-ups where we agreed to become their technology partner and arranged executive oversight for the smaller ones.

During my first time as a technical strategist, we looked at hundreds of start-ups, but funded only about 2%. That's a problem because most start-ups flamed out in a year or two for business reasons, not technical reasons. Given the high mortality rate of start-ups, the only way to end up with winners was to start with a sufficiently large sample, focus on the most promising companies (not concepts), and provide them more oversight than they realized they needed.

In the media fawning over start-ups, it's easy to overlook that the majority fail—sometimes spectacularly. For every wildly successful start-up (founders are wealthy), there are 10 modestly successful (founders are making a living), a hundred barely successful (founders are squeaking by), and ten thousand craters where dreams crashed (founders' net worth evaporated). It's easy to ascribe success to skill, but it often is due to luck.

Highly successful start-ups are the rare survivors of a selection process as brutal as any described by Darwin. Having a technology partner doesn't change how evolution works if the business model and plan aren't well thought out and executed diligently.

Proof of Concept

The most professional satisfaction I've had with strategy has been serving as technical strategist on proof-of-concept projects. As part of the chief innovation office, we made technical innovations easier for customers to try on small scale before committing to large scale.

Proof of concepts (POCs) are quick enough to provide timely feedback and small enough to allow strategists to steer them toward success. The other strategy approaches described earlier are bigger, slower, more expensive, and more complicated. POCs are thus the agile alternative to big strategy.

POCs fill a niche. They don't replace Research or Development. They are worthwhile projects that otherwise would not get done because they have hit a roadblock. Oftentimes, the roadblock is organizational complexity and risk aversion, not financial or technical matters. For instance, executives in different business units cannot agree how to proceed in a way that benefits everyone, especially the customer. Therefore, the Strategy team provided funding that participants had to repay only if the POC proved that the technology works, customers would buy it, and the business could turn a profit with it.

Off-loading risk from business units onto corporate strategy enabled rapid decisions. With coaching, proposal teams could present viable concepts that could be funded in days rather than weeks or months. Funding decisions were made with imperfect information, but it turns out that vital information usually surfaced early. Most regrets that surfaced later could not have been foreseen even with more time because they happened during execution.

POCs are inherently risky, so the goal was a 50% success rate. The actual success rate was higher. (This is another inspiration for the title, *Exceeding the Goal.*) POC duration was dictated mostly by customer, but initial results were expected within 90 days, and final results in a year or less.

Although external VCs doing start-ups and internal strategists doing POCs both provide funding and oversight, there are significant differences:

- External VCs sit on the board of a start-up. Internal strategists have frequent checkpoints, if not daily contact, with the execution team because the POC time scale is much shorter. POCs can't wait a month or a quarter to find out there are problems.
- External VCs invest for 7 to 10 years with multiple rounds of funding. POCs invest for one to two years at most with seldom more than one or two rounds of funding.
- External VCs have an exit strategy for start-ups: Go public or get acquired. Internal strategists have an entrance strategy for POCs: Grow the business through innovation.

The POC program got leverage by feeding back lessons learned into business strategy, product strategy, service strategy, and organization design. POCs reveal what works, what doesn't, and where there are better alternatives.

ADVENTURE: **Diving Catch**

The first proof-of-concept project I led was a last-minute scramble. Our customer was on the verge of selecting a competitor's software solution when we negotiated a 90-day grace period. During that time, we implemented a layered architecture with software running on high-performance hardware. The result was a solution that could analyze 20 times as much data with the same response time as our competition. Size and speed mattered because the solution was for big data analytics, so data volumes would grow rapidly, and our solution's cost scaled linearly rather than exponentially.

The successful project got us back in the game long enough to develop a scalable solution that could handle a thousand times as much data. If not for that project, we would have lost the deal, and the original product would have languished instead of being scaled up because it was limited by the original hardware platform. Thus, an entirely different strategy would have been missed.

Lessons Learned

Different approaches to strategy yield various kinds of feedback and learning. In general, however, Technical Strategy has evolved toward methods that learn faster and pivot sooner.

- Technology research is solutions in search of problems.
- Business opportunities are problems in search of solutions.
- Technology partnerships enable technology sharing.
- Proofs of concept rapidly test innovations, despite roadblocks.

Even low-tech industries and small enterprises are increasingly dependent on Information. Thus, nobody is immune to what happens in the Information field, even if it's not their core business.

Where We Want to Be

"Strategy" has many meanings. The one used here amounts to "How do we get from where we are to where we want to be?" In many organizations, the answer is heavy on finance and marketing, light on operations, and nearly silent on

Information. Yet operations and Information may ultimately determine whether a strategy based on finance and marketing is successful.

Much has been written about strategy, so this chapter does not attempt to summarize the entire field. It instead highlights key business and technical strategy concepts that will be useful when we explore constraints in the Information field.

The remainder of this chapter is an assortment of tools I have found useful when formulating and executing strategy:

- Horizons are about foresight.
- Blue Ocean is about divergence.
- Profit Patterns are about innovation.
- OODA (observe-orient-decide-act) Loops are about agile strategy formation.
- POCs are about agile strategy execution.
- Strategy Traps are mistakes to avoid.
- Strategy Principles are best practices to emulate.

Horizons

As originally presented, the Horizons Model in Figure 3-1 divides every strategy into three parts [Baghai, 2000]. In my experience, however, there's a fourth horizon of equal importance having to do with past business.

- **Horizon 3 (H3)** is the seeds of future business. Ventures are small and embryonic. Many of these ventures will not survive into H2, but they are absolutely necessary.
- **Horizon 2 (H2)** is the emerging stars. Ventures here are fast and entrepreneurial. This is where growth comes from.
- **Horizon 1 (H1)** is the current business. Its products and services account for most cash flow and profit. However, their growth potential is typically low.
- **Horizon 0 (H0)** is the past business. Its products and services are in decline. Perhaps no longer profitable by themselves, they remain in the offerings because customers rely on them, and to discontinue them would jeopardize H1 products and services.

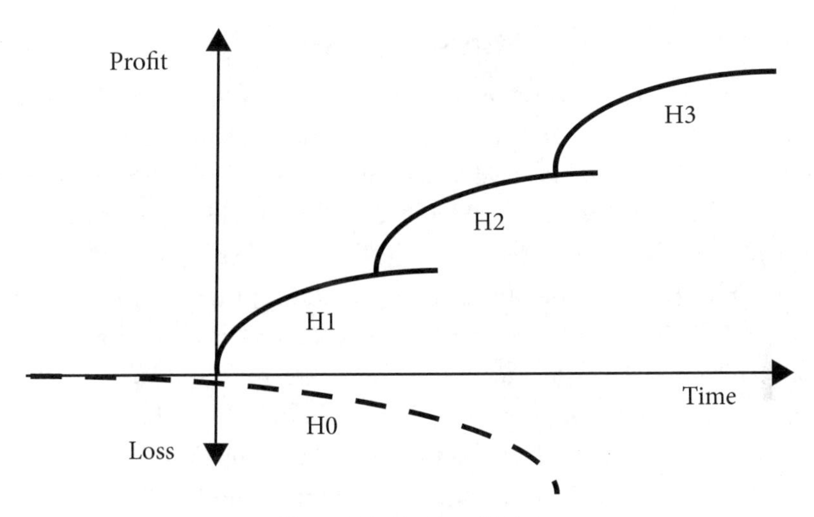

Figure 3-1 Horizons Model.

For example, manufacturers still carry parts for products no longer in production, sometimes for decades. Likewise, enterprise software is typically supported for three to eight years after it's no longer sold. Legacy products and services in H0 do eventually reach end of sales and end of life—but it can be a long tail.

I find it useful to distinguish H0 from H1 for two reasons. First, H0 products and services are not justified solely by their own profitability, but also by their effect on customer loyalty. Second, a large share of the Information budget in going concerns is for operation and maintenance of legacy assets that still deliver genuine business value while simultaneously acting as a drag on strategy.

At the same time an organization is planting seeds in H3, it should be cultivating sprouts in H2, harvesting fruit in H1, and eventually plowing things under in H0. In other words, H3 is for innovation, H2 for effectiveness, H1 for efficiency, and H0 for closure.

Strategy must orchestrate all these horizons in a coordinated manner because decisions affecting each horizon will affect the downstream horizons eventually. The original model used five years as the boundary between H2 and H3. However, in fast-paced industries, such as in the Information field, the actual boundary may be measured in months.

Blue Ocean

A Blue Ocean Strategy creates "uncontested marketspace that makes the competition irrelevant" [Kim, 2005]. For example, circuses were based heavily on animal acts until an entrepreneur decided to present human acrobats, street performers, and distinctive music instead. Old-style circuses couldn't match the novel appeal, and no other entrants were able to duplicate the innovative formula.

This strategy is such a beautiful concept that it's hard to argue against the logic. However, Blue Oceans are rare, and they are incredibly hard to create and sustain on a billion-dollar scale. Automotive companies, once held up as Blue Ocean examples due to their dominance of market segments, no longer have uncontested markets today. Even the best-known Consumer Information companies all have competitors, and they are elite members of the start-up survivor group. To see the full effect, you also need to consider the other 99% of start-ups that perished while navigating what they hoped would be their own Blue Oceans.

For most organizations, Strategy must contend with various shades of Red Oceans where competition ranges from notable to fierce. If you deliver only what customers say they want, you are following the market. To have a Blue Ocean, you must create something that customers don't know they want—and deliver it in a way that competitors can't match.

Profit Patterns

Profit Patterns are strategic changes that shift the business environment and thereby create winners and losers [Moser, 1999]. They are, in a sense, the opposite of Blue Ocean Strategy. BOS makes competition irrelevant. Profit Patterns take competition head on.

Profit Patterns are like winning strategies for game play: You need to know the winning moves, at least to counter them, if not to employ them. If you don't know which Profit Pattern is affecting your industry and your organization, you may not see competition coming in time to react.

Unfortunately, Profit Patterns are increasingly hard to see. It used to be that being the biggest firm in your industry conveyed many benefits: higher efficiency, better talent, customer respect, and so forth. Nowadays, size can be a liability if it means the firm isn't nimble because value migrates rapidly to where innovations are happening.

Here are some common Profit Patterns:

- **Profit Shift.** Once upon a time, most customers were profitable. Today, not all customers are. The bottom third may generate an outright loss. The middle third breaks even. And the upper third generates a profit. For the firm to be profitable overall, that upper third must more than overcome the loss on the bottom third.
- **Value Chain Squeeze.** Within many value chains, not every member necessarily benefits. Value can migrate so that a few members capture the bulk of the profit. Firms in the squeeze must strengthen their position, hop to another value chain, or participate in more than one chain to avoid getting squeezed into a no-profit position.
- **Technology Shifts.** This pattern occurs when technology shifts firms out of their usual positions. Emerging technologies are not just affecting technology vendors, they are affecting technology users. For example, even non-Information products are being fitted with sensors for remote monitoring, maintenance, and repair.

OODA Loop

OODA is an abbreviation of Observe, Orient, Decide, Act. The first three steps are strategy creation. The last step is strategy execution. Among those steps, the second one is arguably most important because it determines viable options.

Speed is the competitive advantage created by OODA. An organization that traverses its OODA loop faster than its competitors will be at least a step ahead if not an entire cycle ahead.

OODA loop originated as a military tactic. U.S. Air Force Colonel John Boyd famously used it to defeat other fighter pilots in less than a minute apiece. OODA was later incorporated into the strategy behind Desert Storm: Destroy the enemy's command and control, then use flanking maneuvers faster than the enemy can react.

In a nutshell, OODA exploits a competitor's mistakes; if your competitors can execute their own OODA loop faster than you can, watch out.

Alignment

Technical Strategy will be a recurring topic in following chapters, so alignment of Enterprise and Technical Strategies provides a context for topics to come. Alignment simply means making sure that whatever gets committed in the Enterprise Strategy is actually feasible in the Technical Strategy and whatever gets done in Technical Strategy directly benefits the Enterprise Strategy.

If the enterprise goal is to increase revenue by 10% within three years, for example, and the Enterprise Strategy is to start a new line of business, the Information infrastructure should be able to support that business. On the other hand, if the information systems aren't capable, and could not be made capable without considerable delay and additional investment, perhaps the business should think twice about pursuing its goal by starting a new line of business. Expanding a current line of business might be less risky, assuming the Information infrastructure is scalable.

At the highest level, Technical Strategy boils down to "make, buy, or sell?" That is, should the enterprise make or buy technology for its own use? And if it sells technology, should it make or buy what it sells? At an intermediate level, Technical Strategy must address multiple facets of Information, including hardware, software, data, network, skills, and projects. And at the lowest level, Technical Strategy concerns specific approaches to developing, purchasing, and selling—and which technologies to avoid or discontinue. Technical Strategy also depends on how much the business depends on technology. Is it an adjunct or core competency? Is it a commodity or specialized?

Strategy Traps

Table 3-1 summarizes Strategy traps [BCG, 2018]. The rows are strategy dimensions:

- **Options.** Existing alternatives versus new alternatives
- **Calibration.** Getting trapped in the past versus searching for better alternatives
- **Resources.** Misjudging current conditions versus overlooking opportunities

Table 3-1 Strategy Traps

		Warnings	Guides
Options	Existing	Narrowing options Incremental improvements Competitive pressure	Explore breadth and depth Open innovation Preemptive responses
	New	Unlimited options No path from current to future Declining innovation	Clear vision Plan for critical mass Able to stop experiments
Calibration	Trapped	Uncertainty in core business Maverick competitors Focus on past	Increased diversity of capabilities Definition of stopping rules Dynamism in talent and technology
	Searching	Low cost exploration Low return on R&D Competitors lead in technology	Feedback from sales to development Reward exploration and exploitation Track R&D contributions
Resources	Misjudged	Rising costs of new markets Loss from commercial operations Bold goals with cash flow	Assessment of competitive conditions Refocus on core operations Stress testing and contingency plans
	Unleveraged	Unexplored opportunities Competitors better positioned Neglecting breakthrough opportunities	Dedication of resources to exploration Role independent from current operations to spot future opportunities
Investment	Small bet	Ambitious goal with small investment Lack of progress on big problem Faith in iterative experimentation	Investment sufficient for big objectives Scalability Organization structure for follow-through
	Big bet	Entrenched attacks hard to repel Big projects consistently fail Faith in silver bullets	Selection from portfolio of initiatives Feedback looks that guide research Flexible organization
Learning	Fixed	Fixed resources in changing situation No noteworthy innovations Metrics and incentives from the past	Balance current and future business Culture of experimentation Learn alternative business models
	Wandering	Imitating competitors Incomplete lessons learned Repeating strategic options	Track lessons learned and financial outcomes Filter out unproductive exploration

Summarized from [BCG2018].

- **Investment.** Making small bets versus making big bets
- **Learning.** Having a fixed strategy versus wandering through strategies

The columns are indicators of traps:

- **Warnings.** Bad practices
- **Guides.** Best practices

For example, imitating competitors is a warning sign of a wandering strategy. On the other hand, fostering a culture of experimentation is a guide sign that the organization is breaking away from a fixed strategy.

Strategy traps also could be called strategy anti-patterns. In software engineering, patterns are effective techniques worthy of emulation rather than reinvention. Anti-patterns are thus ineffective techniques that should be avoided.

Strategy Principles

The problem with strategic plans is they foretell a future that may not exist if customers, competitors, employees, and regulators don't behave as expected. Stuff happens. Static Strategy means everyone works the plan even when faced with a changing environment. Dynamic Strategy means the enterprise pivots rapidly, as in the OODA Loop. It's impossible to plan for all contingencies due to combinatorics, so Sense and Respond is a practical alternative to Predict and Prepare.

Here are Strategy principles derived from the adventures and tools described earlier:

- **Alignment Principle.** Business and Technical Strategy must work together to achieve the enterprise goal, so they cannot be created and executed separately.
- **Blue Ocean Principle.** If you can create an uncontested marketspace that makes the competition irrelevant, do it.
- **Dynamic Strategy Principle.** If a fixed strategy cycle is too inflexible, make the strategy dynamic with tools such as the OODA Loop and Proofs of Concept.
- **Execution Principle.** Even the best of strategies will be tested by issues arising during execution, so make strong execution a priority.

- **Horizons Principle.** Strategy must cover the nascent business (H3), emerging business (H2), current business (H1), and past business (H0).
- **Innovation Principle.** Disrupters are innovators, but incumbents can be innovators, too.
- **OODA Loop Principle.** An organization that traverses its Observe, Orient, Decide, Act Loop faster than its competitors will be ahead of its competition.
- **Profit Patterns Principle.** Strategic changes shift the business environment and thereby create winners and losers.
- **Strategy Traps Principle.** Watch for warning signs of flawed strategy and guide signs of best practices.

Conclusion

Strategy is less about predicting the future and more about shaping it. Effective strategy execution moves innovations from research in Horizon 3 to development in H2 to business as usual in H1 to legacy status in H0.

Conventional Enterprise Strategy is a long loop to get feedback. But there are ways to get feedback from short loops. And it's easier to drop a concept after a short loop without upsetting the entire strategy because investment is low and commitments (product plans and sales plans) are still adjustable.

CHAPTER 4

INFORMATION TECHNOLOGY, COMPUTER SOFTWARE, AND TECHNICAL SERVICES

This book uses Information as the short name for everything in Information Technology, Computer Software, and Technical Services. That covers anything for gathering, organizing, summarizing, storing, retrieving, transmitting/receiving, and processing data, information, or knowledge:

- **Hardware.** Computers, storage, printers, scanners, cameras, tablets, phones, sensors
- **Software.** Operating systems, middleware, applications
- **Networks.** Voice, data, image, video
- **Services.** Professional, scientific, technical

This book also addresses devices and systems that aren't usually considered members of the Information field. They embed sensors, processors, and software. Or they connect to Information devices for diagnosis and maintenance. For example, aircraft, cars, trucks, trains, ships, elevators, navigation systems, parking meters, medical equipment, farm equipment, manufacturing machines, and utility meters are non-Information products that nonetheless depend on Information elements for operation and maintenance.

The word "system" is potentially confusable. An information system is a set of hardware, software, network, and data that work together. An organizational system is a set of people and things that work together. There isn't a viable substitute for "system," so context will have to indicate which type is meant.

Does Information Matter?

The central questions for this book are (1) what can Constraint Management teach Strategy, if anything, about buying, building, consuming, and managing Information and, (2) what can Strategy teach Constraint Management about Information constraints?

- Some say that Information doesn't matter: It's not a competitive differentiator anymore because it's a commodity [Carr, 2003]. The latest version of this commoditization argument says Cloud Computing enables organizations to rent Information capabilities as needed.
- Others say CXO jobs have never been more important because Information is a potential differentiator that evolves far faster than other technologies, it's an efficient way to reduce operating expense, and it's getting ever more complex [McFarlan, 2003].
- And then there's this: "Without software, twenty-first century science would be impossible" [*Science*, 2018].
- Even Constraint Management has something to say about Information: "It's a perpetual bottleneck [Cox, 2012]." Nobody gets everything they want, even though just about everything depends on Information, indirectly if not directly.

This book won't settle the debate because there is no universal answer. Information can be a differentiator, a barrier, or a commodity. It depends on what your circumstances are and what you do with Information. If Information is a commodity, you must manage your return on investment carefully. If Information is a barrier, you must consider alternative strategies. And if Information is a differentiator, you must figure out how to preserve and extend your competitive advantage. So, if you're responsible for Information, or you depend on Information, managing Information constraints should interest you.

ADVENTURE: **The Deep End**

My first job in the Information field was programming simulations on a supercomputer of its day. To say that it felt like I had stumbled in the deep end of the computing pool doesn't begin to describe how I felt. Today we have a name for it: Imposter Syndrome, which is the feeling that you don't know enough to justify having your job, because everyone around you seems to know so much more. And it's still quite common in the Information field among recent college graduates and for some people taking on new roles.

What I didn't realize is that there is so much to know that formal education, technical training, and job experience can't prepare you entirely. Nobody knows everything they need to know, except perhaps in some highly confined domains. A lot of what makes an Information professional effective is the ability to seek out new knowledge, figure things out, and learn from mistakes.

Porting, Refactoring, and Redesign

The simulation code had been written at another university and used to conduct experiments there. My job was to port the code and the data to my university's computer, then refactor the code (make internal changes that are not visible externally) and reorganize the data so that the experimental designs could be table-driven rather than hard-coded. That way, we could run multiple experiments without having to modify the simulator code every time, as had been done at the previous university.

Table-driven logic allowed one simulator version to support multiple experiments simultaneously, which was a lot more efficient. Not only had changing code been laborious, all code changes had to be retested, so it also meant that version control and retrofitting changes to previous versions were burdens.

Research Lab

My previous job, described in the Prologue, had been coordinating operations in a factory, plus making its planning system work. Now I was using that manufacturing experience to do experiments with simulated factories. Domain knowledge like this helps Information professionals figure things out when there's no subject matter expert around, which is why experience matters.

As the simulation experiments got under way, my job expanded to managing the research lab where other researchers were doing their experiments. As more people joined the research program, they needed mentoring to get up to speed on the technology and our methodology. All that preparation paid off when my team used the simulator for some award-winning research. At that point, the deep end didn't feel quite so deep.

Lessons Learned

There is no universal formula for getting started in the Information field. Nevertheless, time has proven that some of my lessons are closer to universal than I realized.

- Imposter Syndrome is normal. Senior professionals should help junior professionals get their feet on the ground before they start climbing the ladder.
- Business/government experience, and the domain knowledge that results, can be as important as Information field experience. That fact is sometimes lost on new recruits who were attracted to the field in the mistaken belief that they wouldn't have to learn about the enterprise's goal and how it functions in pursuit of that goal.
- Inheriting systems and data is common. Starting from a blank slate is uncommon.
- Maintenance and enhancement are valuable. New development is more prestigious, but it's not the only source of value.

The Role of Information

Information can play various roles. This can affect where the constraints are.

- **Information is the business.** Hardware manufacturers, software vendors, resellers, hosting, technical services
- **Information enables the core business.** All industries
- **Information is an adjunct to the core business.** Cloud hosting with spare capacity
- **Information is embedded in products and services.** Many types of equipment

Central Information versus Shadow Information

The centralization-decentralization debate is perpetual, and extends to more than just the Information field. For this book's purposes, however, it doesn't matter where Information management sits in the organization. All that matters is how Information contributes to achieving the enterprise's goal.

- Central Information is owned and operated by the Chief Information Officer.
- Shadow Information is owned and operated by CXOs or line-of-business executives.

The term "shadow" means operating outside enterprise-wide policies. That can expose the enterprise to legal issues, among other things. On the other hand, closer alignment of Information with business unit goals can be a good thing. Thus, some have suggested that Shadow Information will eventually just be called Business Information.

System Types

Information systems are created to serve various purposes:

- **Systems of Record** have been fixtures in business and government for decades. These systems gather and maintain the organization's data.

- **Systems of Insight** are a more recent development. These systems perform analytics to turn quantitative data into actionable information. Or if the data is qualitative (text, voice, image) a cognitive approach turns data into knowledge.
- **Systems of Engagement** are social networking for business.
- **Systems of Innovation** reinvent business processes, products, and services.

Technical Services

A service is something that someone else or something else does for you. "Someone else" means it's a labor-based service. "Something else" means it's a technology-based service. These service alternatives form a continuum from fully manual to fully automated. Professional, Scientific, and Technical Services are covered in depth in *Reaching the Goal* [Ricketts, 2008] and summarized in a later chapter of this book.

ADVENTURE: Distant Rumblings

My second Information job was leading a team that programmed decision support for a national government agency. Today, high-resolution color graphics for data analytics and visualization is commonplace, but back in the day, it was cutting edge.

The hardware cost hundreds of thousands in today's dollars, so it was too expensive to have multiple instances, and overseas data communications were expensive and nowhere near as fast and reliable as today, so we shipped the hardware around the world in crates to hot spots as needed. That presented a dilemma when there were problems in the field, because we had no system back home for replicating problems and testing solutions. It took a lot of imagination to diagnose problems on a computer over 6,000 miles away based on what someone described happening on screen.

Readers who have never known a world without cell phones may not appreciate what communications were like in a world with only land lines. International calls were especially unreliable and expensive.

Desert Trouble

One time I received a frantic call from Egypt. The system wouldn't boot up, and the problem appeared to be in the storage unit. Through a series of questions and answers with our local contact, we were able to determine that although the local power was 120 volts, it was 50 hertz (cycles per second), not the 60 hertz the hardware was designed for. Thus, the disk was rotating too slowly to read or write without error. We solved that problem by having a local machine shop machine a new hub for the disk drive. That was more of a mechanical engineering solution than something a computer scientist might think of, but it was a successful fix.

Jungle Trouble

Another time, I received an even more frantic call from the Philippines. Power was as expected, and all hardware diagnostics showed no problems, so it was probably a software problem. My challenge was determining whether the problem was in the system software, which we could not alter, or in the application software that my team wrote, because we might be able to fix that in the field. To make that determination, I had to talk the local technician through a cold start-up sequence that would isolate the problem.

What made this call challenging was the jungle location, which was so remote that part of the call was a radio link. The radio operator kept an eye on his wristwatch and thumped the table every 60 seconds. Our local technician had to put cash on the table to keep the radio link alive. Years later, I was chief technology officer of a worldwide technical support organization, but never again was there pressure quite like those early days when urgent calls came from a distant desert or jungle.

Lessons Learned

The Information field is not just hardware and software. Support is critical for maintenance and recovery.

"Five 9's" availability—99.999%—is an extremely high service level. It allows less than six minutes of unplanned downtime per year. For

Personal Computing, some downtime is generally not an issue. But for Enterprise Computing, such as financial exchanges and emergency services, even a few minutes feels like a lifetime.

Information Trends

Trends matter because they affect where Information constraints occur. The following sections describe these trends:

- **Dematerialization and Virtualization.** Physical products are becoming virtual
- **Commoditization and Value Migration.** Where profits have gone
- **Moore's Law, Brooks' Law, and Jevons Paradox.** Performance stimulates demand
- **Enterprise Computing versus Personal Computing.** Large scale versus small scale
- **Cloud Computing.** Information on demand
- **Machine Learning, Artificial Intelligence, and Cognitive Computing.** Smart Information
- **Internet of Things.** Sensors everywhere
- **Role Reversal.** The future will be technology-led and people-assisted

Dematerialization and Virtualization

Significant parts of our world are dematerializing. "Software operates in cyberspace, not Newtonian space [Saylor, 2012]." Here are some examples:

- A problem with icing in commercial aircraft engines was fixed, not by reengineering or repairing the engine, but by updating its software.
- When my own car displayed a drivetrain failure message, the fix was not parts replacement but a software update.
- Servers, storage, and networks in data centers are dematerializing into Virtual Machines plus Software Defined Storage and Software Defined Networks.

Also, consider this: Financial Services make up 20% of global gross domestic product, yet only 8% of currency exists as cash. Checks account for less than 2%. The other 90% of money—credit cards, debit cards, retirement accounts, mortgages, insurance cash value, electronic funds transfers—is just digital representations. Trillions of dollars have dematerialized. And cryptocurrencies never were material.

What does dematerialization have to do with Constraint Management? Dematerialized things—money, software, services—may not fit Constraint Management applications created for the material world. Software is much more malleable and reusable than physical products. Services are not managed the same way as physical products.

Commoditization and Value Migration

Strategic products and services from Information providers are meant to start as differentiators, but over time, they become commodities as trade secrets are reverse-engineered, patents expire, and innovations supersede older technology. Meanwhile, Information buyers justify their implementations with cost saving, revenue generation, and regulatory compliance.

Information profits have migrated from hardware products to labor-based services to technology-based services, so Information providers and buyers have reacted accordingly. Want hardware? It's available via Cloud Computing. Want software? It's available via Software as a Service. Want data? It's available via Data as a Service. Want network? It's available as Software-Defined Networks.

What do commoditization and value migration have to do with Constraint Management? Technology-based services are not constrained by physics as much as physical products are. And technology-based services are not constrained by human frailties as much as labor-based services are.

Moore's Law, Brooks' Law, and Jevons Paradox

The unit price for computer hardware used to fall by half every 12 to 18 months—and this continued for 50 years! Looked at another way, for a given price, computer performance doubled about every year and a half.

This phenomenon is called Moore's Law, and it meant that computing hardware did not exhibit the same cost behavior as other products. Computing power that cost over one million dollars in 1965 costs less than one dollar in 2015. Global Positioning System (GPS) receivers that originally cost $10,000 are now built into smartphones for a few dollars.

That said, Moore's Law often describes the smallest element of Enterprise Information cost because it has fallen decade after decade. Thus, its significance isn't quite what it was when first observed.

Hardware manufacturers increased chip densities, cranked up clock speeds, and added processor cores until the laws of physics intervened. So, unless scientists and engineers figure out new ways to make computer chips denser and faster, which is unlikely, Moore's Law is over. It's more likely that engineers will add more functions to create specialized chips, or they will design chips to include multiple "chiplets."

Computer software, however, does not and never has followed Moore's Law because software is handcrafted, not manufactured. The laws of physics aren't what govern software productivity. Software productivity is affected by many things, but it ultimately comes down to having smart people write code, and it's hard to make that happen faster every year. Indeed, Brooks' Law says adding people to a late software project makes it later.

Jevons Paradox notes that making any technology more efficient decreases the amount needed to satisfy current demand, yet demand increases. Hardware sales have borne that out consistently. However, software has increased its resource consumption as hardware price-performance has improved, thereby maintaining a rough balance between what hardware supplies and what software uses. Therefore, now that Moore's Law has effectively ended, the pressure is on developers to improve software performance, as well as their own productivity.

What do Moore's Law, Brooks' Law, and Jevons Paradox have to do with Constraint Management? Most Information devices are not 100% utilized. Far from it. Personal computers and smartphones sit idle about half of every workday, and completely idle overnight. Hardware has become so inexpensive that availability is more important than utilization.

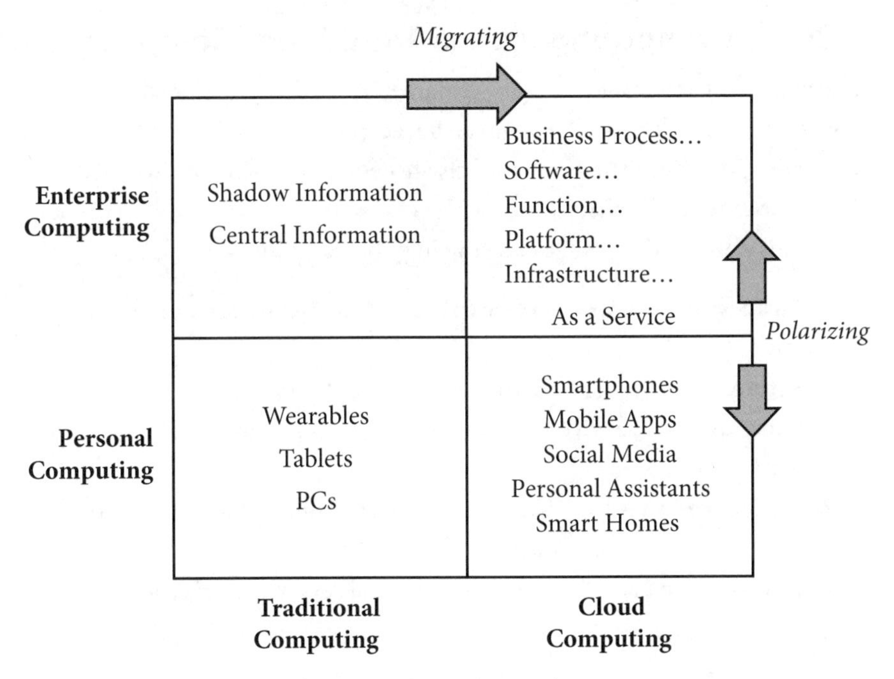

Figure 4-1 Computing quadrants.

Enterprise Computing versus Personal Computing

The business-to-business (B2B) and business-to-consumer (B2C) markets have been around for decades, but the Information market overall is polarizing, as illustrated in Figure 4-1. Enterprise Computing is large scale and getting larger, but hardware and software don't scale linearly. Personal Computing is small scale and getting smaller, so personal computers are being marginalized as smartphones, tablets, wearable devices, and virtual assistants take over tasks previously done with PCs.

What do Enterprise Computing and Personal Computing have to do with Constraint Management? The combination of Enterprise and Personal Computing creates additional challenges for Information managers. Regardless of whether the approach is "bring your own device" (BYOD) or "company-owned, personally enabled" (COPE), consumer products often need additional software to meet an enterprise's requirements for security, reliability, and compatibility.

Cloud Computing versus Traditional Computing

Fundamentally, Cloud Computing is virtual servers and virtual storage on a virtual network, such that all those elements can be reconfigured readily. Multi-tenancy is the illusion that those virtual components are dedicated when the physical devices they run on are actually shared.

Clouds can provide a range of technology-based services:

- **Business Process as a Service (BPaaS).** Complete processes such as "order to cash"
- **Software as a Service (SaaS).** Complete applications
- **Functions as a Service (FaaS).** Tiny programs running as though without a server
- **Platform as a Service (PaaS).** A virtual environment for application development and operation
- **Infrastructure as a Service (IaaS).** Virtual servers, storage, and network

Clouds can take various forms:

- **Public.** In a Cloud center, but for public use
- **Private.** In a Cloud center, but only for one company's use
- **Dedicated.** In a company's data center for its own use
- **Hybrid.** A mix of public, private, and dedicated

From the Information buyer's perspective, public Clouds are rented rather than owned, and the charges are for what's used (that is, "by the drink"). Thus, operating expense replaces capital expense. Furthermore, the provider manages the services its cloud provides, taking this responsibility off the buyer. Thus, the Ownership Economy based on product purchases is transitioning into a Consumption Economy based on rented software.

Cloud Computing can support both Personal and Enterprise Computing, though the methods are different. Personal Cloud Computing is consumer friendly. Enterprise Cloud Computing is business friendly. Beyond functionality and usability, what differentiates Cloud Computing from Traditional Computing is the business model: rent versus buy. As illustrated in Figure 4-1, some Traditional Computing is migrating to Cloud Computing, but that does not mean Traditional Computing will disappear any more than Personal Computing will disappear.

Reasons to keep Traditional Computing include:

- Security
- Response time
- Disaster recovery
- Regulations prohibiting cross-border data flow

What does Cloud Computing have to do with Constraint Management? Clouds are elastic. Servers, storage, and network can be scaled in or out on demand. For Information buyers in industries with widely variable demand patterns, such as retailing, not having to maintain capacity for peak demand is attractive. Thus, constraint capacity may be flexible rather than fixed.

Machine Learning, Artificial Intelligence, and Cognitive Computing

Systems that can behave in ways beyond their original programming include:

- **Machine Learning.** Discoveries from complex data (for example, facial or voice recognition)
- **Artificial Intelligence.** Perform tasks that usually require human intelligence (for example, speech recognition, image recognition, language translation)
- **Cognitive Computing.** Reason about a domain (for example, explain why a medical treatment might be appropriate for certain symptoms)

Such systems generate insights that cannot be gained from conventional Information systems.

What do Machine Learning, Artificial Intelligence, and Cognitive Computing have to do with Constraint Management? Machine Learning discovers underlying patterns in complex data sets. Artificial Intelligence is trained with large samples of speech, images, videos, language, etc. Cognitive Computing gains knowledge from large samples of text or from knowledge fed in directly from subject matter experts. Thus, these external sources can be a constraint apart from the computer systems themselves.

Internet of Things

The Internet of Things consists of sensors on devices in consumer and enterprise settings that send data to another location. For example, vibration sensors can detect a faulty bearing, thereby shutting down the machine, triggering an alert for the operator, and notifying the maintenance department. Or sensors on security cameras can alert security while storing video in the Cloud. Wearable devices, such as exercise monitors, and implantable devices, such as pacemakers, also fall in this category.

Internet-connected devices are easy to build, but those devices will be attacked by hackers unless they are designed and implemented with security features such as encryption and strong passwords. And an IoT device that has been compromised can be misused to attack non-IoT devices elsewhere on the Internet.

What does this have to do with Constraint Management? IoT devices are contributing to the tsunami of data generated every day, which puts demands on networks and storage.

Role Reversal

The last trend for discussion here is subtler than those discussed earlier. It doesn't have a catchy name because it's happening gradually and without fanfare. Organizations are shifting from being people-led and technology-assisted to being technology-led and people-assisted. Sounds simple enough, but the implications for job roles and employment are profound.

In the past, a manager would use technology, such as a spreadsheet, to make decisions. And workers would use technology, such as a computer numerical control (CNC) machine, to manufacture products.

In the future, technology will make routine decisions and notify someone when the inputs or the resulting decision are outside normal parameters. In business, automated stock trading already does this because human decision-makers are orders of magnitude too slow for high-frequency trading. In factories, robots and 3D printers will know how to build certain products based on their design specifications, without being programmed, but they will request human assistance when something unexpected happens, such as a design that exceeds the robot's range of motion or tolerances too tight for the 3D printer. In personal transportation, autonomous vehicles require human drivers to take over under a shrinking set of circumstances.

Information providers are enabling this trend. Systems Integrators used to pull together disparate hardware and software into an integrated solution for a business problem. Those providers are evolving into Services Integrators who build software that provides an application programming interface (API) to whoever wants to construct a solution. And those services may come and go far more rapidly than integrated systems did.

The remainder of this chapter covers some Information topics that aren't trends, but nevertheless could affect where Information constraints are found. Those topics include:

- Fantasy Information
- Technical Debt
- Obsolescence
- Legacy Systems
- Leapfrogging

Fantasy Information

Some things that used to be laborious are now automated. Some things that used to be unaffordable are now affordable. Some things that used to be impossible are now possible. Nevertheless, Information projects still fail when they exceed what's feasible.

There are boundaries. Not everything that can be imagined can be built for financial, staffing, technical, and regulatory reasons. Not everything that can be built will perform adequately, because some problems are truly hard. Not everything that performs adequately will produce the desired result, because systems are complex. Not everything that produces a desired result will actually solve a business problem. In other words, not everything that can be done with Information should be done.

Nonetheless, CIOs and CTOs often get more requests than can be accommodated because manager and consumer expectations for Information are high.

Dark humor for such circumstances goes like this: "Tell me what you want, and I'll show you how you can live without it (because you don't have the data, funds, people, or patience to make it succeed)."

Technical Debt

Technical debt is the price paid later for something done (or not done) today. For example, when a developer builds something quick and dirty today, the system may crash later when the load on it increases. Or a quick fix to one minor bug may create more severe bugs and a maintenance headache later. (Hardware and software problems are called bugs because a moth accidently disabled one of the earliest computers.)

System architecture is riddled with decisions about requirements affecting availability, performance, security, maintainability, etc. Systems of Record, Systems of Engagement, and Systems of Insight may be on different Information platforms and therefore require different skill sets. These complexities create opportunities for technical debt to accumulate.

Technical debt can vary widely in scope as well as depth. It may affect just one program, multiple programs within one system, multiple systems, or even multiple organizations. The more a design decision or an implementation is shared or reused, the wider the potential impact. For example, an ill-advised implementation in a library routine will have broader impact than a routine that's not shared.

Obsolescence

Obsolescence comes in several forms:

- **Technical obsolescence** happens when a better technology that performs the same function comes along.
- **Financial obsolescence** happens when another technology is significantly more affordable—but the functionality often is different.
- **Practical obsolescence** happens when the benefits exceed the cost to replace older technology with newer technology.

Switching cost means it's sometimes better to stick with an old technology that's good enough.

Note that Information buyers and Information providers may view support cost quite differently. The more back-level versions still in use, the longer the support tail for providers. On the other hand, upgrading technologies can be so expensive for buyers that they delay until they are several versions behind. However, this happens less with Cloud Computing when upgrades are handled automatically by the provider.

Legacy Systems

A conservative definition of Legacy Systems is that they are any older generation of technology. A radical definition of Legacy Systems is that they are anything running in production, even if built with modern technology. Regardless of which definition is adopted, the largest portion of Information budgets may go to keeping Legacy Systems running. This is because the business would collapse without them, because that's where transactions are processed and business rules are enforced.

Ironically, the more successful a technology, the more likely it is to remain in use until it tips past the point where it's easy to replace. There are several remedies:

- **Legacy replacement** rips out the old and inserts the new in a big bang. This can be like ripping off a bandage, but it's the shortest distance from old to new.
- **Legacy migration** slides out the old and slips in the new, piece by piece. An approach called the "strangler pattern" gradually and imperceptibly replaces Legacy Systems.
- **Legacy transformation** modernizes only the roughest edges of Legacy Systems. Rewriting the most defect-prone code and adding a few critical functions may be all it takes to breathe some new life into an old system.
- **Legacy inheritance** is reassignment of data from a Legacy System being retired. Shepherding data between systems can be as big a job as replacing old code, especially if the new system has a significantly different data model.

As new technologies become mainstream and then legacy, the variety of technologies in active use grows.

Leapfrogging

Leapfrogging is essentially the opposite of the Legacy Systems problem. Organizations that have not adopted a legacy technology can jump directly into a modern technology. This has happened with Cloud Computing, smartphones, and mobile apps in developing countries. But it's a one-time leap.

Once a new technology has been widely or deeply adopted, legacy status is inevitable unless that technology can be retired and replaced regularly. That's an enviable accomplishment by smartphone and video game vendors that has not been replicated to the same degree in other corners of the Information field.

Information Principles

This chapter has covered some Information principles, and there are more to come in the following chapters. The list here is not a complete list because the Information field is too broad for that. But it should be enough as a mental warm-up for topics to come.

Here are Information principles derived from the adventures and tools described in this chapter:

- **Commoditization Principle.** As Information products and services become commodities, value migrates.
- **Dematerialization Principle.** Physical products are becoming ever more reliant on software.
- **Legacy Principle.** Highly effective Information Systems outlive the technology they were built on/with.
- **Longevity Principle.** Highly effective Information Systems outlast their creators.
- **Polarization Principle.** Personal Computing and Enterprise Computing have disparate requirements, so those markets are diverging.
- **Role Reversal Principle.** Organizations are shifting from being people-led and technology-assisted to being technology-led and people-assisted.
- **Technical Debt Principle.** Technical decisions made now but paid for later tend to accumulate.

- **Variety Principle.** The leading edge of Information moves faster than the trailing edge, so technologies in use become more diverse.
- **Virtualization Principle.** Physical devices are being reconfigured into virtual devices.

Conclusion

What the Information field has accomplished in its relatively brief history is considerable. Within my own lifetime, a million-fold improvement in price-performance is unmatched by any other technology. Nothing else even comes close.

Hardware is especially prone to rapid commoditization, then obsolescence, while software and services can be long-lived. Technical Strategy therefore becomes a balancing act between the wish for things on the leading edge, the burden of things on the trailing edge, and the value of all things in between.

Information is a force multiplier because fewer people can do more when they have effective information. However, Information trends are toward much higher complexity. On that point, Peter G. Neumann warned, "Complex systems break in complex ways [Neumann, 2012]."

CHAPTER 5

CONSTRAINT MANAGEMENT

Theory of Constraints was named by Eli Goldratt, the scientist turned management guru who started it. However, the "theory" is generally misunderstood outside of academic circles, so this book refers to TOC by its practical nickname: Constraint Management.

To get started, consider this scenario. A system produces items in five sequential steps. The number of items steps #1 through #5 could do per hour are 11, 12, 10, 13, and 12, respectively. Utilization of those steps (portion of time they are doing work) is 100%, 95%, 110%, 70%, and 80%. There are items waiting ahead of most steps, but the amount ahead of step #3 is huge and a substantial portion of orders are completed late.

How would you improve what this system produces? A conventional approach is to push more work into the system in order to drive more utilization at steps #2, #4, and #5.

In contrast, the Constraint Management approach is to manage the system based on its constraint, which is the step that determines what the system overall can produce. In this scenario, it's step #3. That step has over 100% utilization because it's beyond normal capacity. Step #1 has 100% utilization only because no other step is impeding it. However, steps #1 and #2 produce more than #3; that's why so much work is backed up ahead of step #3. And because the constraint can't keep up with orders amongst the chaos, some orders are late.

Thus, to manage the constraint, it's necessary to do something that defies conventional wisdom: Accept lower utilization on non-constraints to maximize utilization on the constraint. The only way to get this system to produce more is to get the constraint to do more by reducing its overload. Although this is a simple scenario, the same principles apply to real systems.

Weakest Link Analogy

Every system has a constraint. Otherwise, enterprises could grow without bounds.

Constraint Management was invented to manage systems that consist of a series of dependent events: A precedes B, which precedes C, and so forth. That's the way factories and supply chains work. In factories, raw materials are turned into parts, which are turned into subassemblies, which are turned into finished goods. In supply chains, products flow from manufacturer to wholesaler to retailer to consumer or from manufacturer to distributor to business customers.

Like the weakest link in a chain, systems consisting of a series of dependent events usually have one element that cannot produce as much as the other elements. Constraint Management recognizes that constraint as an operational control point. If managers and workers focus their attention on getting the most out of that constraint, the rest of the system will not require constant attention.

Constraint Management can be applied to systems with interdependent events (A, B, and C all depend on each other) and independent events (A, B, and C do not depend on each other), but the chain analogy doesn't fit as readily. In the latter system types, events happen in parallel, the sequence isn't fixed, and roadblocks and inspiration can occur anytime. These are complex adaptive systems if they learn from experience.

Here are some Information field examples:

- **Dependent events.** Manufacturing computers or constructing data centers are serial tasks.
- **Interdependent events.** Requirements, design, coding, and testing are related tasks, but not as serial as project plans portray them.
- **Independent events.** Calls handled by a help desk are usually unrelated unless they are in reaction to a wide outage (common cause).

Flowing Water Analogy

To further illustrate core concepts behind Constraint Management, here's an often-used analogy. Imagine a series of tanks arranged so water flows by gravity, without pressurization, from one tank to the next. The pipes connecting consecutive tanks have different diameters. The constraint is the narrowest pipe.

Picture someone pouring buckets of water into the first tank. No matter how much water pours into the first tank—and it comes in spurts—what flows out the other end of this system is whatever the pipe with the narrowest diameter passes through. The levels in some upstream tanks will rise when input to the first tank exceeds the constraint and water backs up. Levels in tanks downstream will empty occasionally because they can shed more water than the constraint supplies them.

All water in tanks ahead of the narrowest pipe is the constraint buffer. If the upstream tanks aren't empty, a disruption doesn't immediately affect the constraint, hence the buffering effect. But if the buffer runs dry, the constraint runs dry, too. In order to maximize what the system produces, that buffer should not run dry.

To increase overall flow through the system, the narrowest pipe must be widened. Increasing the width of any other pipe has no effect on overall flow. And trying to maximize flow through all of the pipes—100% utilization—will just overflow the upstream tanks because the constraint cannot handle that much flow.

This tank-and-pipe system is analogous to a factory because both are dependent-event systems. Some of the Information field is like that. For instance, running a series of computer programs to process data in pipeline fashion resembles a factory. The Information field even calls that "production."

Attempts to create software factories failed, however, because software is developed in a complex adaptive system with a mix of dependent, interdependent, and independent events. As noted above, things happen in parallel, the sequence isn't fixed, and roadblocks and inspiration can occur anytime. Just finding the constraint can be difficult.

A better analogy for Information is a sports team. Players practice and compete together. They obey rules or suffer penalties. Coaches help them act as a team by following a playbook. They have a shared goal, and many things happen at once. Where is the constraint in a sports team? There must be one, because teams don't win every game. That's the question we will be answering for the Information field in the following chapters of this book.

If we already know that standard Constraint Management solutions aren't always a perfect fit for the Information field, why bother trying to manage Information constraints? Focus. If we can find constraints instead of trying to control everything, managing those constraints will improve the organization's performance.

ADVENTURE: New Territory

When I was a professional development executive in a large technology services company, the executive I worked for was intrigued by Constraint Management. He handed me a copy of *The Goal* [Goldratt, 2014]. After a pause, he said, “Why can’t we implement its lessons here?” After reading the book, I was intrigued too, but I had misgivings.

My experience in both Manufacturing and Services meant I saw a chasm between those business models that seemed too wide to bridge. For instance, Constraint Management is great at managing physical inventory, but there is no inventory in labor-based services. You can’t deliver a service before a customer shows up. The closest analog in Manufacturing is just-in-time delivery, but even that requires physical inventory.

Getting Our Feet Wet

When my workmates and I met Eli Goldratt in person, he showed us various Constraint Management solutions for product-based businesses, but none of them seemed to apply to services. In our labor-based services business, we had issues with chronic overstaffing when sales forecasts over-predicted actual demand. But Constraint Management solutions said nothing about staffing, apart from being sure to have enough operators for the factory constraint.

So, at Eli’s urging, we put a couple dozen of our executives and senior staff through Constraint Management training, which includes a thinking process. At the end of that course, we decided to tackle our staffing problem with a solution called Replenishment. It didn’t feel like a good fit, but it was the best we could find as a starting point.

Gathering the Faithful

To expand our acquaintance with Constraint Management experts, we hosted the Founder’s Conference for the professional organization. We did not, however, become a member ourselves or pursue certification because we weren't offering Constraint Management services.

Law of the Hammer

Rather than build a Constraint Management solution ourselves, we investigated a consultancy that also offered software. As we explained our problem, the partner kept assuring us that their software was the solution. But it became apparent that his consulting focused on Manufacturing and Distribution, and he didn't really understand large-scale Technical Services.

The harder we pushed back, the more he reasserted his position, and he warned us against trying to implement a solution ourselves. After a polite goodbye, we were reminded of the Law of the Hammer, which says, "If all you have is a hammer, the entire world looks like a nail."

Adapting for Services

With the training, conference, and consultant taken care of, we turned our attention back to the internal issue that led us to investigate Constraint Management in the first place. The dilemma we faced was that a solution designed to manage physical inventory can't manage staffing in a labor-based services business for one simple reason: goods flow predominantly one way through a supply chain, but people always flow both ways in a services business. That is, people are assigned to a task or project, and when their work is complete, they return for assignment to another task or project. Because the flow is circular, the constraint must be managed differently.

Having seen that Constraint Management was a hard sell externally, we took a different tack internally. To get buy-in, we created a simulator and used it to demonstrate alternative staffing approaches. Managers could plug in their own decisions and see that Constraint Management produced better results. In effect, they were proving it to themselves. The solution did staffing on demand rather than to forecasts, which was radical at the time. Details are explained in *Reaching the Goal* [Ricketts,

2008]. Within a few months, overstaffing subsided. That solution is still in use many years later.

Lessons Learned

Lessons learned include:

- Getting started with Constraint Management was worthwhile, though painful at times.
- Experts can suffer tunnel vision.
- Simulation can be a useful R&D tool as well as an adjunct to a live implementation, but effective simulations are hard to get right the first time. We had to throw the first one away and start over with just a weekend to recover.
- Getting buy-in is harder when you're convincing someone to be a pioneer rather than a follower.
- Implementing Constraint Management delivers tangible business value.

The Goal

Constraint Management begins and ends with the goal. What are we trying to achieve? Are we achieving it?

Traditionally, Constraint Management has assumed that the goal of for-profit organizations is to make money now and in the future. This is a condensed version of the Horizons Model. It doesn't distinguish H2 from H3, or H0 from H1, but recognizing the now-versus-future duality of goals is essential when reconciling operational productivity against strategic positioning.

Furthermore, an organization can make money more than one way. Turning a profit on products or services is the most obvious method. But stock appreciation is another method common with technology start-ups and disrupters in other industries. Making money from intellectual capital is yet another method.

The goal for non-profit and government organizations may be defined in "goal units" now and future. For instance, a public clinic might have the goal of treating 100 patients per day this year and 150 in three years.

Constraint Management works well when the goal is clear, but not as well when the goal is ambiguous or conflicted. Thus, a prerequisite for Constraint Management is getting agreement on the goal.

Moreover, the goal—to make money or achieve goal units—depends on a host of necessary conditions, sometimes called critical success factors. Employees, customers, business partners, owners, and other stakeholders all need to feel that they are being treated fairly. For instance, optimizing share price via cheaper labor will be a failed strategy when critical skills are lost. If any stakeholders feel they're not in a win-win relationship, the goal won't be achievable eventually, if not immediately. Thus, critical success factors contribute to achieving the goal.

Even Eli Goldratt himself in his later years questioned whether the goal is really to make money. If living a good life is the ultimate goal, then making money is just one of several necessary conditions.

Systems

The word "system" is potentially confusable. An information system is a set of hardware, software, network, and data that work together. An organizational system is a set of people and things that work together. Of course, organizational systems use information systems. There isn't a viable substitute for the word "system," so context will have to indicate which type is meant.

A system has properties that its elements don't have. For example, you can fly on an airplane, but not on its parts. Thus, a system is not the sum of its elements, it's the sum of their interactions.

The boundaries of a system affect where the constraint is. Therefore, separate observers may not agree on the boundaries or the constraints. For example, aircraft, airlines, airfreight, airports, airmail, and air traffic control are different systems with different boundaries and constraints even though they share some elements and operate in the same industry.

For Constraint Management purposes, system boundaries are based on flow because a constraint restricts flow. Materials and parts flow through manufacturing. Products flow through sales and procurement. Dollars flow through accounting and finance. But the flow may span organizational boundaries by crossing departments or by crossing between organizations. In other words, system boundaries and organizational boundaries don't always coincide, which is a source of conflict.

Viewed at the executive level, an organization is a system of systems driven by subordinate goals. If the subordinate goals conflict with each other or if they don't align with the enterprise goal, that's a problem because the systems may oscillate back and forth as each strives to achieve a different goal. Thus, goal setting and conflict resolution are big parts of an executive's job.

Flows

Systems create flows. Constraint Management strives for balanced flow. It does this by modulating the system so that its overall workload does not overload the constraint. With balanced flow, a disruption anywhere except the constraint may stay local, while the system overall continues to function, despite this instability.

In contrast, Lean Manufacturing strives for balanced capacity. It does this by constructing a system so that every element has the same capacity. With balanced capacity, however, a disruption anywhere becomes a disruption everywhere. Thus, balanced capacity is only practical in stable systems.

Velocity is the rate of flow. Performance improvement means increasing velocity by giving the constraint more capacity or by reducing overload on current capacity. However, attempts to improve performance sometimes have unintended consequences. For example, when lanes are added to a congested highway, traffic often slows down due to induced demand. That is, the increase in capacity attracts an even larger increase in traffic. In business, however, an increase in demand is usually welcomed.

Just as a system has two goals, now and future, systems have two flows. The current-value flow is described above. The strategic-initiatives flow is changes that will establish the future system. Those initiatives can include performance improvement, as well as new products, marketing campaigns, and so forth.

The Information Constraints part of this book will cover Information flows. But it's worth noting here that those flows can have characteristics distinct from physical flows. For example, data can be transmitted around the world in less than a second, but transporting physical products takes much longer.

Limits

In common usage, a constraint is anything that limits what a system can produce. In Constraint Management, however, a limit and a constraint are not the same. A limit does indeed restrict what a system produces, but a limit cannot be used to manage the system. For example, if a company has a no-overtime policy, its factory will produce only during the day shift. How much it produces during that shift is not governed by that limit, so the system cannot be effectively managed via overtime policy.

A constraint, on the other hand, not only restricts what a system can produce, it also can be used to manage the system because it's a productive element of that system. Suppose we find the least-productive machine in a factory. Making sure that machine always has work will maximize what the factory overall produces. If that machine runs out of work or is down for repair, it effectively causes time to be lost to the whole factory. Thus, how much the system produces is governed by that constraint, and it provides a control point for management.

Why make a distinction between limits and constraints? Managers frequently must remove internal limits before they get to a constraint they can manage. For example, if removing the no-overtime policy increases daily capacity to the point that the lowest-producing machine then is an active constraint, Constraint Management can optimize production. Otherwise, overtime policy remains the limiting factor.

Note, however, that many external limits have been established to prevent the unbridled pursuit of profit from endangering health, safety, and economics. Medical, environmental, and financial regulations are responses to sickness, death, and abuse. Regulations do limit what organizations produce, but they also prevent undesirable consequences. Therefore, some limits are beneficial and should not be evaded.

Unlike constraints in a factory, which are physical, limits are often intangible. For example, rules, procedures, regulations, laws, and policies may all impose limits on a system. But some constraints are intangible too. For instance, the constraint in a project is the longest subset of all tasks.

At one time, Constraint Management considered policies to be constraints, and they are discussed that way in *Reaching the Goal* [Ricketts, 2008]. However,

Goldratt rescinded that position. Therefore, the certification organization now does not consider policies to be constraints, at least not for operations management purposes.

Constraints

Constraint Management didn't invent constraints, of course. Liebig's Law of the Minimum originated in agriculture, and it says crop growth is controlled not by the total amount of resources available, but by the scarcest resource. Constraint Management took that concept, applied it to Manufacturing and Distribution, and used it to improve productivity.

If the constraint restricts what a system can produce, it's an internal constraint. If the market will not buy all that a system can produce, it's an external constraint. However, if customers aren't buying because the sales process is broken or understaffed, that's still an internal constraint. Likewise, if suppliers aren't shipping because the procurement process is broken or understaffed, that's also an internal constraint.

A natural constraint—one that happens by chance rather than by design—can occur anywhere in a system. When searching for the natural constraint, however, it pays to examine convergence points and integration points.

The Heat Treat department is often a convergence point because parts from multiple departments cycle through there repeatedly. Multiple inflows to a single department are likely to use all its capacity and create a backlog of work in process.

Final Assembly is an integration point because parts from multiple departments come together, literally. In the Information field, doing daily software builds and integration testing is roughly analogous to final assembly in manufacturing.

A guiding principle is to position the constraint where it supports the strategy because a strategic constraint leads to better goal achievement than an arbitrary constraint. Positioning might require acquisition of additional capacity to elevate the current constraint until the intended element becomes the new constraint.

Although Constraint Management says a system generally has only one operations constraint at a time, it acknowledges that the constraint can change over time. At the operations level, for instance, one system can have an internal constraint during peak times and an external constraint during slack times. At the strategic level, the current constraint can be different from the future constraints, which is consistent with the Horizons Model. Ideally, however, the constraint does not move unexpectedly.

If each system has one constraint at a time, then most of the elements in a system are non-constraints. That means they have capacity to spare, which can be used to sprint. Sprinting is used to catch up after an outage elsewhere in the system. Near-constraints are least able to sprint because they have the least spare capacity above what the constraint itself has.

Systems within an organization have their own local constraints, most of which aren't the enterprise constraint. But those local constraints can be of interest to someone working to improve the performance of those systems. They may not limit what the enterprise produces, but they do consume resources that might be an unnecessary drag. However, local optimizations do not add up to global optimization.

Buffers

A buffer decouples elements of a system so that a disruption in one does not necessarily affect another. In a production system, a typical place for a buffer is ahead of the constraint. That constraint buffer holds enough work to keep the constraint busy while repairs are made upstream, if necessary, but it has a second purpose. Buffer management happens when the constraint buffer level triggers a controlled release of new work into the system so that the system doesn't get choked with work in process.

Similarly, finished goods inventory is a buffer between a factory and the market for its goods. New orders for stock products can be fulfilled immediately from that inventory, and when the inventory falls low enough, a new job is released into manufacturing to replenish it.

Buffers come in several types:

- **Stock buffer.** The buffers described above are stock buffers because they contain inventory units of one kind.
- **Space buffer.** A physical area after the constraint to hold inventory of any kind. When the space is full, the constraint stops producing because there is no place to put its output.

- **Time buffer.** Time allowed for contingencies. Typical location for a time buffer is in a project plan or in production lead time for a make-to-order product.
- **Cash buffer.** Funds allowed for contingencies.
- **Skill buffer.** Human capabilities that are not readily available on the job market can be hired and trained in advance of demand.
- **Capacity buffer.** Spare capacity on the internal constraint keeps an external constraint from emerging when there is a spike in demand.

Buffering is a familiar concept in the Information field because it's been used throughout human history to decouple devices that work at different speeds. For example, computers are faster than storage, which is faster than networks, which are faster than people. A buffer keeps the faster device from flooding the slower one or the slower device from starving the faster one. In the Information field, such buffers may be called a cache or a staging table, but they are still buffers.

Constraint Management Solutions

With an appreciation for the goal, systems, flows, limits, constraints, and buffers, we have the tools to understand standard Constraint Management solutions. The original versions were designed for Manufacturing and Supply Chains, but extended versions are available for Services.

Constraint Management solutions answer these questions:

- **Production.** How can production be optimized? How can a service or a business process be optimized?
- **Distribution.** How can inventory in a supply chain be optimized? How can staffing in services be optimized?
- **Projects.** How can we finish faster and more predictably? How can we do more projects with the same resources?

As you read the following sections, be aware that whole books have been written about these topics. This book just covers the highlights, not the details of each subject.

Production

It sounds counterintuitive, but the route to higher production overall is lower production everywhere it is not needed. Consider this scenario for an internal constraint. Work is pushed into the factory to get high utilization on every machine, but because the constraint can't handle the load, work in process backs up everywhere and orders are filled late. However, when buffer management is applied, and the release of new jobs is choked, work in process goes down and productivity goes up, as measured by orders filled on time. In other words, work is pulled through the factory by demand.

The scenario just described is one where excess work in process is impeding production. However, if work in process has been "over-Leaned," there is effectively no buffer, which means every disruption anywhere in the system likely wastes time on the constraint. In that scenario, building a buffer (increasing work in process) is necessary to make the constraint an effective internal control point.

If the constraint is external, in the market, finished goods inventory becomes the constraint buffer, but buffer management otherwise works the same way. This scenario is increasingly common as excess capacity has accumulated worldwide. As much as 70% of manufacturers do not have an internal production constraint, yet buffer management is still valuable because it prevents overproduction.

When producing goods, buffer management triggers the release of jobs into production when the buffer level falls below a designated threshold.

The buffer has three zones:

- Green means no action.
- Yellow means prepare for action.
- Red means refill the buffer.

For services, the production solution works differently. Because there is no inventory in services, there is no constraint buffer of work in process. Instead, service levels are used to adjust capacity to match demand. For example, during peak periods in a call center, as wait time rises, more agents can be brought in to handle calls—or inbound calls can be diverted to a call-back queue, which smooths demand. During slack periods, when there are fewer calls to answer, some agents can be assigned other tasks or sent home.

When delivering services, Buffer Management triggers capacity changes when the buffer level falls or rises beyond thresholds.

Again, the buffer has several zones:

- High red means cut capacity.
- High yellow means prepare for action.
- Green means no action.
- Low yellow means prepare for action.
- Low red means add capacity.

Figure 5-1 is a vignette comparing conventional and Constraint Management approaches to production of goods and services:

- **Conventional Production of Goods** has relatively balanced capacity and pushes materials into production to achieve utilization, which results in a disorderly process because work-in-process accumulates. Nobody knows or cares about the constraint. Work is pushed through the system.
- **Constraint Management Production of Goods** has unbalanced capacity and pulls materials into production based on demand, which results in an orderly flow with minimal work in process. The constraint is the narrowest region in the pipeline. One stock buffer is ahead of the constraint. Another holds finished goods. Work is pulled through the system by buffer levels and due dates.
- **Conventional Production of Services** plans capacity and pushes service requests into the process, which results in variable service levels and under- or overstaffing. Nobody knows or cares where the constraint is. Work is pushed through the process.
- **Constraint Management Production of Services** uses service levels to adjust capacity on demand, which pulls service requests through the process and results in consistent service levels without staffing issues. The constraint is the narrowest point in the pipeline. Capacity is adjusted to attain target service levels. The skill buffer enables flexible capacity. Work is pulled through the process.

Can the production solution be used for Information work? Yes. In addition to call centers, it can be used to manage business processes and their Information

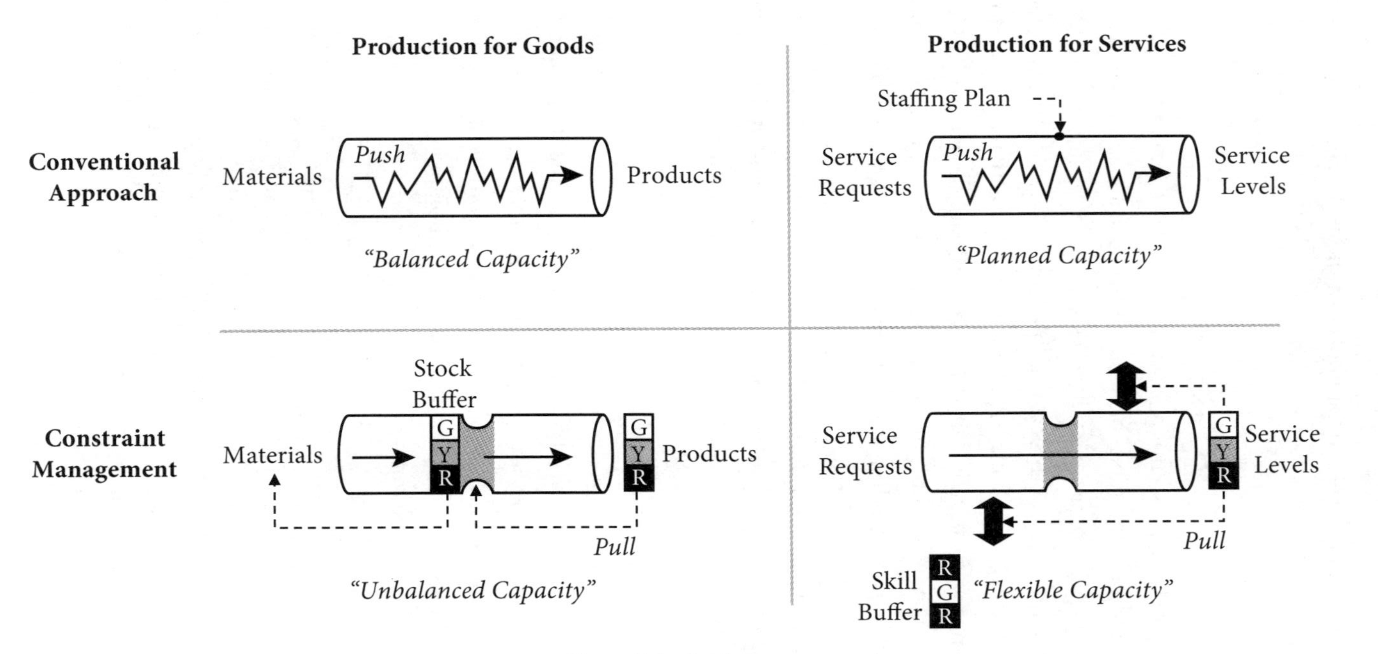

Figure 5-1 Production.

support. And changes to Information support can move the constraint to a more strategic location in the process.

Distribution

The Constraint Management solution for distribution also runs against conventional wisdom. Traditionally, goods have been pushed as far and as fast as possible through supply chains because retail stores are where consumers buy. Stock is ordered in fixed economic quantities at variable reorder times dictated by sales.

The distribution solution, called Replenishment, turns that around. Stock is ordered at fixed intervals—usually daily—in variable quantities based on that day's sales. Shipping costs may be a little higher for frequent shipments of small quantities, but the savings from avoiding stockouts and overstocks more than compensate.

Just enough stock for a few days' sales is held at each retail location, and the rest is held in a central warehouse because aggregated demand at the warehouse is much smoother than sales at any retail store. Stock flows only to stores actually making sales.

Thus, the traditional approach pushes inventory through the supply chain, but old merchandise eventually must be discounted to make way for new merchandise. The Replenishment approach pulls inventory through the supply chain, and because it only flows where sales are being made, retail stores do less discounting.

When distributing goods, Buffer Management triggers reordering of inventory when the buffer level falls below a designated threshold.

A typical buffer has three zones:

- Green means no action.
- Yellow means prepare for action.
- Red means replenish.

For services, the Replenishment solution works differently because the flow is workers, not goods. Unlike the flow of goods, which is generally one way, the flow of workers is circular. They come back from assignments when their work is complete to get reassigned. Staffing is therefore adjusted up or down based on demand for services.

The staff buffer has several zones:

- High red means reduce staff, preferably through attrition or by halting overtime.
- High yellow means prepare for action.
- Green means no action.
- Low yellow means prepare for action.
- Low red means increase staff through overtime or hiring from the job market.

Figure 5-2 is a vignette comparing conventional and Constraint Management approaches to distribution of goods and services:

- **Conventional Distribution of Goods** pushes inventory to retail outlets, which results in overstocks and stock-outs.
- **Constraint Management Distribution of Goods** holds inventory centrally and replenishes on demand, which prevents overstocks and stock-outs. The stock buffer level triggers replenishment based on aggregate demand.
- **Conventional Distribution of Services** hires to plan and pushes people onto projects, which may result in under- or overstaffing.
- **Constraint Management Distribution of Services** hires on demand according to the pull for resources by project, which prevents over- and understaffing. The skill buffer level triggers capacity increase or decrease based on demand.

Can Replenishment be used for Information work? Yes. The original version works for distribution of Information products and the extended version works for Information staffing.

Projects

The Constraint Management solution for projects, Critical Chain, has a similar name to an older method, Critical Path. But they are quite different because Critical Chain is designed to overcome the shortcomings of Critical Path that condemn projects to finish late and over budget.

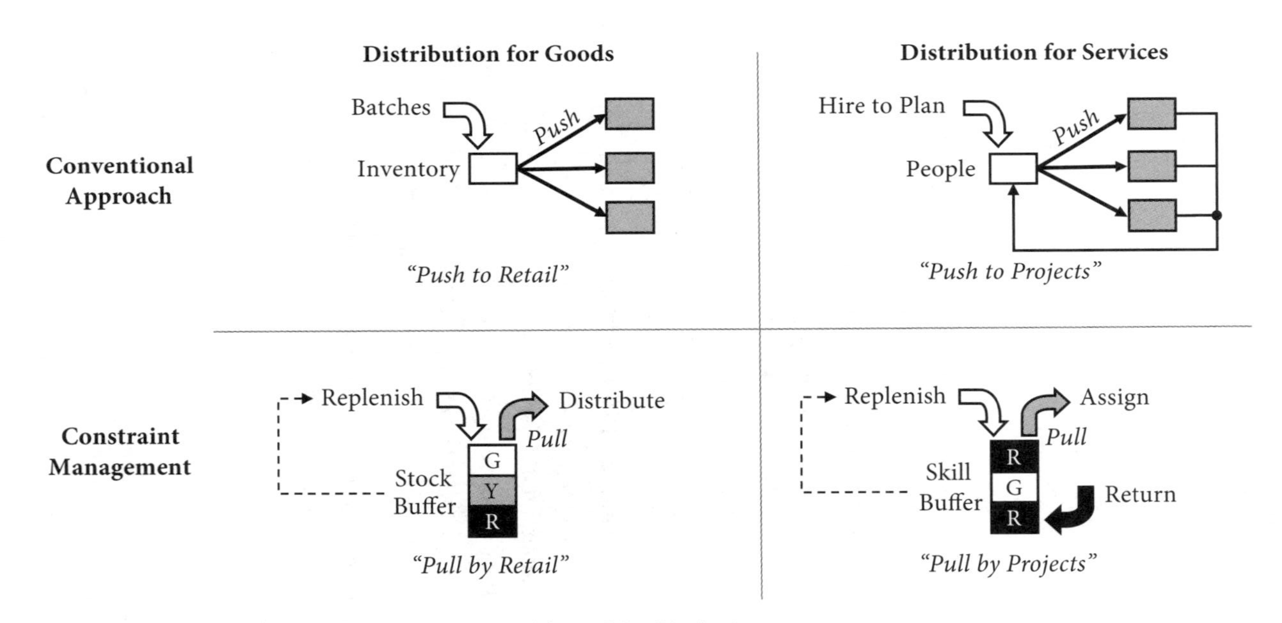

Figure 5-2 Distribution.

A project is a set of tasks that deliver a specific outcome, such as engineering a new product or constructing a bridge. Tasks are arranged in the plan by precedence: If task A produces an outcome that B depends on, A precedes B in the plan. Thus, both methods start with the same tasks in the same order. But how projects are estimated, executed, and measured is different.

A Critical Path task is estimated with 90% confidence that it will be completed on time, while a Critical Chain task is estimated with only 50% confidence. Thus, Critical Path estimates embed contingency, while Critical Chain task estimates without contingency are much shorter for the same scope of work. However, Critical Chain places contingency into a time buffer at the end where it protects the entire project rather than individual tasks.

The Critical Chain and Critical Path are both the longest path through their respective networks of tasks. But because the tasks are estimated differently, the Critical Chain may not include the same subset of tasks that the Critical Path does. Because Critical Chain needs less contingency to have 90% confidence in on-time project completion, Critical Chain plans generally have shorter overall duration.

Critical Chain estimates tasks differently because it executes projects differently. Critical Path not only allows multitasking, it is too often encouraged. Because anyone can work on more than one task at once and early task completions are discouraged, late task completions are common, and they accumulate. On the other hand, Critical Chain enforces the Relay Race work rule: Each person works on just one task at a time, and they hand it off as soon as possible to receive the next task. Although late task completions do occur, they do not accumulate as readily because they are offset by early task completions.

Critical Path measures progress by the percent of tasks completed, which is misleading because only tasks on the critical path determine on-time project completion. Critical Chain measures penetration of the project buffer.

The buffer has three zones:

- Green means no action.
- Yellow means prepare for action.
- Red means expedite whichever tasks are causing the buffer penetration.

For example, if 50% of the scheduled duration has passed and late tasks have only penetrated 30% into the project buffer, it's in the green zone. But if buffer penetration is 70%, it's in the red zone.

Figure 5-3 is a vignette comparing Critical Path and Critical Chain project management methods:

- **Critical Path** embeds contingency in every task, tolerates multi-tasking, and pushes for on-time task completion on the assumption that local optimization adds up to on-time project completion. Nevertheless, CP projects are frequently late. The commitment is to complete the project by the end date on the last task, so any late task completion on the critical path endangers the entire project.
- **Critical Chain** puts contingency in the project buffer, enforces Relay Race work rules, counts on some early task completions to offset late task completion, monitors project buffer penetration, and thereby pulls the entire project to on-time completion with shorter overall duration. The commitment is to complete the project by the date at the end of the project time

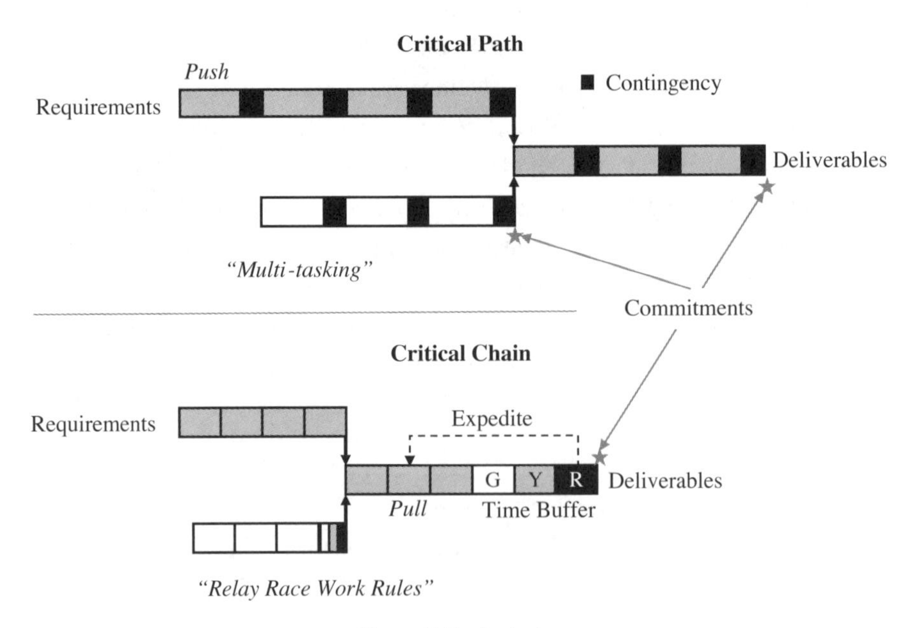

Figure 5-3 Projects.

buffer, not the last task, so late task completions are not necessarily cause for alarm. When the buffer level falls into the red zone, however, expediting selected tasks seeks to get back on schedule.

Can Critical Chain be used for Information projects? Yes. Anywhere a project might be managed with Critical Path Method, it can be managed instead with Critical Chain. For example, constructing a data center or delivering an Information consulting project can be done with Critical Chain. This does not mean, however, that every Information project is a candidate for Critical Chain. But that's a topic for later chapters. Moreover, most organizations do multiple projects at once, so that is also another topic for later.

Sense and Respond

Constraint Management solutions exemplify Sense and Respond instead of Predict and Prepare. In other words, Constraint Management works on demand instead of to forecast.

Conventional wisdom says decision-makers should seek more-accurate forecasts. Constraint Management offers an alternative that eliminates forecasts altogether.

The problem with forecasting is many models will fit historical data, but only one of them predicts the future. Finding that one model is always an elusive objective. That's why, instead of managing to a forecast, Constraint Management solutions use current data to react rapidly when a buffer level strays into a red zone.

Focusing Steps

The Focusing Steps drive performance improvement:

1. **Identify** the constraint—if it's not in the right place, move it.
2. **Exploit** the constraint—ensure it is not inhibited.
3. **Subordinate** everything else—don't let anything overload the constraint.
4. **Elevate** the constraint—increase its capacity.
5. **Repeat** these steps—don't stall after each improvement.

It is important to perform these steps in order. The natural impulse is to jump ahead to step #4, Elevation, but that just amplifies the waste and confusion already in the system. Step #3, Subordination, drains the waste and settles the confusion first.

Constraint Management recognizes that every improvement is a change, but not every change is an improvement. Thus, improving a non-constraint does little or nothing to help an enterprise reach its goal.

Natural constraints are not always in a good place. Moving the constraint means increasing an arbitrary constraint's capacity while decreasing the intended constraint's capacity until it becomes a strategic constraint you can manage.

Unfortunately, the Focusing Steps do not distinguish natural constraints from strategic constraints. That subject is covered later. When in doubt, however, an internal constraint should not be optimized to the point that it jeopardizes the market, which will quickly emerge as the enterprise constraint if customers are neglected.

Conflict Resolution

Performance improvement can lead to conflict, and so can differences in system goals. For example, an Information products team wants to sell as quickly as possible to minimize cost of sales, but an Information services team wants to bill as many hours as possible because that's what generates services revenue. Even though their goals are different, both products and services are needed to solve the customer's problem, and both contribute to the enterprise goal.

Conflicts are good if they challenge people to find and correct false assumptions. They are bad if people do not accept logic and evidence, and thereby remain trapped in the conflict.

Constraint Management has a conflict diagramming technique called an Evaporating Cloud, or just a "cloud" for short. As seen in a previous chapter, Cloud Computing enables capabilities to be acquired as needed, without necessarily buying hardware or software—and those are also called a "cloud." Note, however, that these are totally different concepts that coincidently share the same short name. To avoid confusion, this book uses "cloud" only to refer to Cloud Computing.

Constraint Management has guidelines for how to deal with conflict—and how not to:

- **Dominance** is a common method. Someone gets what they want while others get substantially less, if anything. Of course, this invites attacks on the dominant position, so conflicts persist. This is the win-lose scenario, also called a zero-sum game because existing assets or benefits are reallocated.
- **Compromise** is another common method. Everybody gets something they want, but nobody gets everything they want. That means nobody's entirely satisfied, so the conflict will likely resurface. This is the lose-lose scenario, also called a negative-sum game because existing assets or benefits are reduced.
- **Resolution** is the preferred method. Everybody gets exactly what they need—or better—even if they couldn't see it in the beginning. That means everybody is satisfied and should be less prone to future conflict. This is the win-win scenario, also called a positive-sum game because new assets or benefits are generated.
- **No deal** is another possibility. The best alternative to a negotiated agreement may be no loss and no gain, or just opportunity cost instead of overt loss.

The "game" label comes from Game Theory, which applies math and psychology to strategy. Constraint Management favors conflict resolution, of course. This is accomplished through the buy-in process.

Buy-in

Too many attempts to motivate others fail because they begin with a solution rather than the problem. Hence, the early steps of the buy-in process are critical:

1. Agree on problem.
2. Agree on direction of solution.
3. Agree that solution solves problem.
4. Agree that solution will not lead to any significant negative effects.
5. Agree on the way to overcome any obstacles that block implementation.
6. Agree to implement.

For non-trivial problems, the buy-in process takes more than one meeting over more than one day. It takes that long for participants to think through the steps, so rushing it doesn't help.

The buy-in process includes some counterintuitive aspects. For example, at first look, encouraging people to identify negative effects seems to be the opposite of gaining buy-in. However, when people offer up their own methods for overcoming negative effects, their buy-in becomes stronger than if those negative effects had not been surfaced.

Each step in the buy-in process has a corresponding level of resistance:

1. Disagreement about the problem.
2. Disagreement about the direction of the solution.
3. Lack of faith that the solution will yield significant results.
4. There are too many side effects.
5. The solution is too hard to implement.
6. Unspoken fear.

Of course, resistance occurs before acceptance, so it can guide the buy-in process.

Strategy

The Constraint Management approach to Strategy is to figure out what a customer needs and fulfill that need in a manner that makes competition irrelevant. This is accomplished by reducing each customer's limitations by helping them manage their own constraints. For example, by implementing solutions for production, distribution, or projects, an organization can promise faster, more reliable delivery—and perhaps earn a premium price for it.

Thus, Constraint Management strives to present an "unrefusable offer" to customers—and an "unfathomable strategy" to competitors. An unfathomable strategy happens when competitors remain convinced that nobody else can really deliver faster or more reliably than they already do, never mind get a higher price.

Strategy and Tactics Trees lay out the steps to implement a solution. Standard S&T Trees are undifferentiated, so they must be customized for each organization.

There are issues with every approach to Strategy, and the S&T method is no exception:

- S&T Trees are an organizing tool, not an idea-generation tool.
- Executing a strategic initiative can take years.
- Some strategic decisions do not depend on dramatic process improvement.
- There are no alternatives or decisions in S&T Trees.
- A strategic initiative that's not flexible is vulnerable to unforeseen events.

The S&T method has proven effective in capital-intensive industries, especially where capacity changes happen in relatively large increments and competitors' relatively slow reaction time opens a sustainable first-mover competitive advantage. Building another factory or opening another store are not small decisions. Even adding one large machine to a factory or one major department to a store can require significant investment and planning.

The S&T method is harder to apply in organizations that are knowledge-based, technology-enabled, or innovation-driven because competitors react rapidly. Capacity changes can happen in modest increments and fast-follower responses make first-mover advantages hard to sustain. For example, mobile phone apps have proliferated because they are readily matched by competitors. Consequently, an unrefusable offer and unfathomable strategy won't fly when customers already have fast, reliable delivery and competitors are quick to catch on to innovations.

There is no standard S&T Tree for the Information field because there is no Constraint Management solution specifically for it. However, if the Information field uses any of the standard solutions, their S&T Trees can be followed.

Technology

The Constraint Management perspective on technology is it's beneficial if it diminishes a constraint or limit. Not all technology does that, however, so Constraint Management poses these questions [Goldratt, 2000]:

1. What is the power of the technology?
2. What current constraint or limit does the technology eliminate or vastly reduce?

3. What policies, norms, measurements, and behaviors are used today to bypass the previously mentioned constraint or limit?
4. What policies, norms, and behaviors should be used once the technology is in place?
5. Do the new rules require any change in the way we use the technology?
6. How can the change be brought about?

By evaluating technology based on its ability to diminish constraints or limits, Constraint Management reminds us that the Information field sometimes gets more complicated than it needs to be. If Information is supporting local optimization of systems rather than global optimization of the enterprise constraint, it may be pursuing cost reduction while neglecting revenue generation and strategic positioning.

Decision-making

When making operational decisions, such as evaluating a new product that would consume capacity on an internal constraint, Constraint Management uses Throughput Accounting. *Reaching the Goal* summarizes it for goods, services, and software [Ricketts, 2008].

When making strategic decisions, such as acquiring another company, Constraint Management uses Throughput Economics to compare alternatives [Schragenheim, 2019]. An appendix to this book explores it for the Information field.

Here are the fundamentals:

- **Revenue:** R = money generated by sales, interest, royalties, etc.
- **Totally Variable Costs:** TVC = money spent on materials and parts
- **Throughput:** T = R – TVC
- **Investment:** I = money spent on land, buildings, machinery, inventory, and technology
- **Operating Expense:** OE = money spent turning I into T
- **Net Profit:** NP = T – OE
- **Return on Investment:** ROI = NP / I

To be viable, a proposed technology must move the needle on NP or ROI. That in turn requires improvement in T, I, or OE.

Product cost, a staple of Cost Accounting, is prohibited in Throughput Accounting because the allocation of TVC distorts decision-making. As described in the Prologue, products that are actually profitable may appear unprofitable when product costs are compared.

Cost cutting, another staple of Cost Accounting, is not favored in Throughput Accounting because once TVC is removed, OE typically does not vary with production or sales volumes. Furthermore, attempts to cut costs too often endanger the constraint. Throughput Accounting therefore favors increasing T instead of decreasing OE.

Principles

A good way to appreciate Constraint Management is to recognize the principles embodied in its solutions. Here are Constraint Management principles derived from the adventures and solutions explained earlier:

- **Aggregation Principle.** Inventory and resources are best held centrally because that is where consumption varies least.
- **Buffer Principle.** Strategically placed units or time for contingencies minimize the spread of disruptions across an entire system.
- **Capacity Principle.** Unbalanced capacity is more productive because the constraint can be used to manage the entire system.
- **Chain Principle.** A system can produce only as much as its weakest link, the constraint, will allow.
- **Conflict Principle.** Win-win solutions are not only possible, they are the best way to prevent conflicts from recurring.
- **Decision-making Principle.** The effect of a proposed change can be assessed by comparing its net profit and return on investment to the baseline.
- **Demand Principle.** Forecasts are usually wrong, but reacting to actual demand is usually right.
- **Flow Principle.** Managing the constraint controls flow through an entire system.

- **Focus Principle.** An entire system can be managed by focusing on the constraint.
- **Multitasking Principle.** Trying to concentrate on more than one task at once decreases performance on all of those tasks.
- **Optimization Principle.** Local optimizations do not add up to global optimization.
- **Pull Principle.** Rather than pushing work through a system, setting up procedures that pull work increase the total amount of work done.
- **Relay Race Principle.** Instead of multitasking, completing one task before moving on to another accelerates projects.
- **Replenishment Principle.** Restocking from a central location leads to fewer stockouts and overstocks than pushing stock to retail locations.
- **Strategy and Tactics Principle.** The sequence of tasks necessary to implement a strategic initiative can be drawn in a tree diagram.
- **Technology Principle.** Technology can reduce a limitation if correctly applied.
- **Time Principle.** Time lost on the constraint is time lost to the entire system.
- **Utilization Principle.** The only utilization that matters is on the constraint.

Here are principles that form the Focusing Steps for performance improvement:

- **Goal Principle.** Setting an unambiguous goal can be considered step zero.
- **Constraint Principle.** The first step is to identify the constraint.
- **Exploitation Principle.** The second step is to use the constraint to its fullest.
- **Subordination Principle.** The third step is to ensure that non-constraints do not generate extra work for the constraint.
- **Elevation Principle.** The fourth step is to increase the constraint's capacity.
- **Improvement Principle.** The fifth step is to repeat the previous steps to generate continuous improvement.

Information Constraints

Can Information be the enterprise constraint? Yes. When an enterprise gets its revenue from hardware, software, network, or services, and it can't deliver what customers want to buy, it may have an Information constraint. That's common in start-ups, but not as much in established Information providers. They may simply have a constraint elsewhere in the business, such as sales or procurement.

If Information isn't the core business, Information probably isn't the enterprise constraint because it is not what restricts what the firm can produce. For instance, a mining company can still extract ore and an airline can still fly planes when their payroll systems are down.

Of course, Information can be a chronic problem. And if your area of responsibility includes Information, you may be intensely interested in where the local Information constraint is because managing it is a necessary condition for getting the most out of your enterprise constraint. For instance, if a mining company cannot extract ore or an airline cannot fly planes because their workforce scheduling systems are down, those Information problems are wasting time on the enterprise constraint.

Nevertheless, a growing number of non-Information products and services rely so heavily on Information that it is literally an integral component of the core business. When Information is down, the business is down. Warehouses increasingly rely on hardware in the form of cameras, sensors, and robots—as much or more than human workers. Much of the value in automobiles today is software—not steel, aluminum, glass, and rubber. Most of the world's money is in digital data—not coins and paper currency. In industries like these, Information may not be the enterprise constraint, but it surely makes for a bad day when it goes sideways.

The following section of this book will further explore Information constraints and strategies for:

- Hardware
- Software
- Data
- Knowledge
- Networks
- Architecture

- Skills
- Methodology
- Projects
- Processes
- Portfolios
- Services

Although Information may not be the enterprise constraint, each of those domains may harbor a local constraint that executives and managers could use to improve Information performance.

Conclusion

Constraint Management deals well with uncertainty and conflict, but it cannot deal with disorder and ambiguity. If a system is irretrievably chaotic, Constraint Management won't stop the chaos, and it can't manage a system where the constraint hops frequently. Furthermore, if the system exhibits goal ambiguity, that must be sorted out before Constraint Management can improve performance. That's why standard Constraint Management solutions are more often used by established organizations than by start-ups—unless their offering is products or services for Constraint Management.

Constraint Management solutions are used to manage current constraints, which correspond to H1 and H0 in the Horizons Model. Strategy is for managing future constraints: H2 and H3. Both are necessary for organizations to survive and thrive.

Flow constraints pertain to just one system. Strategic constraints pertain to a system of systems. With Constraint Management, there's less micromanagement and more strategizing.

Constraint Management gets buy-in by showing how changes targeted at solving core problems eliminate multiple pain points. In that way, Constraint Management gets leverage by making modest changes with big benefits.

PART 2

TECHNOLOGY

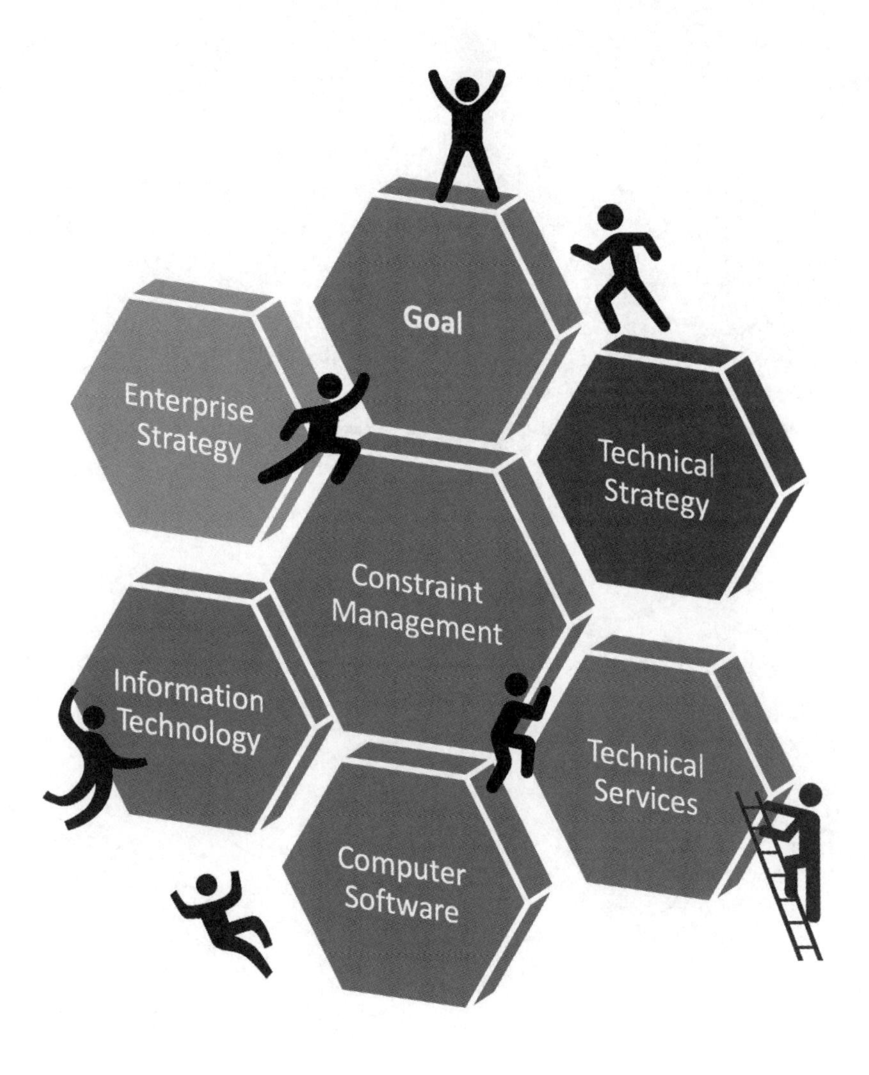

CHAPTER 6

HARDWARE

Of all the chapters about the Information field, hardware is the best place to start examining Constraint Management because it's a perfect fit. Hardware providers do engineering, manufacturing, and distribution of products, which are Constraint Management's original domains.

Hardware also requires installation, operation, maintenance, and repair, which are Technical Services, not products. Fortunately, there are extended Constraint Management solutions for services, too.

How buyers acquire hardware has changed. Moore's Law and commoditization of products has put downward pressure on prices, so the market has polarized toward Personal Computing and Enterprise Computing. But Shadow Information means hardware buyers are scattered across organizations. And Cloud Computing means hardware ownership is not the only option.

What is hardware? It might seem that hardware is physical products, and that's not far off. But hardware runs a wide spectrum of software, and some of that software is either built into the hardware or it's written in close coordination with the hardware manufacturer because the hardware and software are tightly coupled. From a user perspective, that software is mostly if not entirely invisible, so it warrants some explanation.

Layers

Computing systems are composed of layers, described next from top to bottom. In those layers, hardware is at the bottom, so be patient.

- **Applications.** Software for users or developers.
  - Software for users includes word processing, spreadsheets, presentations, accounting, scheduling, email, etc. Some user applications are

standard packages from software vendors, while other applications are custom software just for the organization that creates it.

  - Software for developers includes editors for writing code, compilers and interpreters that convert human-readable source code into machine-executable code, and development environments that also enable diagramming and testing. Developers use this software to write applications for users and software at lower layers.

- **Middleware.** Software that provides common services to application software, such as a Database Management System, a Transaction Processing System, Message Queuing, or a video game engine. ("Middle" means between applications and the operating system.)
- **Operating System.** Software that manages hardware and software resources, such as input/output, main memory, and storage. For example, when you create a file, the OS determines where and how that file is stored.
- **Hypervisor** (also called the virtualization layer). Software that abstracts physical memory, storage, and processors so that they act as virtual devices to layers above. For example, multiple virtual machines can run simultaneously and independently on a single physical machine.
- **Firmware.** Software in read-only memory, which is persistent even when powered off. It makes hardware components work together to accomplish the system design.
- **Hardware.** Physical devices, such as central processing units, graphic processing units, input/output processors, memory chips, servers, storage, network, printers, scanners, etc.

In addition to developing the hardware and firmware, the hardware team may have a hand in developing software at the other levels. Or that software may be developed independently.

ADVENTURE: **Disappearing Act**

When I led a team that was developing data visualization applications for a government agency, all work suddenly ceased because sections of source code had begun disappearing as programmers were editing it. That's right. Code was suddenly and irretrievably gone right in front of their eyes. It was as though the computer had turned into an evil magician.

The programs looked normal, except a few hundred characters starting and ending in the middle of lines would be replaced by random characters. The location of those blocks was itself random. The time between occurrences was also random.

No Clues

Of course, we set about replicating the problem. Who was doing what when it first occurred? Same question for the second occurrence. And for the third. The pattern that emerged was . . . there was no pattern.

Hence, we proceeded to investigate potential causes:

- Hardware problem? No. Diagnostics showed no faults.
- System software problem? No. System tested good.
- Environmental problem? Static electricity? No. Overheating? No.
- Power problem? No. 120V, 60Hz.
- Water problem? No leaks. Humidity normal.
- Data problem? No. The issue happened during development, not execution.
- Storage problem? No. Plenty of space available.
- Network problem? No. Not network connected.
- Bug in the development environment? No. Problem occurred even when idle.
- Human error? No. Problem occurred even when no one had hands on keyboard.
- Encryption run wild? No. Nothing on the system did encryption.

- Did restoration from backup solve the problem? Nope. Still random.
- Someone sabotaging the project? No. I saw the problem happen with my own eyes.

Superstitions

In vexing situations like this, it's human nature to grasp for explanations. As the team brainstormed, some of the hypotheticals bordered on superstition. Maybe the hardware was worn out and having sporadic failures, except when the diagnostics were running. Maybe the diagnostics themselves were corrupted. Maybe the keyboard was defective and sending random strings. Maybe it was an electromagnetic pulse from the physics building. Maybe the code was self-destructing, like *Mission: Impossible*. Maybe we were losing our minds. The truth, however, was simple.

Lucky Break

Sometimes it pays to take a break. As I ambled down the hall while lost in thought to get a cup of coffee, through an office door slightly ajar I glimpsed a space heater. Full stop. Reverse. Knock, knock. Is this office on the same circuit as our development lab? Oh, yes.

First step, kick on the heater and watch the voltage sag cause memory refresh cycles to be missed, which converted code into random characters at random locations over random intervals. Second step, confiscate the heater. Final step, upgrade power conditioning in the lab.

Problem solved. Everyone, back to work.

Lessons Learned

Lessons learned include:

- When you think you have eliminated all potential causes of faults, circle back.
- Look outside the walls of your area, both figuratively and literally.
- Don't give in to superstitious explanations.

This latter point has served me especially well when working with people outside the Information field. Rather than describe their computer problem, they are likely to begin with a superstitious explanation of why it won't do what they want it to do or why it does something they didn't expect. It often takes patience to separate the core problem from symptoms.

Performance

Personal Computing is familiar because it's ubiquitous, not just in our personal lives, but also at work. Enterprise Computing, on the other hand, is largely unseen because it's literally locked away in computing centers. Therefore, here are some differences that may not be obvious to folks outside Enterprise Computing:

- Enterprise Computing is the opposite of Personal Computing because enterprise solutions are intentionally impersonal.
- Eighty percent of the world's corporate data originates or resides on enterprise-grade technology.
- Enterprise technology typically lasts 20 years, while personal technology is technically obsolete within about 2 years.
- Enterprise technology gets a lot of service because so much depends on it, while personal technology manufacturers hope to minimize repairs.
- Enterprise technology is built for higher availability, and service providers pay penalties when it does not meet the Service-Level Agreement.
- Enterprise technology is built for high performance, sometimes with specialty hardware, while personal technology is more often built from commodity hardware.

Availability

Availability is determined by:

- **Reliability.** Mean Time Between Failures (MTBF)
- **Maintainability.** Mean Time to Repair (MTTR)

High reliability is achieved with redundant components and hot failover, which means when a particular hardware unit fails, another takes over immediately and without disruption. In a triple-redundant system, the support team isn't notified until the second component fails and the system is running without redundancy.

Availability matters more than utilization in contexts where response time matters. Police, fire, and ambulance services are classic examples.

Availability also matters when outages have severe impact. For example, when the U.S. Internal Revenue Service implemented new hardware for master files, the old system could not communicate with the new hardware, and the entire system went down on Tax Day. Because many tax returns are filed online at the last moment, the IRS had to extend the deadline.

Compatibility

Backward compatibility has been a distinguishing feature of Enterprise Computing for decades. It means you can run software written for an old computer on a new computer because the manufacturer designed it to work that way. Enterprises have such large software portfolios that it would be prohibitively expensive to rewrite and retest all that software when upgrading the hardware that it runs on.

Backward compatibility is not as common on consumer products such as personal computers, which may not run the latest operating system or applications. Backward compatibility is even shorter on smartphones, which get automatic software updates for only a few years.

Security

Much of the concern about security—or lack thereof—has historically focused on software. However, the discovery of hardware bugs has shifted attention because hardware bugs are difficult, if not impossible, to fix once manufacturing is complete. One type of bug could allow an attacker to read decrypted data that they should not be able to access during processing. More nefarious than an accidental bug is the intentional insertion of tiny spy chips that compromise server security.

Only 2% of data in commercial data centers is encrypted. Breaches of personal data succeed because only about 4% is encrypted. However, the most robust enterprise computers can encrypt data both at rest (in memory or storage) and in motion (moving between processors, memory, and storage), so this is a solvable problem.

Storage

Manufacturers create many types of storage devices because enterprises store data in a hierarchy that trades cost, capacity, speed, and volatility at each level. For example, main memory, also called RAM for random-access memory, is very expensive, very small, very fast, and very volatile. When power is lost, so is the data. That's why you restart your PC after a power outage.

Solid-state drives (SSD) and hard disk drives (HDD) are cheaper, larger, slower, and non-volatile. The data is still there when the power comes back. Magnetic tapes are much cheaper, much larger, much slower, non-volatile—and easily stored at another site as a backup against disaster. Thus, the less-frequently data is needed, the further down that hierarchy it is transferred. Even though users don't see it, tape is still the dominant storage medium for backing up data on the Cloud.

Smartphones and Cloud Computing

Smartphones are displacing personal computers, and Cloud Computing is challenging Traditional Computing. That's not to say PCs are disappearing, but the smartphone market is up, while the PC market is down. Likewise, Traditional Computing isn't disappearing, but selected applications are being migrated to Cloud.

Another difference between smartphones and PCs is the design of processor chips. In a PC, a multi-core chip is likely to have several identical cores, while a smartphone is likely to have cores with different capabilities. Some of those cores are for heavy workloads, which deplete the battery. Others are for light workloads, which conserve the battery.

Virtual Hardware

Cloud Computing relies on virtualization. That trend covers all types of hardware:

- **Virtual Memory** makes main memory seem bigger than it really is by moving pages between physical memory and temporary storage elsewhere, such as on an SSD.
- **Virtual Machines** make an entire server seem like it is multiple computers.
- **Server-less Apps** have just the bare minimum of software interfaces to run. The hardware is essentially invisible because the Cloud provider manages it.
- **Software Defined Storage** manages data independent of the actual hardware.
- **Software Defined Network** enables rapid reconfiguration of communications.

Robots

Robots do highly repetitive, dangerous tasks on assembly lines, such as welding and painting. There are over one million industrial robots worldwide, and they are growing by 100,000 per year. That partly explains why manufacturing revenue is up while employment is down. Robots also do repetitive tasks in data centers by retrieving, mounting, dismounting, and storing tape cartridges.

Not all robots are physical, however. Robotic Process Automation performs business processes, as well as Information operations. For example, RPA is software that can watch the operators' consoles in data centers and act to avoid outages. This allows human operators to focus on alerts that RPA doesn't recognize or know how to handle. And in residences, personal assistant devices control lights, music, and security based on spoken instructions. Thus, even some robots are dematerializing.

Internet of Things

The Internet of Things (IoT) refers to hardware with sensors and actuators that engage in machine-to-machine (M2M) communication. For instance, pumps, valves, meters, gauges, thermostats, and lights can be IoT devices.

IoT devices can also have compute power. Processors smaller than a grain of salt and costing less than 10 cents to manufacture are headed out of research and into production. They don't have a lot of processing power by today's standards, but their diminutive size and miniscule cost mean they will show up in single-use objects like shipping labels that track temperature and boarding passes that track location.

Time

Time on a human scale we take for granted. A few minutes here or there often makes no difference. But time on a hardware scale is different. A one-second difference in Global Positioning can be fatal because it translates to about a third of a mile or half a kilometer in positional error. A one-second difference in stock trading is worth over a hundred million dollars.

Hardware generally uses Coordinated Universal Time (UTC), which is the same worldwide. Then programmers use library routines to convert UTC to local time because the rules for time zones are complex. Some zones, for instance, differ by 30 or 45 minutes, not 60 minutes. Moreover, China recognizes a single time zone even though geographically it spans 5 time zones, while France has a dozen time zones, and the United States has 11. Furthermore, the start and end dates for Daylight Saving Time are not consistent even within countries.

Remote Control

Remote control is an alternative to fully autonomous vehicles. R&D is underway for drone ships. With no sailors on board, those ships will be operated from a central control room that looks like a ship's bridge.

When I worked in a strategy group, our software group offered a data analytics solution that ran on personal computers. As data analytics turned into big data analytics, the software solution was constrained by the memory and processing power of PCs. Even a top-of-the-line PC just wasn't enough anymore because the volume of data was growing exponentially. Therefore, it was getting harder to find needles in haystacks and disambiguate apparent duplicates.

Some of our customers were considering another vendor's solution, which they claimed could handle twice as much data. In response, I funded a proof of concept that would demonstrate within a few months that we could more than meet that volume.

We're Gonna Need a Bigger Boat

During the POC project, however, we could not port the entire solution to midrange computing hardware because it relied on the graphics capabilities of PCs. What we did do, however, was rearchitect the solution so that the user experience stayed on the PC while the back-end processing was off-loaded to scalable midrange hardware that ran database software and analytics software on storage tuned to this kind of big data analytics. Customers could run their analyses on shared hardware or acquire their own.

The POC project was successful. We demonstrated that our minimum hardware configuration could handle not just twice the original volume, but 10 times our competitor's volume, and still return results in seconds. Six months later we completed a product release that could handle thousands of times the original volume.

Lessons Learned

The workload on successful IT solutions grows. Sometimes it grows faster than the original hardware platform capability. Fortunately, if the problem can't be contained, the solution can be scaled up or out.

Hardware can be the key to rapid scalability. One method is scaling out (horizontally) by adding more devices like the original, which is typical in Cloud Computing. The method we used in this adventure was scaling up (vertically) by moving back-end workload to an Enterprise Computing platform. We made that architectural decision because it was expeditious, highly scalable, and it leveraged the existing front-end, so no additional user training was required.

The proof-of-concept project saved the day. Without it, our initial demonstration would have merely met rather than exceeded the goal.

Research and Development

Production of hardware begins with Research and Development. Pushing the frontiers of chip design requires materials research because silicon has already been pushed about as far as it can go. Computer-aided design (CAD) tools specifically for computer hardware are used to validate, optimize, and test design alternatives. Unfortunately, hardware can have design defects as well as manufacturing defects. By the time hardware defects come to light, they may have been present in millions of devices for years. As Atkin's Law of Demonstrations reminds us, "When hardware is working perfectly, the really important visitors don't show up."

Once the functions are right and sufficiently reliable, the design must be tweaked so devices can be manufactured and serviced. Serviceability is more important in Enterprise Computing than in Personal Computing because enterprise devices can last for decades, while consumer devices are more likely to be replaced than repaired.

While hardware is manufactured in highly repetitive processes, software is developed in processes that are only slightly repetitive. The tasks may be the same, but the execution of those steps is different for each software product. That is, the purpose of hardware manufacturing is to churn out many identical units, but the purpose of software development is to churn out only unique products, which can then be copied.

Manufacturing

High-tech Manufacturing is one of 18 industries in the Manufacturing sector. In addition to computers, it manufactures storage, printers, scanners, cameras, sensors, and network elements, such as fiber optic cables. It also manufactures chips and circuit boards that go into non-IT products, such as automobiles and aircraft.

High-Tech Manufacturing is different because it relies so much on technology to build technology. Chip fabrication plants, for instance, use hermetically sealed containers for silicon wafers because the tolerances are so tight that contamination is likely even when workers wear masks and bunny suits. Moreover, High-Tech Manufacturing steps are automated because the tolerances are too small for manual work.

Nevertheless, defects happen. Chips with manufacturing defects may be sold with lower speed ratings because they will not perform reliably at higher speeds. However, processor and memory chips may be designed with more circuits than needed so that defective ones can be disabled, and the chip will still perform as designed.

Distribution

Distribution of Personal Computing devices relies heavily on retailing to consumers, but a variety of channels to businesses. Sales to businesses can be direct, retail, or wholesale.

Distribution of Enterprise Computing devices relies more on direct sales by manufacturers and their business partners. Those sales are more likely to include services for installation, maintenance, and repair in addition to products.

Constraint Management

Can hardware be the enterprise constraint? Yes, a machine can be the enterprise constraint in any manufacturing company. And that machine can be a computer. However, computers are an unlikely place for an enterprise constraint because compute power gets cheaper every year and is therefore readily expandable.

As mentioned at the start of this chapter, hardware is a good place to start examining Constraint Management for the Information field because it's a perfect fit:

- **Project application** was invented for engineering, so it works fine for hardware R&D and system software development.
- **Production application** was invented for manufacturing, so it works fine for computer hardware. An extended form of this application works for Technical Services.
- **Distribution application** was invented for supply chains, so it works fine for hardware.

To see how Constraint Management applies, consider the following profiles from multiple management viewpoints of systems procuring, producing, or servicing hardware. This is just an illustration, not a complete picture. Each profile is a different view of interlocking systems.

Profile: Engineering Manager

System	Research and Development
Goal	Competitive price-performance of hardware
Flow	Ideas → experiments → prototypes → specifications
Application	**Critical Chain**
Constraint	Critical Chain
Buffer	Project (time) buffer
Limits	Sufficient skilled scientists and engineers; H2/H3 investment

The R&D team may be separate or integrated into product development. For consumer products, design itself ("look and feel") can be a major differentiator.

Critical Chain was invented for engineering projects. Its shorter duration and higher predictability are advantages when speed and dependability matter.

Profile: Production Manager

System	High-Tech Manufacturing
Goal	On-time completion
Flow	Materials → parts → subassemblies → products
Application	**Production for Goods** (Releases orders based on availability of the constraint)
Constraint	Probably external, in the market If internal, machine or person with least productive capacity
Buffer	If external, finished goods inventory If internal, work-in-process ahead of the constraint
Limits	Sufficient skilled technicians; H0/H1 investment

Manufacturing is Constraint Management's home turf. However, internal constraints are less prevalent than they were when CM was invented.

Profile: Sales Manager

System	Sales
Goal	Revenue
Flow	Prospects → sales orders
Application	**Sales**
Constraint	Seller activity
Buffer	Sales in process
Limits	Finished goods inventory

Constraint Management does not have a standard application for sales, but it does discuss retail sales, which is how most hardware for Personal Computing is sold. For wholesalers and Enterprise Computing, however, the dominant sales method is direct sales. And if Cloud Computing applies, those sales are to builders of Cloud Computing rather than to buyers of it.

The conventional view of the direct sales process is a funnel-shaped pipeline. At each activity in that pipeline, the number of potential sales orders is reduced by eliminating customers who won't buy. Somewhere in those activities, however, is a constraint on sales. Oftentimes, it is non-sales activities that limit the time that sellers interact with customers.

Thus, a Constraint Management approach is to apply the Focusing Steps, which start with identifying the constrained activity. Then a buffer of work-in-process optimizes that constraint.

Profile: Buyer

System	Procurement
Goal	Competitive price-performance of hardware
Flow	Requirements → request for proposal → order → delivery
Application	**Accounting**
Constraint	Budget
Buffer	Dollars
Limits	Lead time. Preferred vendors. Services may be separate.

For enterprise hardware, the buyer is probably in a procurement department and working from requirements written by the CIO office. But for small hardware, the buyer may be in a line of business that has its own Shadow Information team.

There is no standard Constraint Management application for procurement. The usual advice is to use Throughput Accounting (see the Constraint Management chapter) to ensure that the purchase contributes to net profit and return on investment. However, the release date can affect the timing of orders. That is, if the next generation of hardware offers better price-performance, as they generally do, waiting may be the sensible decision.

Whoever is driving the purchase also needs to keep the Horizons Model in mind. Underinvesting in hardware for H0/H1 endangers current profit and ROI. Underinvesting in hardware for H2/H3 endangers future profit and ROI.

Profile: Distribution Manager

System	Distribution
Goal	On-time delivery
Flow	Parts and products in warehouse → retailers, installers, customers
Application	**Replenishment for Goods**
Constraint	Inventory
Buffer	Stock
Limits	Sufficient warehouse space and transportation

Distribution is also Constraint Management's home turf. However, brick-and-mortar retail is arguably less prevalent now than when it was invented. Online retail is already more disposed toward holding inventory centrally, but the Replenishment application brings buffer management instead of economic order quantities and the large batches they favor.

Profile: Service Manager

System	Technical Services
Goal	Service level
Flow	Service requests → installation, operation, maintenance
Application	**Production for Services** (adjusts capacity based on service levels)
Constraint:	Skilled technicians
Buffer:	People on the bench, overtime, contractors, new hires
Limits:	Sufficient inventory of parts and replacement products

In the hardware realm, Technical Services don't get the respect they deserve. Preventive maintenance fixes hardware before it breaks. Repairs fix it afterward. And technical support answers "How to?" calls.

System of Systems

A system of systems means management decisions and worker actions in one system can have repercussions elsewhere. Figure 6-1 illustrates how these profiles interact:

- **Engineering** turns requirements into deliverables: hardware design, system software.
- **Production** turns the hardware design, materials, and system software into products.
- **Sales** generate orders.
- **Distribution** replenishes finished goods in the warehouse based on consumption.
- **Services** achieve service levels by completing service requests.

Thick arrows indicate flows of designs, hardware, software, or orders. For instance, hardware designs flow from engineering so production knows what and how to produce products. Distribution of hardware initiates installation, operation, and maintenance.

Thin dashed arrows indicate flows of signals (or information). For instance, when an engineering project buffer is in the red zone, that's a signal to expedite selected tasks. Or when a service level drops into a red zone, that's a signal to adjust service capacity.

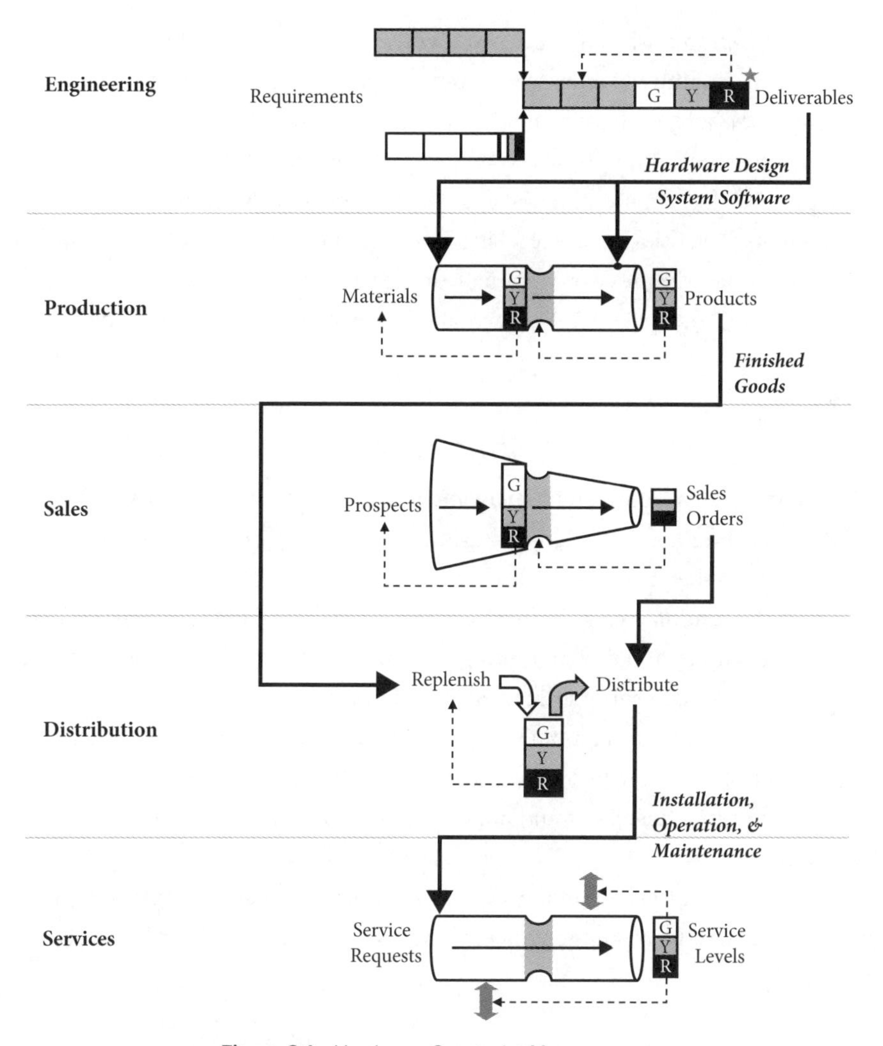

Figure 6-1 Hardware Constraint Management.

Optimizing these systems individually may not optimize them collectively. For example, over-engineering can delay the hardware revenue stream, drive manufacturing cost up, and impede post-sale services. Over-production generates goods that cannot be sold immediately and that may have to be discounted or scrapped eventually. Overzealous sales entice customers to order products that aren't available in a timely manner or that don't meet customer requirements.

Over-distribution pushes inventory out where sales vary most. Over-delivery of services can squander inventory by replacing parts when the root cause of problems lies elsewhere, such as insufficient training or carelessness.

Therefore, even though each system has its own operational constraint, at the strategic level they should function as a system of systems pursuing one enterprise goal. A buffer level or a service level in a red zone may trigger action in an adjacent system—in the extreme case, rippling from services all the way back to engineering. Moreover, strategic initiatives may alter not just how individual systems work, but how they work together.

Conclusion

It's no surprise that Constraint Management applications fit hardware well because they address Engineering, Manufacturing, and Distribution. But hardware is just one element in the Information field.

As virtualization layers obscure hardware and Cloud Computing makes hardware ownership optional, it's tempting to conclude that hardware doesn't matter. However, out of sight should not mean out of mind.

When hardware has an outage or workload exceeds hardware capacity, the importance of hardware becomes indisputable. Modern commerce, communications, science, and defense would not be possible without computing and telecom hardware.

Furthermore, the laws of physics and economics make continuous improvement a challenge. Computer engineers have a saying: "It's called hardware because it's hard."

CHAPTER 7

SOFTWARE

One major difference between computer hardware and software is explained by this old humor:

- Where does old hardware go? Onto the scrap heap.
- Where does old software go? Into production every day.

As seen in the previous chapter, hardware is a perfect fit for Constraint Management because creating hardware uses the high-tech version of Engineering, Manufacturing, and Distribution. Software, however, is not as good a fit because it's intangible, unique, and replicable—therefore, not at all like hardware.

Inventory management is a central concern of traditional Constraint Management, but software inventory is an entirely different animal, so nobody manages software the same way as materials, parts, and finished goods. Nevertheless, Constraint Management principles, if not its applications, are still worth considering. To do that, however, we will need an appreciation for the wide variety of software, because it can't all be managed the same way.

What is software? Start by recalling the computing system layers from the Hardware chapter: Applications, Middleware, Operating System, Hypervisor, Firmware, and Hardware. Five of six layers are software.

This chapter expands on that, but it focuses more on what software is than on how it's developed. The "how?" will come later.

Strange Familiarity

There are few aspects of modern life not touched by, or wholly dependent on, software. You probably have software running in your hands, pocket, or purse right now. Your health, safety, comfort, and finances depend to a remarkable extent on software.

We are so accustomed to things being done by software that we've been desensitized to its strangeness. It's intangible. It's reusable. It's flexible. It's complex. It's omnipresent.

"Software" is a misnomer because softness is a physical property, yet software is not physical. It is an intellectual creation.

Unlike manufactured products, software does not wear out. It becomes obsolete due to a misfit between requirements and implementation, or incompatibility with the hardware it runs on or with other software.

Software is arguably the most complex entity ever built by humankind. If you consider spreadsheets, websites, and smartphone apps to be software, and you should, only the most primitive organizations use no software.

Code

In a nutshell, software is a set of computer instructions controlled by decisions based on data and user input. Data goes in. Calculations and comparisons are performed. Decisions are made. Repeat until done. Information comes out.

Why is software called code? Hardware is built to act on symbols called operation codes. Software is called code because its "op codes" tell the hardware which operations to perform, such as comparing strings of text or adding numbers or printing output.

The code that computers run is called executable code. The code that humans read and write is called source code, so named because it's the source of executable code.

All source code is written in languages with tighter rules about syntax, grammar, and naming than natural languages, and there are a lot of programming languages to choose from. (More about that later.)

To get from source to executable code, there must be a conversion. If high performance is not required, some languages are interpreted into executable code at run time, which means the same instructions may be converted repeatedly. But if high performance is required, the source code is assembled or compiled before run time, which means the conversion is done just once. Then, at run time, the code does more useful computation and less code conversion.

Code modules or programs are sets of instructions that implement algorithms (step-by-step procedures) according to a design (purposeful arrangement of code and data). For example, if the design specifies an ordered list of items, an algorithm implemented in code can sort the data.

Reusable software designs are called patterns, and they consist of data structures as well as algorithms. For instance, the bridge pattern decouples two software implementations so they can be modified independently, such as when software written by separate organizations uses different character sets or different communication protocols.

Anti-patterns are attempted solutions to recurring problems that create more problems than they solve. For instance, hard coding references to hardware, operating systems, middleware, or data storage binds that software unnecessarily, which makes it less portable to other environments.

Of course, software can be used for nefarious purposes as well as noble ones. Dark patterns trick users into buying or registering for things that they did not intend to.

Software Industry

The software industry was born in the late 1960s when computing systems were unbundled. Prior to unbundling, software just came with hardware, and so did services in the form of onsite system engineers to keep it all running. But that was years before Personal Computing and the Internet.

By today's standards, computer memory was tiny, processors were slow, and storage was expensive. Things that take a fraction of a second today took hours back then. We've come a long way, but excess leads to sloppiness. Software bloat has soaked up gains in hardware performance.

The software industry still exists because some companies' products are entirely software. But how software is sold and delivered is quite different today. Video games are an example. Furthermore, because software is embedded in so many physical products and labor-based services, it's getting harder to find companies that aren't a software company to some degree.

Software Origins

Software originates several ways other than the software industry:

- **Packaged software.** Purchased from vendors
- **Embedded software.** Incorporated into physical products
- **Custom software.** Created by in-house developers or by contractors
- **End-user software.** Created by employees who are not professional developers
- **Open-source software.** Contributed by individuals or organizations
- **Software as a Service (SaaS).** Used over the web

Open-source software is not zero cost. Somebody has to create it, and it can be hard to generate enough revenue to make ongoing improvements. However, SaaS successfully monetizes open source along with custom software.

Software as a Service

SaaS is more self-correcting than on-premise software products because its evolution is continually driven by usage. SaaS also can be rolled out in waves, with corrections before mainstream rollout.

SaaS can have a user interface, such as social media apps provide, for direct use. But SaaS can also have Application Program Interfaces for use by other software. APIs can be:

- **Private.** Used only by one organization
- **Protected.** Used by suppliers, customers, and business partners
- **Public.** Used by anyone on the open web

Microservices

APIs enable microservices, which are loosely coupled code modules that typically perform just one service apiece. For instance, one microservice might access a database, while another generates a graph from the data.

Although microservices are typically simpler than monolithic applications that implement equivalent functionality in fewer, larger programs, some complexity in microservices architecture moves from code to interfaces—and there are a lot of interfaces.

Additional complexity arises in distributed applications from what happens when a microservice is not available or is performing poorly. Is the application completely broken or can it continue to function in some degraded state? Once the wayward microservice is restored, does the application have any clean-up or catch-up work to do, perhaps because transactions were queued rather than processed? If the provider of the microservice and API is a third party, what happens if the firm goes out of business or an individual stops doing it as a hobby?

Cross-enterprise dependencies create vulnerabilities that cannot be mitigated in-house. For example, when the third-party software that provides weight-and-balance information failed, all flights on multiple airlines were grounded because they couldn't get takeoff clearance.

Furthermore, in an end-to-end process, availability of a series of microservices is less than individual microservices. For example, 99.9% × 99.9% × 99.9% × 99.9% = 99.6%, or 1.5 days of downtime per year. If every programmer seeks maximum reuse of existing code, dependency chains can become lengthy, and that creates vulnerabilities to mischief as well as neglect.

Functional Requirements

Software is meant to fulfill Functional Requirements, which define what the software is supposed to do. For example, one use case can be to withdraw an item from inventory as soon as a customer puts it into their online shopping cart. An alternative use case can be to withdraw the item only when the customer checks out and payment is confirmed.

The former requirement prevents other customers from buying the same item, but it requires the item to be put back into inventory when customers abandon their shopping carts. The latter requirement doesn't prevent an oversold condition, but abandoned shopping carts can simply disappear without skewing inventory data. However, an online retailer might specify both requirements: one for peak periods and the other for slack periods.

Good user interface design follows the Principle of Least Astonishment: software should do what users expect it to do. Surprises, such as hidden side effects, are a sign that the design should be improved.

A Functional Requirement is a "must have," while a feature is a "nice to have." For instance, every text editor and word processor must have a function to save the results. If it shows a progress bar during lengthy saves, that's a feature because the save function can be implemented without it.

Three major challenges during software development are:

- **Requirements ambiguity** at the start. Is something a function or a feature?
- **Change requests** and **scope creep** at any point.
- **Integration** of software modules or microservices at the end.

In Personal Computing, consumers often don't know what they want until they see it because it's a fashion choice as much as a function choice and the requirements are often straightforward. Glass-backed smartphones originally enabled wireless charging, but soon became fashionable, too.

In Enterprise Computing, users can more often state what they want because it's mostly a function choice, and there are often many complex requirements. When users are subject matter experts, however, there's always danger in over-engineering requirements. In Constraint Management terms, a requirement should not exist unless it optimizes a constraint, but that guidance is routinely violated because nobody knows where the constraint is or why its management matters.

Developmental Requirements

Software also fulfills Developmental Requirements, which define system attributes from the developer perspective. Here are some common Developmental Requirements:

- **Affordability.** Satisfy the market on price to license and cost to operate.
- **Globalization.** Support multiple languages and character sets.
- **Maintainability.** Design and code are easy to understand, modify, and test.
- **Multi-tenancy.** Allow more than one client to use same software simultaneously.
- **Portability.** Do not depend on specific hardware or software.

Operational Requirements

Software should also fulfill Operational Requirements, which define system behavior from the operator perspective. In Personal Computing, users are the operator, but in Enterprise Computing, operation is a separate role unless combined with development.

For example, high availability means the software can resume if interrupted, but continuous availability means there should be no interruptions. For developers, those similar-sounding requirements lead to distinctive designs.

Here are some common Operational Requirements:

- **Accessibility.** Usable by people with disabilities.
- **Auditability.** Make operations observable.
- **Configurability.** Make capabilities selectable.
- **Efficiency.** Minimize energy usage.
- **Privacy.** Do not disclose personal information.
- **Scalability.** Increase and decrease capacity automatically on demand.
- **Security.** Do not allow unauthorized access.
- **Usability.** Make easy to understand and manipulate.

"Working as designed" refers to what developers implemented. "Working as desired" refers to what users wanted. They do not coincide when requirements are missing, erroneous, or ambiguous.

Never Used versus Unexpected Use

Clients who study their custom applications say that 90% of the code is never used or is seldom used. For instance, some code only runs at year end. Other code pertains to products and services no longer offered. In a separate study, over 95% of a popular software development environment was determined to be irrelevant. It could therefore be eliminated without seriously impairing the developers who use it.

Software tends to implement old business rules, which makes those rules hard to change. Moreover, software increases in complexity over time unless there's a concerted effort to simplify it. Simplifying software is hard, but it's often easier than simplifying the business, which is extremely hard.

A key challenge in manufacturing is to stop overproduction, which Constraint Management accomplishes with its production application. The equivalent challenge in software is gold plating, which is creation of nonessential functions or features.

Sometimes software is used in ways that its developers did not anticipate. For instance, some aviation software computed altitude in negative fathoms because it reused code from a naval software library.

Software Engineering

Software engineering means taking a disciplined approach to software. The harder the requirements, the more discipline matters. The greater the temptation for gold plating, the more discipline matters. The larger the risk, the more discipline matters.

Software engineering is the only engineering field where the subject is entirely a human creation and the entities being engineered are intangible. Some people question whether software engineering is really engineering because it has no generally accepted metrics, estimates, or benchmarks. Every other engineering field can calculate factors like loads and safety margins, but software engineering cannot because software is not governed by physical laws like gravity, momentum, and thermodynamics. Thus, it's ironic that software does calculations for all the other engineering fields.

There are no universal metrics in software engineering because the problems are complex, with nonlinear relationships between many, many variables. And there is a lot of uncertainty in requirements and skills. Coding is always an endeavor to create something new, because if it already existed in code, it would be far easier just to reuse it.

Mature engineering disciplines are based on centuries of trial and error. Bridges and buildings collapsed in the early days. Test pilots are respected because so many early aircraft crashed. The seeds of software engineering as we know it today were sown in the 1970s. So, under the most generous interpretation, the software engineering field is about 50 years old. Unfortunately, the tools that software engineers use haven't changed much since the 1980s. Developers still write code one line at a time, sometimes using a simple editor tool, other times using an

Integrated Development Environment. Historians will look back and say we were in the infancy of software engineering.

Metrics

Metrics are quantifiable measures of software attributes or the process used to create software. Many metrics have been defined, but too few are used.

Here are some major classes of software metrics:

- **Size.** The amount of code or requirements
- **Complexity.** Number of logic paths in code or coupling between modules
- **Defect.** Invalid requirements or deviations from valid requirements

Those may not seem like profound measures, but some software teams resist all efforts to gather metrics about their process, deliverables, and capabilities. But without metrics, software development is a craft, not engineering.

Size

Size of existing code can be measured automatically. A common method is to count millions of lines of code. For such metrics to have any validity, however, the code counting rules must be explicit and consistent. For instance, are non-executable lines counted, such as blanks, comments, declarations, and braces?

Before developers or managers reading this go ballistic, I know MLOC is a terrible metric. But it's better than no metric. If your organization has adopted a better metric, bravo! Use it.

Suppose one MLOC was printed with 40 lines per page and 400 pages per book. The resulting 60 volumes would fill a couple bookshelves. Additional MLOC would require more bookcases. Now imagine your job is to find and fix mistakes in those volumes, where one mistake can be as small as a single letter or word. Even if you narrow it down to one shelf and then one volume, you still have 400 pages to investigate. That's one reason why debugging code is hard.

For all its faults, this metric is a way of making at least rough comparisons:

- Pacemaker: 1 MLOC
- Video game: 3 MLOC

- Web browser: 5 MLOC
- Passenger jet: 10 MLOC
- Office software (spreadsheets, presentations, documents): 25 MLOC
- Operating system: 50 MLOC
- Automobile: 100 MLOC

That said, incentives that reward code production have serious unintended consequences. The best code is as short and simple as possible. Writing more code than necessary creates an immediate debugging challenge and a long-term maintenance headache.

An altogether different sizing approach is to quantify requirements. This is done by manually examining user stories or design documents before code is written, and quantifying work effort in T-shirt sizes: S, M, L, XL. That approach works fine for individual requirements on small projects, but it doesn't scale up to large projects or portfolios.

Just as MLOC metrics are criticized for not correlating with requirements, manually quantifying requirements has its own challenges. It is labor intensive, so it's expensive to do on a large scale. Also, people doing the counting don't always agree, so inter-rater reliability may be questionable.

Rough size comparisons don't have to be across domains. No matter how you measure size, if developers are asked to develop a project larger than any previous success or in less time than any previous success, alarm bells ought to go off because the risk of failure is high on projects outside the previous success zone. Organizational capabilities should grow and stretch, but post-mortem analysis shows that overly ambitious projects are often doomed from the outset.

Complexity

One view of complexity is the number of independent logic paths through code. For existing source code, this can be measured automatically by counting decision keywords. This complexity metric correlates with size, but more importantly, it correlates with testability. As the number of decision statements increases, the number of test cases necessary for complete coverage rises markedly.

Another view of complexity is the number of interfaces, parameters, and return or status codes. For existing source code, this can also be measured automatically by examining calls between modules. In this context, return and status codes are values that the called module uses to communicate success or failure back to the calling module. Like the code complexity metric, this design complexity metric correlates with testability.

Defects

A common definition of a software defect is a deviation from requirements. For instance, choosing the wrong algorithm is a design defect. Implementing an algorithm incorrectly is a coding defect. These are also called bugs. But requirements themselves can be missing or incorrect or conflicting, thus creating a requirements defect.

Software defects can be complex, but even simple defects can cause major problems. For instance, a system that accepts incoming calls for emergency services (911 in the U.S.) assigned a unique number to each call—until it ran out of numbers in what developers call a field overflow error. When the system crashed, thousands of people nationwide could not summon police, firefighters, or ambulances.

In a separate incident, legacy code that had been unused for nine years was inadvertently reactivated. The financial services firm then lost 465 million dollars in faulty trades, effectively bankrupting itself.

Reports of software defects are easy to find:

- Uncontrolled automobile acceleration was traced to extremely convoluted code.
- Trying to divide by zero disabled a Navy ship.
- A bug in magnetic resonance imaging (MRI) software jeopardized years of research.
- An online retailer credited customer credit cards if they ordered a negative quantity.
- A missing parameter file governing its engines led to a fatal cargo plane crash.
- Bugs caused robots to miss structural welds and instead weld auto doors shut.

Despite the headlines, most software defects never result in failure, and those that do often have negligible effect. That's fortunate because defect-free software is impossible. Nevertheless, every developer wrestles with the question of how much testing and debugging is enough.

Brian Kernighan famously said, "Debugging is twice as hard as writing code in the first place. Therefore, if you write code as cleverly as possible, you are, by definition, not smart enough to debug it."

Here are some common defect types:

- **Edge cases.** Zero balance is usually a special case.
- **Boundary conditions.** The line between actions, such as authorize or reject payment.
- **Off-by-one errors.** In some languages, an array with dimension 9 has 10 elements because the lower bound is at element 0.
- **Inequality bugs.** Does "Up to X" mean "≤ X" or "< X"?
- **Unit errors.** Failure to convert between Imperial units and the Metric System.

The most dangerous defects, however, fall in these categories:

- **Integration bugs.** Modules not interacting as expected.
- **Resource management.** Improper creation, use, transfer, or destruction of objects.
- **Porous defenses.** Poorly implemented security.

The stealthiest defects cannot be detected by testing because they occur only at run time under unforeseeable conditions:

- **Race condition.** Software interferes with other software by accessing or modifying a shared resource concurrently.
- **Deadlock.** Locking shared resources results in two or more software instances halting.

This explains why a restart gets things running again, though the previous results may be invalid. Rather than letting processes run indefinitely, knowing that they could eventually get into trouble, another approach is to restart them automatically and randomly. This forces developers to make the restart process quick and robust, though it means that users or consumers sometimes see brief lapses in availability.

Estimating

Before funding is approved for software, sponsors want an estimate of schedule, budget, staffing, and benefits, such as revenue earned, efficiencies created, regulatory compliance achieved, or goodwill generated. Without software metrics, the only way to create estimates is by educated guesses.

Productivity rates turn size, complexity, and defects metrics into estimates. Productivity can be expressed in hours per MLOC or function, subject to complexity and requirements, plus assuming certain skill levels.

Getting productivity rates means either gathering productivity data internally or using published data. It's common, however, to find software teams that have no historical data about their own productivity, so it's hard to establish a baseline for improvement.

Software development suffers diseconomy of scale because design, coding, and testing activities do not scale linearly. A system that's twice as big or twice as complex takes more than twice the schedule and twice the budget to build. And the more anyone tries to compress the schedule and budget, the more likely they are to compromise the requirements. Ironically, software deployment gains economy of scale when development, operations, and maintenance cost can be spread across more users because unit costs decline.

Without metrics, however, it's hard to know when someone is committing to an infeasible project. Fail fast is a philosophy that says find out what works through rapid experimentation, but that is disastrous on projects of size and complexity well beyond what an organization has successfully completed before. There is much to be learned by benchmarking an enterprise's own successes and failures.

Benchmarks

Benchmarking can compare software or software projects within one organization or across organizations if consistent metrics are available. When estimates for a proposed project reveal conditions beyond any previously successful project, that's a warning sign. When those metrics reveal that no other organization has been successful at that magnitude, that's an even bigger warning sign.

Benchmarks can, however, be misunderstood and misused.

- **Benchmarking Fallacy.** The industry average says nothing about limits.
- **Benchmarking Trap.** Beating the industry average is not always a good thing.

The most valuable benchmarks indicate the limits of performance, not average performance. Pursuing the industry average cost engages in a race to the bottom by squeezing out innovation, rather than a race to the top by finding a competitive edge.

The Information field and Constraint Management have complementary principles on this subject:

- **Metrics Principle.** Metrics are the foundation of estimating and benchmarking, and are essential for predicting whether a proposed project is feasible.
- **Measurement Principle.** Measures drive behavior, and mismeasures drive misbehavior.

ADVENTURE: Viewing the Invisible

The longest stretch of my career, in the Information field, began with research and development of software products. The company was a software start-up that had been bought by a large consulting firm which was attempting to diversify. But the new parent mostly left the software group intact and separate. As part of their growth strategy, they hired me as research manager because I had consulted with them previously on their software metrics.

Visualizing Software Metrics

The primary business was software reengineering. We sold tools to rewrite Legacy Software to make it more readable, testable, and maintainable. However, the benefits were hard to visualize. Reading lengthy computer code takes considerable mental effort, and management decision-makers weren't programmers themselves, so reading code was out of the question.

Therefore, as a supplemental business, the firm developed a tool for calculating software metrics. It measured dozens of characteristics, including these:

- **Size.** Exceptionally large programs are hard to understand and modify without error.
- **Complexity.** The more logic paths through the code, the more test cases needed.
- **Defects.** Bad practices, such as obscure coding styles, are prone to bugs.

Even with metrics, the benefits of reengineering were hard to appreciate without some visualization. Thus, I developed a prototype. Its three-dimensional color graphics could be rotated and viewed from any angle, and animated simulation showed how reengineering improved each program.

For a typical software portfolio, about 50% would start in the red zone and another 25% in the yellow zone due to their large size, high complexity, or low quality. During automated reengineering, however, about 90% would migrate into the green zone. Reengineering could improve quality and simplify code, but it could only reduce size by eliminating dead code: parts of a program that could not be reached by any logic path. However, with some additional tooling, even the excessively large programs could be refactored into smaller programs in the green zone.

Sometimes You Don't Know What You Got 'Til It's Gone

We were on the verge of productizing the visualizer/simulator when our parent company ran into a cash constraint. Shortly thereafter, our software group was sold to another software company with complementary products.

During my exit interview, I turned over source code for the prototype to the incoming vice president. Because I had developed the research prototype on my own personal computer, which I was taking with me, it wasn't on the same computing platform or under the same source code control as the firm's software products. However, he didn't seem to care. He tossed the storage media and documentation casually onto a pile behind his desk. Hence, I packed up my office and moved over to my new job across town.

Some weeks later I got a call from my previous manager. He explained that they were looking for new revenue opportunities, and he had been directed to inquire whether I still had a copy of the prototype, which I did not, of course, because it was their intellectual property and I had wiped my storage. To my amazement and disappointment, the VP had lost the only instance of the prototype, and searches of his office had come up empty-handed.

Thus, the new owners didn't know what they had until it was gone. This software asset that would have been trivial to safeguard was impractical to re-create because nobody remaining in R&D had the necessary knowledge to do so, and the documentation had disappeared along with the source code.

Lessons Learned

Lessons learned include:

- Metrics identify software most in need of reengineering.
- Metrics quantify improvements from reengineering.
- Visualizing metrics is a lot more persuasive than comparing numbers.
- Software assets from H3 may get lost on the way to H2 for inexplicable reasons.

Life Cycle

The software life cycle includes stages that may be distinct or blurred:

1. **Research.** Experimentation with prototypes
2. **Development.** Initial creation of production software
3. **Announcement.** Customers order product
4. **General Availability.** Running in production
   - **Maintenance.** Repairs to defects/bugs
   - **Enhancement.** Addition of new functionality
   - **Reengineering.** Improvements to coding, but not function
   - **Refactoring.** Improvements to design or operation, but not function
5. **Migration.** Movement to successor software or hardware
6. **End of Life.** Withdrawal from the market or production
7. **End of Service.** Withdrawal from support

Each phase of the life cycle produces deliverables. It's often assumed that code is the largest deliverable, and that is true for small projects. But the larger the project, the more documentation needed as a percent of all deliverables. Documentation includes requirement lists, project estimates, budgets, schedules, staffing reports, status reports, issue reports, design diagrams, test plans, test results, certifications, etc.

Testing

An assortment of testing activities can be embedded in life-cycle stages, though not all kinds of tests are done every time:

- **Unit test.** Developer checks individual code modules for defects.
- **Integration test.** Testers check that code modules work correctly together.
- **Regression test.** Testers re-run unit and integration tests to detect new defects attributable to changes to existing code.
- **System test.** Testers check a fully integrated product or custom application with test data.
- **End-to-end test.** Testers check the flow of processing with production data.

- **Alpha test.** Developers verify usability.
- **Beta test.** A subset of users or customers provides feedback.
- **Gamma test.** Testers verify that safety requirements are met.
- **Acceptance test.** Users or customers check that all use cases are satisfied.

Work Breakdown Structure

One way to classify software is by work breakdown structure. Some software has stricter requirements, so more tasks are performed to meet those requirements.

- **Military** software has the most tasks, and it undergoes extensive testing.
- **Systems** software has fewer, but still a lot of tasks.
- **Application** software has even fewer tasks.
- **User-developed** software has the fewest tasks, and it may undergo no testing.

Because more tasks require more labor, this list forms a hierarchy from most- to least-expensive software.

The F-35 warplane has almost 10 million lines of code running on the plane itself, and another 20 MLOC in support on the ground. Because the software literally flies an airplane that is aerodynamically unstable, its reliability must be superb. Thus, this is the most expensive weapon built to date.

User-developed software has obvious benefits but hidden risks. The biggest benefit is productivity: Spreadsheets can do some types of data preparation and analysis much faster than alternatives. Thus, users prepare data for import into Enterprise Systems, and perform analysis on data exported from Enterprise Systems using spreadsheets.

Unfortunately, most spreadsheets—and virtually all complex spreadsheets—contain errors. An erroneous formula contributed to the six billion dollar loss known as the London Whale. And 20% of scientific papers on genetics contain errors from automatic data type conversions: mistaking a gene ID for a date, for instance.

Spreadsheet formulas can be self-documenting when each key parameter is stored in a named range. Unfortunately, too many spreadsheets have constants embedded in thousands of formulas, which makes testing, verification, and modification unnecessarily difficult.

Languages

Another way to classify software is by the language used to write it. Hundreds of programming languages have been invented, and clever people keep inventing more. Consequently, most systems today are written in multiple languages, each suited to a specific domain, thereby making coding a polyglot world.

Explaining the wide variety of languages is unnecessary for this book. What is useful, however, is to understand language levels because higher-level languages do the same amount of work with less code and are therefore more productive. For example, spreadsheets are popular because they are over 10 times more productive than the languages used by professional programmers.

- **Level 50.** Spreadsheet languages
- **Level 25.** Query languages
- **Level 10.** Statistical languages
- **Level 8.** Database languages
- **Level 7.** Simulation languages
- **Level 6.** Object-oriented languages
- **Level 4.** Report generators
- **Level 3.** Procedure-oriented languages
- **Level 1.** Machine or assembly language

Dark humor among software engineers asks: What's the most common programming language? Profanity.

Object-Oriented Programming

Common problems with procedure-oriented languages include:

- **High coupling.** Code can have side effects that break other code.
- **Low cohesion.** Code can do multiple things poorly instead of one thing well.
- **Type specific.** Code can only process one type of data (number, text, image, etc.).

Rather than separating data from procedures, as procedure-oriented languages do, object-oriented languages allow developers to define "objects" which combine data and procedures known as "methods." For example, if customer accounts are defined as objects, invoking the method to update a specific account's balance keeps the details of how that happens neatly hidden away so tasks like currency conversion are handled correctly.

Here is how object-oriented languages address the problems mentioned earlier:

- **Encapsulation.** Objects hide their implementation from other objects, so fewer side effects.
- **Polymorphism.** Objects enforce separation of concerns, so do one thing well.
- **Inheritance.** Objects can inherit methods (code) from other objects, so the hierarchy can handle more than one data type.

There is a lot more to object-oriented programming than this, but for non-developers, it's worth knowing that programming languages and best practices have evolved considerably. Thus, object-oriented languages dominate Modern Systems while procedure-oriented languages are still common in Legacy Systems.

Internationalization

Accommodating a variety of natural languages is a hard problem. Natural languages require different character sets to accommodate accents, among other things. Some languages depend on gender. Some languages are written right to left. Expressing the same thought in different languages can require text of widely varying lengths, which is problematic for laying out text fields and control buttons. Even within one language, the same words are innocuous to some people and offensive to others.

Internationalization also requires date formatting: YYYY-MM-DD for standard international format, but DD.MM.YYYY for Europe, MM/DD/YYYY for the U.S., and Julian dates (9999999) for astronomers. It also requires number formatting: comma for decimal point in Europe, period for decimal point in the U.S.

Promotion

In common use, "promotion" means a better job or an enticement to buy. In software, however, promotion means moving code to a higher readiness status. The following is a typical sequence:

1. Development
2. Test
3. Pre-production
4. Production

Each state is implemented in a separate storage area called an environment. And different teams may control each environment. For example, the pre-production environment may be virtually identical to the production environment, except the integration team controls pre-production, while the operations team controls production. (With the advent of DevOps, these boundaries are blurred. More on that later.)

Though the terminology can be confusing, a development environment is used to start subsequent stages of the life cycle: Maintenance, Enhancement, etc. Then the updated code flows through the other environments as usual.

Operations

Operations includes production and a lot more:

- **Installation.** Putting software into an environment
- **Configuration.** Setting parameters for input, output, storage, backups, etc.
- **Upgrades.** Like installation, but with migration from previous versions
- **Backups.** Copying software and data for restoration later
- **Recovery.** Restoring software and data after an unplanned outage
- **Restart.** Resuming production

Legacy Systems

Modern Systems get publicity, while Legacy Systems are a subject nobody likes to talk about. Yet Legacy Systems dominate many Enterprise Computing budgets because all successful software eventually becomes a legacy. Sections of legacy code may have been written decades apart. Indeed, some code is older than the people now maintaining it.

Just one language accounts for 220 billion lines of code worldwide. Most financial transactions are processed using that language, which first came to prominence in the 1960s and peaked in usage during the 1980s [Asay, 2019]. But some languages invented recently are approaching legacy status.

Legacy Systems are:

- Past the development stage
- Running in production
- Based on older technology
- Delivering significant value
- Expensive and risky to replace
- Sometimes blamed for limiting the organization

The legacy problem is not unique to the Information field, of course. Cities stop building subways when their operation and maintenance consume the budget and there is nothing left for expansion of infrastructure.

One of the strange aspects of Enterprise Computing, however, is sponsors and users expect it to work flawlessly, cost less, and last indefinitely. That's ironic because no other product elicits such lofty expectations. Physical products come with warranties because they break or wear out—and nobody is surprised when they do.

Developers prefer to write new code instead of debugging old code because writing code is easier than reading code. On the other hand, sponsors and managers are likely to insist that legacy code be maintained rather than rewritten due to risk. When replacing Legacy Systems, the minimum viable product is often all requirements met by the Legacy System, which makes an incremental approach hard to pull off unless it's done covertly via the strangler pattern.

If your organization has been around a while and you think the Legacy Systems problem doesn't apply, do an inventory of software applications, walk through your software development area, and peruse the list of backlogged change requests. Then ask why more new systems aren't being implemented.

Application Understanding and Impact Analysis

Whenever a new requirement must be implemented as an enhancement or a bug fixed during maintenance, software engineers must find the right places to make changes. Sometimes a change has side effects that will break code in the same application or in other applications, and it's better to know that in advance.

Application Understanding tools read source code and other software artifacts to determine the following:

- Which programs call other programs
- Which programs create, read, update, or delete data
- Which screens/portals display data
- Which reports print data
- The sequence of events

Impact Analysis tools show all the code and artifacts affected by a proposed change. For example, to lengthen a data field, changes may have to be made to database tables, programs, screens, reports, and documentation.

Sometimes multiple changes must be implemented in sequence. Sometimes they must be implemented simultaneously. Sometimes they must be synchronized with a business process modification or with a physical change, such as warehouse layout or truck route.

The number of software artifacts in the portfolio of a large enterprise is typically quite large—hundreds of thousands or more. Just the source code modules may number in the tens of thousands. However, the number of dependencies between those modules is some large multiple of the count: the number of potential direct connections between n modules is $n^*(n - 1)/2$. Consequently, a diagram of all active connections looks like a furball. For such a diagram to be comprehensible, only a small subset of artifacts and their dependencies can be diagramed at once.

ADVENTURE: **Inheritance Isn't Always a Gift**

When I was an executive consultant, we were engaged to create Technical Strategy for a large company's Legacy Systems. Their software portfolio had been neglected to the point that it was beginning to affect the CIO budget negatively. They had the usual assortment of licensed software packages and custom applications, but they were spending a lot more on Legacy Systems than desired. For instance, they had six general ledger systems.

Kicking It Down the Road

Duplicate systems happened because growth through acquisition poses a challenge to the acquirer. Such deals are often justified in part by the savings from eliminating duplications. However, the effort required to migrate an acquired company from their legacy software to the new parent's software was deemed unaffordable during the acquisitions because so much data would have to be converted and so many users retrained. They couldn't just map the data fields and consolidate databases because the account codes and activity codes were mostly incompatible. Data would have to be validated, extracted, transformed, and reloaded during migration. Therefore, migrations had been deferred for years in anticipation of a steady-state phase to occur at some unspecified future time.

The problem with kicking the can down the road on legacy migration is the problem only gets worse. Consolidating onto one system would require investment that added no value, but running six systems instead of one was roughly six times the expense, and the migration cost itself kept climbing as the data volumes grew. What they thought was a sound Enterprise Strategy was based on a terrible Technical Strategy.

I never could confirm what led them into this predicament, though the grapevine said the lines of business dominated executive decision-making. The CIO was never seen. His subordinate would carry messages in and out of his office, so communication was sporadic, terse, and generally unhelpful.

From Surplus to Scarcity

The Legacy System problem is at its peak when source code is lost. Don't laugh. It happens more than you think. This client had created a custom report generator (software that puts headings above columns and subtotals below them, among other things) and made sure it was used by every application in its portfolio. That's the good news because it simplified their applications. The bad news is they lost the source code, which eliminated all repairs and enhancements to the report writer.

They were looking for a technical service capable of reverse engineering the source code for a small price. Unfortunately, the work couldn't be done for a small price because there's no economical way to revert executable code into source code. It's like trying to unscramble a million eggs because the names that have meaning for programmers are missing in the executable code and there are many optimizations in executable code that aren't present in the corresponding source code. Furthermore, the risk of disrupting the client's entire portfolio of software production jobs was extremely high.

I wish I could say that this adventure had a happy ending, but it didn't. There was no miracle fix. As far as I know, the client is still running duplicate packages and using a report generator frozen in time. Maybe the CIO is still barricaded in his office.

Lessons Learned

Lessons learned include:

- Great Enterprise Strategy can lead to terrible Technical Strategy.
- Legacy System problems do not improve with age.
- Dropping duplicate software is easy. Consolidating the data is hard.
- There is no painless way to re-create lost source code.
- These are hard lessons—and the CIO reaction can be to shoot the messenger.

Technical Debt

Technical debt is the price paid later for something done (or not done) today. Oftentimes, debt comes from taking a shortcut. For instance, some quick-and-dirty code may work fine on a small scale but will stagger as the load increases because the data structure and search algorithm are too slow. Somebody is eventually going to have to redesign and reimplement that functionality. Software engineers have a saying that "nothing is as permanent as a temporary fix."

Another kind of technical debt comes from obfuscation. Naming things so that other developers will not be confused is harder than it sounds. For instance, what does a variable named **customer** contain? Name? Address? Balance due? What does a function named **cdr** do? Martin Fowler wrote: "Any fool can write code that a computer can understand. Good programmers write code that humans can understand."

Yet another kind of technical debt comes from inconsistency. Database management systems and web browsers from different vendors may be mostly consistent, yet have just enough unique commands and syntax to inhibit portability.

Code isn't the only place that technical debt accumulates. Comments and user documentation suffer technical debt too, unless they are maintained. When comments are left behind during refactoring, those comments may be incorrect or misleading.

Over the entire life cycle of software, maintenance is often the biggest cumulative expense, but maintainability is rarely a formal requirement. Software engineers think about it, of course, but sponsors, users, and development managers may not because they're focused on getting the project done. If technical debt accumulates, the result can be software that's so obscure that new team members can't understand it and so brittle that old team members don't want to change it.

ADVENTURE: Technical Debt Comes Due

The concept of technical debt originated with software design and coding. Oftentimes, only developers see technical debt directly, but sometimes the symptoms appear when users and the public start seeing strange things. Consider these events:

- Senior citizens got notices to attend kindergarten.
- Medical devices displayed incorrect years.

- Credit cards were rejected before their expiration date.
- Mortgage calculations failed.

What's So Hard About Dates?

These were early symptoms of the infamous Year 2000 (Y2K) problem, also known as the Millennium Bug. What was going on? Developers had for decades been building systems that stored the year portion of dates in two digits because storage was extremely expensive by today's standards. Thus, conserving storage space by using just two digits to represent years in trillions of records created enormous savings when those systems were new.

Unfortunately, as those systems aged and the dates in their data spanned a century boundary, two-digit years caused date arithmetic and date comparison errors. Accordingly:

- Kindergarten notices were mistakenly sent because software calculated that the recipients were five years old instead of 105.
- Medical devices displayed incorrect dates because their internal clocks didn't roll over correctly from year to year.
- Credit cards were rejected early because 00 is less than 97 even though 2000 is greater than 1997.
- Mortgage calculations failed in the 1970s when 30-year mortgages extended into the 2000s.

Why is this distant technical issue relevant today? It was worldwide technical debt with a nonnegotiable due date, so it mobilized the Information field as never before or since. Furthermore, it revealed the need for comprehensive application inventories, portfolio analysis, code remediation, data conversion, cross-system integration testing, disaster recovery planning, and vetting of suppliers and business partners to ensure they were exercising due diligence. It wasn't just a technical problem. It was a management problem.

No Good Deed Goes Unpunished

As an executive consultant, one of my biggest clients was in a regulated industry. That client immediately recognized the potential severity of the problems that would ensue if date arithmetic and comparisons went awry. They established a war room and set about aggressively fixing code, data, and manual procedures.

You might think that the regulators would be delighted by this proactive approach, but they weren't, at least not at first. The client's VP and I were summoned to a regulatory oversight meeting where we would defend our technical approach. Mind you, the client had already spent millions of dollars on this, and the project was nearly complete, but the regulators were late in appreciating the technical issues—and hinting that the work might have to be redone their preferred way. In other words, no pressure.

Of course, if something can go wrong, it will at the worst possible moment. About 90 minutes into our joint presentation, the presentation software crashed so hard that it corrupted the presentation file itself. The only way it could have been worse is if my laptop had set itself on fire, which they have been known to do occasionally. We adjourned for a break in the action, rebooted the laptop, restored the file from backup, and regained our composure before we resumed the meeting. Fortunately, when we showed them our results, the regulators were persuaded that our technical decisions were justified, and the client did not have to suffer a do-over.

History Repeats Itself

There were few serious problems during the run-up or when clocks ticked over from 1999 to 2000. Many people engaged in solving this problem remember that New Year's Eve as blissfully quiet because so much preparation had gone well. Conversely, some pundits and journalists have since claimed that the problem was overblown and the entire remediation effort unnecessary because few problems happened. But that logic is backward. It's like saying that money spent on airline safety is wasted because so few planes crash.

So, we're done, right? No, it's not over yet. One of the remediation methods used was "date windowing," which left two-digit years in files but expanded them on the fly in software to four-digit years. Although date calculations and comparisons then work correctly, the date range is still limited to 100 years. If the full range of dates approaches that limit, those applications must be remediated again.

Could it happen again? Uh, yeah. The Year 2038 problem can already be seen in Google Calendar on some smartphones when it wraps from January 2038 back to December 1901. Similarly, Unix time will overflow on January 19, 2038, thereby wrapping dates back to January 1, 1970. Thus, date calculations and comparisons spanning 2038 will have errors like those that spanned 2000 with two-digit years. It's a solvable problem, but one with eerie echoes of the past. Unfortunately, expanding the date-time field will be impossible in some embedded devices, so they will have to be replaced. Fortunately, early signs are that lessons were learned, and the Year 2038 problem will be solved before it becomes a crisis.

The Japanese calendar, based on the reign of their Emperor, switched to a new era on May 1, 2019, so that remediation was already underway as this book was written. The previous switch happened in 1989, but a lot of new software has been written in 30 years.

Older Global Positioning System devices count up to 1,024 weeks before their counter rolls over to zero. This means that every 20 years, including 2019, the firmware in those devices must be updated to keep positioning calculations accurate. Nevertheless, several flights were grounded because the aircrafts' flight management and navigation software had not been updated. Fortunately, newer GPS devices have a larger field, so they aren't vulnerable.

There will never be another worldwide effort to remediate technical debt, right? Well, preparations for the European Union's General Data Protection Regulations are estimated to have cost multinational firms about eight billion dollars. And it remains to be seen how much the fines and remediation will cost.

Lessons Learned

Lessons learned include:

- Unlike an outage, where alarms go off, technical debt can sneak up over decades.
- Technical debt is not just a developer concern. It can affect users and the public.
- Technical debt can have worldwide impact.
- Resolving technical debt may be subject to regulatory oversight.
- When successful remediation results in “no problem,” skeptics and pundits will claim the issue was overblown.

Constraint Management

Can software be the enterprise constraint? Yes. Software that cannot scale will eventually constrain whatever business process it enables. For example, retail systems on Black Friday and Cyber Monday must be scalable enough to handle peak demand or management should expect decreased performance.

Software can also be a constraint if software is the product and the market is waiting for a new product or updated release of the current product. Shipping late or with disappointing capabilities has damaged or sunk more than one software vendor.

Even if software is not the enterprise constraint, there are always constraints on the development and use of software. Taking a Constraint Management approach means:

- **Project application.** Can be used for software research, development, and post-production activities, such as maintenance, enhancement, and refactoring. However, Critical Chain is not the dominant project management method for software. (More about that in the Methods chapter.)
- **Production application.** Can be used for enterprise software execution, which is typically handled by an operations team. And the labor-based services version of this application can be used for software support.

To see how Constraint Management applies, consider the following profiles from multiple management viewpoints of software creation and use. This is just an illustration, not a complete picture. Each profile is a different view of interlocking systems.

Profile: Research Manager

System	Research
Goal	Business and technical innovations in software
Flow	Ideas → experiments → prototypes → specifications
Application	**Critical Chain** (or alternate method)
Constraint	Critical Chain
Buffer	Project (time) buffer
Limits	Sufficient skilled researchers; H3 investment

The Research team can be separate or integrated with Development. Innovations are limited by the amount of H3 investment and availability of skills, however.

If an alternate project management method is used, the buffer may be different or nonexistent. For example, Critical Path Method does not manage buffers. Likewise, the trigger for expediting will differ. Critical Chain uses only buffer penetration, while Critical Path uses late task completion and missed milestones.

Profile: Development Manager

System	Development
Goal	Competitive price-function-performance of software
Flow	Requirements and defects → code and tests → Development environment
Application	**Critical Chain** (or alternate method)
Constraint	Critical Chain
Buffer	Project (time) buffer
Limits	Sufficient skilled software engineers; H2 investment

Dark humor says developers turn caffeine into code. They actually turn requirements into code, of course, but sometimes caffeine is helpful.

Development is limited by H2 investment. Striking the right balance between fixing defects and developing new code is a perennial management challenge.

Deliverables may include packaged software, custom software, or Software as a Service. However, Developmental and Operational Requirements for those alternatives will differ substantially even when the Functional Requirements are the same.

After the first release of software goes into production, the development manager handles subsequent maintenance and enhancement, too. Maintenance fixes bugs. Enhancement addresses new or changed requirements.

Profile: Operations Manager

System	Operations
Goal	Service-Level Agreement (SLA)
Flow	Hardware, software, and data → Production environment
Application	**Production for Services** (Adjusts capacity based on service levels)
Constraint	Hardware and software capacity
Buffer	Sprint capacity and skills
Limits	Sufficient skilled operators; H1 investment

During normal operations, the production constraint is often a particular piece of hardware or software. For example, a typical daily update job takes four hours because the storage devices cannot operate faster. Finding or establishing a stable internal constraint and using it to manage production is rare because the loads placed on systems are highly unpredictable.

Sometimes, however, the limiting factor is a person, such as when a daily update job cannot proceed until an operator allocates more storage. Because that limitation is transient, it's typically not the production constraint for scheduling purposes.

Service-Level Agreements (SLAs) specify performance targets, such as minutes of unplanned downtime and time to restore normal operations. SLAs can also specify cost targets, which often translates into number of operations staff. Operations are thus limited by H1 investment. Skilled operators can diagnose and correct operations problems faster because they've seen the same or similar problems before.

Profile: Support Manager

System	Support
Goal	Service-Level Agreement (SLA)
Flow	Features, functions, and defects → product roadmap
Application	**Production for Services** (Adjusts capacity based on service levels)
Constraint	Service capacity
Buffer	Spare parts and skills
Limits	Sufficient skilled service technicians; H0/H1 investment

Support includes help desks to answer user questions and repair services to get broken hardware and software working again. Sometimes diagnosis and repair can be done remotely. When it requires a site visit, however, the SLAs take that into account by allowing for travel time based on distance.

Software support often relies on logs and dumps in addition to user reports of trouble. Logs are records of software activities so engineers can figure out what was happening before trouble occurred. Dumps are listings of system status when it halted. In the PC realm, dumps are sometimes called the Blue Screen of Death because the system has to be restarted.

Support is constrained by skilled resources, which are limited by H1 investment. It may also be limited by H0 investment if legacy hardware or software are involved.

Profile: User

System	Business process
Goal	Business performance
Flow	Business needs → requirements → acceptance testing
Application	**Critical Chain** (or alternate method)
Constraint	Critical Chain
Buffer	Project (time) buffer, dollars, SMEs
Limits	Budget, subject matter expert availability.

Users seek software that enables business performance. Thus, users participate in software projects by expressing business needs as requirements, and later by testing whether the software satisfies those requirements.

In addition to the constraint embodied in the project plan, users have limits on budget and staffing. Many development projects therefore prioritize requirements based on available budget and business plans. For example, if an advertising campaign depends on software, the launch date for advertising may dictate the target completion date for the software. Some development projects also have a scheduling limit because users are not available during busy seasons or holiday periods.

Product Managers coordinate research, development, operations, and support. The Product Manager profile will be covered in a later chapter.

System of Systems

A system of systems means management decisions and worker actions in one system can have repercussions elsewhere. Figure 7-1 illustrates how these profiles interact:

- **Research** turns ideas into deliverables in the form of prototypes and specifications.
- **Development** turns requirements into deliverables consisting of packaged software, custom software, or Software as a Service.
- **Operations** runs software, which turns data and user inputs into information.
- **Maintenance** fixes bugs and **Enhancement** creates new functionality, both of which are delivered as software updates.
- **Support** achieves service levels by completing service requests.

Research, Development, Maintenance, and Enhancement are shown as Critical Chain projects. Alternative software project management methods for software will be covered in a later chapter.

Software updates may be accumulated into annual, quarterly, or monthly releases. Or with alternative project management methods, the deliverables may be moved into production continuously.

Packaged software, custom software, and SaaS are substantially different projects with different deliverables. Packages come with vendor services and support. Custom software means the cost is not recovered over many customers. SaaS can

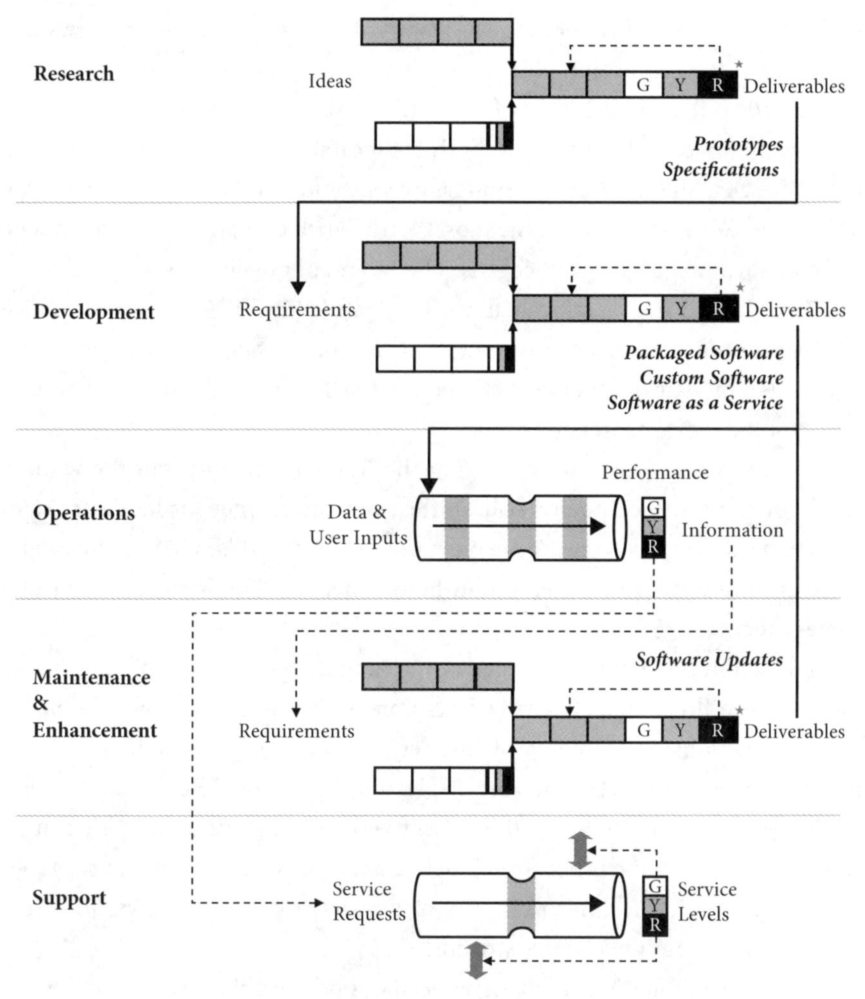

Figure 7-1 Software Constraint Management.

be single-tenant, but multi-tenancy is increasing. Subscription models instead of licenses are becoming more common, as is self-service for all types of software.

In the figure, Operations has multiple grey bands to imply that software will often encounter hardware or software limits during execution. Systems built to scale out and in can adjust capacity automatically. (Inner workings of scale out/in are beyond what can be illustrated in a vignette.) When systems do not scale automatically, that shows up instead as performance variance.

Thick arrows indicate flows of software. For instance, software flows into Operations from Development and from Maintenance and Enhancement. Compared to the smooth flow of work-in-process in a manufacturing setting because the steps are highly and uniformly repeatable, the flow of software is turbulent because the steps are not uniformly repeatable and there are many, many unknowns. For instance, even the most clearly written requirements documents still contain ambiguities, gaps, contradictions, and impossibilities.

Thin dashed arrows indicate flows of signals (or information). For instance, when performance statistics are in the red zone, that's a signal for Maintenance to update the software or for Support to modify the hardware environment, perhaps by allocating more memory or storage.

The sales function is omitted from the figure for brevity, but the vignette would be the same as the one seen in the previous chapter for hardware sales. Indeed, hardware and software are often purchased together. On the other hand, when custom software is developed in-house, there is a buy-in process instead of a sales function.

As seen in the previous chapter, optimizing these systems individually may not optimize them collectively. For example, skewing the H2 software budget toward Functional Requirements while neglecting Developmental Requirements can lead to technical debt. Likewise, neglecting Operational Requirements leads to outages and other SLA violations. On the other hand, underinvesting in H3 research can starve the innovation pipeline and future revenue streams. Margin pressure on periodic software license fees is driving vendors toward services which generate annuity revenue streams.

Thus, even though each system has its own operational constraint, at the strategic level they should function as a system of systems pursuing one enterprise goal. Unlike Manufacturing, however, where a buffer level or a service level in a red zone may trigger action in an adjacent system, the signals of an imbalance in software are neither overt nor immediate. Rather than adjustment in operations, strategic decisions and initiatives alter how the systems work together.

Conclusion

Why is software so difficult? All the simple programs have already been written.

The flow of work through Software organizations is considerably more turbulent than the flow through Manufacturing organizations. Although there is feedback and feedforward through the various systems creating, operating, and using software, the inter-system adjustments can occur on time scales much longer or shorter than in Manufacturing. For example, a red-zone buffer in distribution can trigger daily replenishment and weekly manufacturing, while red-zone buffers in software R&D can take months to resolve. On the other hand, an SLA may drive software outage resolution in minutes.

When it comes to Constraint Management for software, the constraint—the thing that limits what a system can produce—tends to fall in one of three areas. First, there's the set of tasks that determine project completion dates. Second, there are the skilled people who perform those tasks. Finally, there is the operations environment, including hardware, that dictates how rapidly business processes automated by software can run.

The complexity of software development and software deliverables is rising. One sign that a software organization employs software engineers rather than coders is their use of metrics and benchmarks to plan and perform software activities.

Every successful software endeavor eventually yields Legacy Systems, yet they are too often neglected. Thus, technical debt is underappreciated by people outside software organizations. Indeed, lost source code is a dirty little secret in more than a few software organizations.

Software already has a one trillion dollar impact on the U.S. economy. Value is migrating to Software as a Service because it's stickier than hardware or labor-based services. Once implemented, it's hard for enterprises to give up software.

What's the difference between computing hardware and software?

- If you use hardware long enough, it breaks.
- If you use software long enough, it works.

CHAPTER 8

DATA

It's no secret that data is growing exponentially. By some estimates, 80% of data in use today was generated in just the last three years. Some of that growth is due to smartphones, which capture and replay large image, audio, and video files. Data growth is also due to the Internet of Things, which generates data continuously.

The digitization trend continues apace with paperless billing, automatic payments, and electronic records in lieu of paper. Likewise, dematerialization throws off additional data as enterprises virtualize previously physical products.

Nevertheless, apart from frustration with personal data breaches and outrage about abuses of social media, people outside the Information field may not have much understanding of what enterprises do to manage data. That's what this chapter is about.

Data, Information, and Knowledge

Data, information, and knowledge are related but separate concepts. Data is a representation of facts or a recording of events. Computation turns data into actionable information. Experience and interpretation turn information into knowledge.

For example, a current building temperature of 78°F (26°C) is data. Computing that the temperature has been rising all day is actionable information because working conditions are becoming uncomfortable. Thus, this information can be used to decide to turn the air conditioning to a cooler setting. However, experience leads to knowing that the temperature will subside as occupants leave work, equipment turns off, and the sun sets. That knowledge can be used to decide not to change the air conditioning setting because the current temperature is at a peak in the normal daily cycle and the building will naturally cool overnight.

This chapter focuses on data and its computation into information. Knowledge is covered in a separate chapter because the technology and the use cases are different.

Analytics, covered here, applies to structured data, such as names and numbers. Cognitive Computing, covered in the Knowledge chapter, applies to unstructured data, such as text and images.

Misinformation, Disinformation, Malinformation

While this chapter focuses on how data is turned into legitimate information, it's also appropriate to acknowledge that the same technologies can be used for nefarious purposes:

- **Misinformation.** False information unknowingly shared
- **Disinformation.** False information knowingly used to deceive or mislead
- **Malinformation.** False information knowingly used to cause harm

Data at Rest, Data in Motion, Data in Process

Data at rest refers to data in storage. If it's stored on a disk drive, it will actually be spinning, but that is still considered "at rest" because that data isn't going anywhere besides around and around.

Data in motion refers to data in transit between devices at one site or between systems at separate sites. Local area networks and wide area networks are built differently, but they both convey data.

Data in process refers to data in computation. That's when comparisons, calculations, and other manipulations happen.

Data is most vulnerable while in process because it cannot be encrypted there. While at rest or in motion it can easily be encrypted. Thus, highly secure systems minimize the time that data is unencrypted for processing.

The news that some computer chips are vulnerable to malware which could steal data while in process is especially troubling. The number of systems built with those chips is large, and there is no painless fix because the fault is in the hardware chips. In a nutshell, the problem is that malware running on the hardware can see data elsewhere in the hardware that the malware should not have access to. Thus, highly secure hardware runs software in separate address spaces so that this look aside is impossible.

Data Types

Some data types are common. For example:

- **Unsigned integer.** No negative values
- **Signed integer.** Positive, zero, and negative values
- **Floating point.** Numbers with fractional values
- **Date.** Gregorian, Julian, or other calendars
- **String.** Text characters
- **Binary.** Readable by devices, but not by humans
- **Boolean.** True or false
- **Pointer.** Reference to a value stored elsewhere

Some uncommon data types serve specific purposes. For instance, the Soundex code allows names to be indexed by sound rather than spelling. Also, imaginary numbers are used in mathematics and science.

Data Structures

Data elements of diverse types can be formed into data structures. For example:

- **Master record.** Data about an entity, such as a product or a customer
- **Transaction record.** Data about a change to entities, such as a product sale to a customer
- **Indexed array.** Table of elements referenced by integers
- **Associative array.** Columns of elements referenced by strings
- **Queue.** Waiting line (first in, first out)
- **Stack.** Like a pile of dinner plates (last in, first out)
- **Tree.** Elements are connected in a hierarchy
- **Graph.** Elements are selectively connected in non-hierarchical fashion
- **Hash.** Calculations map from keys to values
- **Metadata.** Data about other data (for example, run time of an audio or video file)

Formulas and Expressions

In spreadsheets, comparisons and calculations are written as formulas. In programming languages, expressions are the equivalent of formulas. Both use logic and calculations to return values.

Wildcard symbols are used for approximate string matching, also called fuzzy search. For example, the asterisk in ***.doc** will find every filename with the **doc** extension. Regular expressions can be more elaborate, however. For instance, **^3[47][0-9]{13}$** will find credit card numbers starting with 34 or 37 and ending with any 13 additional digits.

ADVENTURE: Customer Service

How enterprises engage with their customers, suppliers, employees, and business partners has evolved. For instance, call centers handling inbound or outbound phone calls have evolved into contact centers that also send and receive communications by other means, such as chat sessions, screen sharing, social media, and smartphone apps.

Searching Under the Streetlight

When I was senior researcher, I was assigned to a team doing due diligence. As we sat around a conference table to discuss current performance and improvement goals, the room was ringed by stacks of boxes with metrics such as Average Wait Time and Average Handle Time. Those metrics were generated automatically by the automatic call distributor as it directed calls to agents. When I asked what the long-term trends were, nobody knew because the focus had been on statistics for scheduling agents. When I asked what factors drove the metrics up or down, there were lots of opinions but no facts. Thus, those reams of paper reports were useless.

As you can imagine, I was dismayed by how rudimentary the data was and how many gaps there were. For example, First Call Resolution—solving the caller's issue the first time—was a goal, but there was no data about it because it couldn't be measured automatically. There were over a dozen reasons for contacts: buy a product, make a payment, dispute a bill, schedule service, etc. However, there was no data about which types of contacts were most prevalent or which took the most handle time, again because that couldn't be measured automatically.

This situation reminded me of the old joke about a drunk who lost his keys in a dark place but was looking for them under the streetlight because the light was better there. Apparently, nobody at the centers had thought to gather metrics that could be used to assess and improve performance.

Illuminating the Darkness

Therefore, we conducted our own study. It began with captured basic contact attributes, such as date, time, location, and language. During analysis, these attributes enabled visualization of patterns by day of week, time of day, and geography.

The data also included:

- **Purpose.** Activation, deactivation, payment, dispute, trouble, etc.
- **Resolution.** If not first contact, number of previous contacts
- **References.** Which information systems the agent used for diagnosis and resolution
- **Transfer.** Did agent receive incoming transfer or initiate an outgoing transfer?
- **Escalation.** Hand-off from level 1 agent to level 2 agent or level 3 subject matter expert
- **Satisfaction.** Rating of agent by person being assisted

Leveraging the Tail

During data analysis, we confirmed some suspicions, but more importantly, we made some discoveries. We had suspected that First Call Resolution was good, and we confirmed that it was. We also learned that the client had been routing its most difficult customers to the contact center with its most capable agents, who were unflappable even when faced with egregious provocations and deceptions. On the other hand, we discovered that the client had not attained the service levels that they wanted us to commit to in the Service-Level Agreement. Thus, we negotiated a new baseline plus a performance improvement plan.

We also combed the data for leverage points. We began by ruling out the usual suspects. Was staffing adequate? Yes. Was training adequate? Yes. Was reference material adequate? Yes. Were company policies adequate? Yes.

Eventually, however, we found a leverage point hiding in the upper tail of the distribution of handle time. As is often the case with a variable that is only lower bounded, the distribution was positively skewed. That meant that Average Handle Time, the industry standard metric for performance, was a higher value than the mode (peak of the distribution) and higher than the median (value that splits the distribution in half). Thus, while average handle time was 10 minutes, the median was 8 and the mode was 7.

Furthermore, the distribution contained some extreme outliers—scattered values greater than two standard deviations from the mean and widely separated from the main distribution. Naturally, that led us to examine the logs to see what was different about those contacts. It turned out that the extremely long contacts happened when the caller and the agent stumbled onto a mutual favorite topic such as sports, religion, entertainment, or politics. Those calls went closer to an hour.

Therefore, we ran a simulation to see what would happen if we reduced Handle Time of just the outliers by eliminating only the favorite-topic portions of the conversations. By touching only 1% of contacts, we could reach the goal, and by touching 2% we could exceed the goal. All it took was detection of contacts with extreme handle time and coaching those agents. This simple policy tweak was much more practical than squeezing time out of thousands of contacts that were already brief.

Deflecting Contacts

Another revelation was the large and growing number of calls that were simple requests and inquiries which could be deflected to self-service.

When wait times became lengthy, callers were amenable to self-service via voice menus or a web portal. Careful design of smartphone apps eventually fulfilled simple requests and inquiries instead of calls.

Gaming the System

As valuable as metrics are for diagnosing performance issues and designing improvements, the Measurement Principle always applies. People do more of whatever is used to measure their performance, assign compensation, and evaluate promotions—so enterprises should be careful what they measure.

As customer satisfaction surveys have proliferated, employees and their managers have coached customers that any rating less than 10 out of 10 will be considered a bad rating. If every agent gets a top satisfaction rating because callers have been shamed into keeping concerns to themselves, the enterprise is looking under the streetlight for performance improvement.

Trouble tickets are also prone to gaming. If the goal is for all trouble to be resolved in 48 hours, a common pattern is tickets closed by day two, with new tickets opened on day three. And the cycle may repeat, with tickets closed on day four, and new tickets opened on day five. Again, what's the point of measuring ticket closure if the problems aren't really getting resolved in a timely manner?

Lessons Learned

Lessons learned include:

- Even though data collection is automated and everybody in the industry is doing it that way, that doesn't mean you will be able to find and manage issues with it.
- Outliers can be a leverage point, but you must have raw data, not just averages, to discover outliers.
- Visualization can be more enlightening than numbers and statistics. Showing the skewed distribution of handle time was the first time that management had seen it, and the presence of outliers was indisputable.

- Leverage points are places where a small change has a large beneficial effect. They are considerably more effective than "boil the ocean" alternatives.

Data Administration

Data structures can be transient (processed in main memory only when a program is running) or persistent (stored in external memory even when a program is not running). Persistent data structures are implemented in files and databases.

Data modeling is the design of data elements and data structures according to requirements. For instance, a data model can define separate tables representing entities such as customers, employees, products, and services. The data model also specifies relationships between tables, such as which customers buy each product or which employees deliver each service.

A collection of tables forms a relational database. However, some databases use other data structures, such as graphs or key-value pairs, which are better suited to how that data will be used.

Data administration is the implementation and operation of data models. It includes the creation of tables or other data structures and indexing or striping to improve performance. Indexing allows specific records to be located without scanning an entire data structure. Striping spreads a data structure across multiple storage devices for faster access.

Extract, Transform, Load

Getting data from its sources into a file or database built on a different data model is done by Extract, Transform, and Load. An unintended consequence of mashups that combine data from disparate sources is that data analysts and administrators spend an increasing portion of their time on ETL.

Transformation may be needed to change values from one computer's format to another's because they represent numbers or characters differently. Transformation may also be needed to change values from one business' representation to another's because they use different product codes, date formats, languages, or currencies.

Transformation

Sometimes data transformations are a routine part of operation. For instance, if three-dimensional models use a right-handed Cartesian coordinate system, displaying those models on a two-dimensional screen or paper requires a projection into a left-handed Cartesian coordinate system.

"Handed" means if you extend your thumb and index finger 90 degrees, your thumb points toward the positive x-axis, your index finger points toward the positive y-axis, and your palm points toward the positive z-axis. When right-handed, z points up. When left-handed, z points away from the viewpoint.

If the transformation is a perspective projection, distant objects appear smaller than near objects. And z-values indicate which objects are in front of other objects so hidden elements can be removed. Such transformations are an integral part of 3D business graphics, video games, and virtual world depictions of building/landscape plans.

De-duplication

Some data sources—especially the Internet of Things—naturally generate duplicate data. De-duplication is the elimination of duplicates. Not only is that necessary to avoid over-sampling during analysis, it also simplifies processing and economizes data storage.

Compression

Some data types—especially images, audio, and video—are naturally redundant. That means parts can be eliminated without unduly harming the overall look or sound. The same goes for most numeric and textual data.

- **Loss-less compression** means the original data can be fully restored from compressed data. **ZIP** files are an example.
- **Lossy compression** means the original data cannot be fully restored, but the compressed data is significantly smaller. **MP4** files are an example.

Encryption

Encryption scrambles data to make it unintelligible to unauthorized persons. Of course, encryption is only as secure as the algorithm and the key, but secure options are widely available.

Even though encryption is computationally intensive, hardware encryption can be used when software encryption is too slow or too vulnerable to tampering. Thus, there's no excuse for enterprises not to secure sensitive data via encryption.

Passwords can be encrypted via hashing, but secure methods take an additional step invisible to users. Salting adds an additional random input that makes it much harder for unauthorized persons to break the passwords.

Unfortunately, encryption can be weaponized, too. Ransomware is a form of malware that encrypts data and holds the key hostage until a payoff is made.

Streaming

Streaming occurs when data flows continuously. Stock tickers, weather sensors, flight tracking, radio, and television are examples.

Data velocity is high, and the usefulness of data often declines with age. Is yesterday's weather forecast really of use today? Thus, data analysts may decide not to retain any streaming data, sample the stream, or retain the entire stream.

Logging

A log is a record of activity. Log entries can be about routine events, such as a monthly report, or about unusual events, such as an attempted security breach.

Some computer logs are designed for human consumption. For instance, as software applications run, they write status codes and diagnostics into a log file to aid developers in debugging.

Other computer logs are designed for computer consumption. For instance, database management systems write images of records into a log file when processing updates so that the database can be rolled back to a checkpoint if a problem is encountered. This maintains the integrity of the data by undoing partially completed transactions.

Data Errors

Despite diligent efforts, some data errors are inevitable because neither people nor machines are infallible. Invalid values are a common example. Sometimes, however, the error is missing values among otherwise useful data. Thus, analysts must decide what to do about bad data:

- **Error-end.** Drop bad data because it is not repairable.
- **Error-hold.** Sequester bad data until it can be corrected.
- **Error-defer.** Let bad data go because a correction will be applied later.
- **Error-substitution.** Replace bad data with an alternative, such as the mean.
- **Error-through.** Let bad data go because the error is tolerable.

A third or more of the effort on data analytics projects may be devoted to data cleansing. And that project phase is often severely underestimated.

Some errors are truly random. Many, however, are not. For example, transposition errors such as reversed digits are so common that account numbers often have check digits for error detection. Likewise, the construction of computer hardware includes parity bits that also detect and correct errors. But some data types are prone to specific errors. For instance, precise 12-hour errors (switching AM with PM) are common in time data.

Look at the data on the left in Table 8-1. It shows the number of inventory units going in and out each month, as well as the start and finish levels. Fees are computed as 10 cents per average units on hand each month.

Table 8-1 Data Example

	Start	In	Out	Finish	Fees			Start	In	Out	Finish	Fees
Jan	50	12	13	49	$49.50		**Jan**	50	12	13	49	$49.50
Feb	49	33	18	64	$56.60		**Feb**	49	33	18	64	$56.50
Mar	64	260	19	305	$184.50		**Mar**	64	26	19	71	$67.50
Apr	305	28	24	309	$307.00		**Apr**	71	28	24	75	$73.00
May	309	15	23	301	$305.00		**May**	75	15	23	67	$71.00
Jun	301	5	20	286	$293.50		**Jun**	67	5	20	52	$59.50

Can you find any errors? Most people see that units coming in March are about 10 times the units coming in any other month, so that could be an error. That cell stands out because it's different from the surrounding cells, and people are quite capable of detecting deviations from patterns like this.

In addition to transcription errors, numbers sometimes just get lost in translation. Chefs for an Olympic team accidently ordered 15,000 eggs instead of 1,500.

What else do you see? Most people see that there is a discontinuity in fees during March, but it's consistent with the leap in inventory levels.

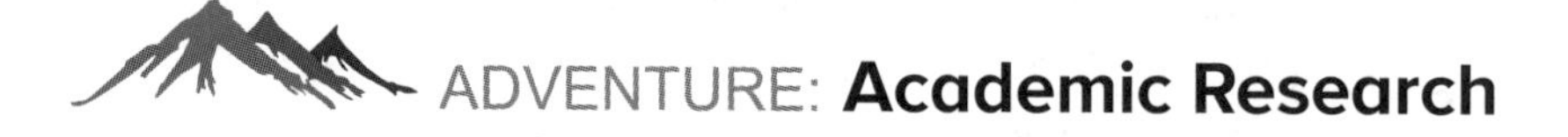

ADVENTURE: **Academic Research**

The second field of my career was a foray into academia. To support myself while I went to graduate school, I wore many hats: I taught computer programming; led teams developing simulation, database, and decision-support systems; I managed a psychology research laboratory; and I conducted experiments as a principal investigator.

My experiments were human-computer interface studies. Subjects acted as production managers in a simulated factory, making decisions about how much to produce of each product. The quality of their decisions was revealed by the profit or loss that the simulated factory made. The experiments compared decision quality under different information scenarios, plus different psychological and demographic variables, to determine if differences were due to the information or to the decision-maker.

A Trivial Discrepancy

The original simulation was written in a programming language that represented numbers as floating point (decimal places) or integer (no decimal places), but not currency (dollars and cents). Consequently, financial reports did not always cross-foot exactly. That is, row totals and column totals would sometimes appear to differ by one cent. In the computer, the actual difference was less than one cent, but automatic rounding would hide this.

Technical Debt Can Be a Matter of Perception

This was not a material discrepancy because one cent would never affect decision-making among alternatives measured in tens to hundreds of thousands of dollars. However, in the post-experiment debriefing, the few subjects who discovered the discrepancy said they could not have faith in a system that could not calculate currency consistently.

Before I got involved, hundreds of subjects had used the reports before anyone mentioned the discrepancy. But it fell to me to deal with it. With a few coding tweaks, I was able to nudge the reports into consistent cross-foots, and I thought that would be the end of it.

Proportional Response

Some languages do have currency data types. However, language migration in this case could not be justified because the original language otherwise perfectly fulfilled requirements. Thus, minor technical debt is often better fixed in place than by major investment that could be better spent elsewhere.

With the fix deployed, no more subjects found cross-footing discrepancies. But the experience revealed how even a trivial discrepancy can affect perceptions.

This is not to say, however, that rounding errors are always trivial. A rounding error in software controlling a rocket led to the destruction of 500 million dollars' worth of uninsured satellites.

Lessons Learned

Lessons learned include:

- Everyone has a materiality threshold, and for a few people even trivial differences matter.
- Distrust of automation can be a sufficient reason to fix discrepancies.

Information Biases

Now look at the data on the right in Table 8-1, where the error has been corrected. What do you see? Most people are content with the correction and move on. Look hard. Did you notice that the fees every month are 10 times larger than expected? Fees in the table are one dollar per unit, not 10 cents per unit. Ninety percent of people shown this bias cannot see it because the erroneous parameter is not visible amongst the data, and all the fees are the same magnitude relative to inventory units—so the bias does not stand out.

When you consider that one erroneous parameter can bias an unlimited number of data elements in this manner, the impact of powers-of-10 biases becomes apparent. And by some estimates, most non-trivial spreadsheets suffer mistakes because they are not tested sufficiently.

Number numbness is the inability to comprehend extremely large or small numbers—or to recognize numbers much larger or smaller than they should be by powers of 10. Implementing reasonableness tests in software is a best practice that's not always followed. For example, a residential customer received a bill from the electric utility for $284,460,000 instead of $284.46, which was off by six zeros. A phone customer was billed €11.7 quadrillion instead of €117.2, which was off by 14 zeros—an amount thousands of times greater than her entire country's annual economic output.

One way of doing a reasonableness test is to designate a floor and a ceiling, then flag any amounts that fall outside that range. Another way is to calculate the standard deviation, which will vary over time, then flag amounts that are obvious outliers. It's not hard for software to detect unreasonable values, so it's baffling when astronomical amounts slip through.

ADVENTURE: **More Academic Research**

As crazy as the off-by-one-cent discrepancy had been, things were about to get crazier. I was about to discover that being off by thousands of dollars may also escape detection.

An Insidious Bug

A major bug in the original simulation had been left uncorrected because fixing it would have invalidated previous experiments. Before I got involved, hundreds of subjects had used some misinformation before anyone recognized the bug. But it fell to me to deal with the bug. That made me wonder how big the misinformation would have to be to be noticeable.

It was an insidious bug because it was a single faulty parameter that generated long streams of numbers, all about the same magnitude, but all different by some powers of 10 from the correct value. In other words, the bug wasn't in the code itself. It was in the parameter file that configured the simulations.

Automation Bias

When debriefed after my experiment, almost all subjects said they just trusted the computer. Only a handful bothered to do calculations by hand to verify what the computer was telling them. This human trait is called Automation Bias and is defined as the tendency to trust automated systems to provide correct information.

Powers-of-Ten Information Biases

Later, while I was a professor, I replicated this study, but extended it to include even larger biases, plus a range of instructions about verifying rather than trusting the computer [Ricketts, 1990]. The experiment confirmed that misinformation biased by larger powers of 10 are not necessarily easier for people to detect. Likewise, even strong instructions to verify information do not lead to significantly greater detection rates.

Thus, I coined the term, Powers-of-Ten Information Biases. Such biases are typically caused by a parameter that's off by some power of 10—it could be greater or smaller than the true parameter—with the effect that entire columns of numbers are about the same magnitude, yet every number is wrong.

People are good at some kinds of pattern analysis, so it's relatively easy to see one anomalous number in a column of numbers, particularly if the anomaly is some number of decimal places different. But it's much harder to recognize when an entire column of numbers is incorrect because a parameter used to calculate that column is wrong.

Lessons Learned

Lessons learned include:

- Powers-of-Ten Information Biases are extremely stealthy.
- Larger biases are not more detectable.
- The Automation Bias of human decision-makers makes bias detection unlikely.
- Instructions to verify information do not overcome Automation Bias.

Analytics

Analytics range from mundane to intricate:

- **Reporting.** Documents events and status
- **Query.** Answers questions
- **Alerts.** Triggers actions
- **Inference.** Indicates differences between population parameters, such as means
- **Simulation.** Shows what could happen
- **Forecasting.** Projects past data into the future
- **Prediction.** Shows the consequences of decisions
- **Optimization.** Computes how to achieve the best outcome

But even mundane analytics are susceptible to data errors and information biases, as illustrated above. When a bot crawled 30,000 published research papers, it found that half had math or statistical issues, and 13% had issues that could alter the conclusions [Resnick, 2016].

Arguably, the biggest analytics challenge is that even random data will always exhibit detectable patterns, so it's hard to distinguish valid patterns from phantom patterns. Many models will fit the data containing a valid pattern, but only one is correct. Science tackles this with replications on the presumption that a valid pattern will show up in multiple studies.

In practice, however, things aren't that simple. There is a bias against publishing replication studies. Furthermore, science suffers the file-drawer problem. That is, studies which don't find statistically significant results generally are not publishable. But when those studies aren't published, they can't be part of the meta-analysis that would show the occasional study finding significant results is probably just due to chance.

Studies of human subjects using inferential statistics often struggle to collect sample sizes large enough to detect anything other than large effects (big differences between groups). On the other hand, with streaming sources, such as sensor data, it is possible to have sample sizes so large that even trivial effects are statistically significant [Ioannidis, 2005]. Unless researchers calculate statistical power, they don't know whether their sample is too small, too big, or just right.

It's also easy to succumb to selection bias. During World War II, bombers returning from missions with holes were subsequently armored in those areas until someone realized that the armor really needed to be in areas without holes because damage in those areas was the reason bombers did not return.

As enterprises use analytics to push the limits of their performance, the world still harbors some rigid limits. Consider healthcare. The average human life span increased at first through reduced infant mortality, and more recently through reduced senior mortality—yet nobody lives past age 116.

ADVENTURE: **One in a Row**

When working on a Strategy team, I funded a proof-of-concept project that implemented an optimization model. The goal was to help the client maximize its profit by pricing a service or offering personalized perks so that orders would make the best use of available capacity, which was perishable and adjustable only in fairly large increments.

If demand was less than supply, capacity would be wasted. If demand exceeded supply, some orders would be refused and the profit foregone. And if the client's bid was not competitive, its customers would choose another service provider.

The client willingly provided historical demand and standard pricing data, though that data had to be cleansed. Pricing via the optimization model meant that each order would subsequently be priced separately based on remaining capacity, and perks might be offered to sway uncommitted customers.

Lost Puzzle Piece

Using simulations, the team demonstrated how a significant increase in profit would result from better pricing and perks. But without actual supply data, the model could only optimize pricing relative to capacity assumptions, and profit would be limited by whatever actual capacity existed.

Within the client, pricing and capacity were managed by separate departments. And the capacity management department balked at providing capacity data. We eventually determined that they feared job loss due to automation, which was an unfounded fear because growth via the optimization model would require more care and feeding than the existing capacity management method.

Half Optimized

So, when it comes to optimizing only demand versus both demand and supply, is there a vast difference? It's half vast.

Lessons Learned

Lessons learned include:

- Analytics with great promise can be derailed by missing data.
- When data ownership is distributed, assembling the parts can be difficult.
- Fear of automation is a powerful inhibitor.

Security

The more valuable data is, the more enterprises should want to secure it, of course. Theft of intellectual property and intangible assets can be even more damaging than theft of physical property.

Perimeter security (firewalls alone) is no longer state of the art because perimeters are too porous. Isolation and access controls within the enterprise, coupled with intrusion detection, are the new frontier. Nevertheless, separation of duties is still a best practice, because then fraud and theft require collusion, which is harder than with a single perpetrator.

Headlines about security breaches tend to focus on exfiltration (theft) of data. But infiltration (tampering) is also a concern because it exposes the enterprise to theft of more than just data. Unfortunately, only about 5% of the data lost through security breaches is encrypted.

Privacy

With the advent of social media and browser tracking, substantial personal information is online voluntarily, yet we still expect a modicum of privacy. Opting out of social media isn't fully effective because shadow profiles are built based on what others post about you. And blocking browser tracking isn't fully effective either because the Internet was designed for openness, not privacy.

When security fails, privacy often suffers. Data breaches have exposed personal information of virtually every adult in the United States, sometimes more than once. And there is no way to undo it. In some cases, the breach was from third parties entrusted to analyze the data.

Nevertheless, some privacy, such as healthcare information, is protected by regulation. The U.S. Health Insurance Portability and Accountability Act (HIPAA) requires that personal data be deidentified before disclosure beyond health management.

The European Union's General Data Protection Regulation (GDPR) mandates the "right to be forgotten." That is, users can not only opt out of personal data collection, they can demand that organizations remove their personal information, and penalties for noncompliance are severe. Furthermore, sovereignty means that data cannot leave the country.

Retention, Backup, Recovery, Archiving, Disposal

Security and privacy notwithstanding, there are many legitimate reasons for enterprises to collect and retain data. Obviously, records are needed for basic business operations. In addition, laws and regulations require records.

Data retention at enterprise scale is harder than it sounds, however:

- **Fault tolerance.** The ability of a system to continue operation when components fail, perhaps at reduced performance and some data loss
- **Hot standby.** A secondary system that takes over immediately when a primary system fails, without reduced performance or data loss
- **Backup.** Copies of data stored on site with primary data or off site so that the copies are unlikely to be lost during a disaster at the primary site
- **Recovery.** Restoration of primary data from a backup copy
- **Archiving.** Migration of aged data from fast, convenient, expensive storage media to slower, less convenient, less expensive media
- **Disposal.** Deletion or erasure of data, sometimes with physical destruction of storage media

RAID drives are a widely used method of achieving fault tolerance for data in active use. RAID stands for Redundant Array of Inexpensive Disks because it combines several physical drives into one logical unit. Then data can be spread across those drives in a manner that allows it to be reconstructed if one, sometimes two, of the drives fail. They can be Hard Disk Drives (rotating) or Solid State Drives (chips).

Tape drives are a widely used method of backup because the media is inexpensive enough that multiple generations of backup are affordable. Although the media is inexpensive, the administrative overhead of physical storage is not free. Therefore, backup media eventually gets reused. The moon landing tapes were lost to accidental reuse, which overwrote the previous data.

The longer that media sits in a vault, the more it deteriorates physically. For instance, tapes sag and stretch. Billions of compact discs (CDs) are silently becoming unreadable.

The longer that media sits in a vault, the closer it comes to digital obsolescence. Floppy disks, for instance, are antiques in computer years. Finding a working floppy disk drive nowadays can be an expedition. In addition to physical media becoming obsolete, so do file and database formats. For example, early spreadsheet applications have been discontinued. Fortunately, their files can still be read by other applications, but it's risky to assume that will continue indefinitely.

Recovery means replacing lost or damaged primary data from the backup copy. If there is no backup, recovery may mean attempting to restore a lost version. Whether that recovery succeeds depends on whether the primary data was simply misplaced, deleted, or wiped (overwritten to make it unreadable).

Archiving means putting data into indefinite, if not perpetual, storage. For reasons mentioned earlier, however, digital archives require maintenance of the hardware and preservation of the software as well as the media. The Internet is not an archive: Content comes and goes without warning.

Disposal refers to more than deleting or wiping data. Some businesses and some regulations, such as GDPR, require verifiable data destruction. The media itself may have to be shredded or otherwise made physically unrecoverable in front of witnesses who attest to the destruction in a log.

Technical Debt

In addition to design and code, data is another source of technical debt. If lost data is recoverable, the cost and time can be prohibitive. If obsolete data is convertible by ETL, that time and effort should be built into migration plans.

Just as losing source code can be catastrophic, so can losing data. The sequel to a popular computer-animated children's movie was accidently deleted, and the backup was bad. Luckily, someone had an unofficial copy because she worked from home occasionally. Once restored, the movie went on to gross around 500 million dollars. I hope she got a bonus check.

In another incident, an owner claimed to have accidently deleted his entire company's software and data, plus his hosted websites, plus all onsite and remote backups, with one errant command. The perpetrator/victim later claimed it was a hoax, but the command in question could have done what was claimed, so skeptics suspect his revised story was a face-saving effort.

And then there's the social media company that lost 50 million songs that had been uploaded over 12 years. According to news media reports, server migration did not go well. Similarly, a photo-sharing company decided that it had been too generous in its offer of free storage, so it reduced the limit from 200,000 to 1,000 photos per user and deleted millions of photos in the transition. Thus, while some citizens and lawmakers press "the right to be forgotten," users lament the loss of data they thought was safely tucked away.

Despite valiant efforts to maintain old hardware and software and media, hardware eventually breaks, software isn't portable, and media wears out. In urgent cases, hardware emulation may be the only option. Emulation means running obsolete software on current hardware that has been programmed to imitate missing hardware. Emulation may thus enable access to obsolete data, but at a high price.

Open Data

This chapter has so far described proprietary data. There is a movement, however, to make some data open and freely accessible by anyone. Of course, personal data must be removed or deidentified or aggregated, but that's not necessarily a roadblock.

Governments, for instance, can post relevant statistics or post the raw data and hope that someone will analyze it. For instance, some apps have used open data to superimpose housing, income, transportation, and crime statistics on maps.

DataOps

This chapter has so far mostly described data administration as something done on behalf of users. There is another movement to have users curate their own data.

This is already happening in Shadow Information corners, but DataOps brings it out of the shadows and into shared enterprise data. DataOps might not be practical for Systems of Record with their complex requirements, but for Systems of Engagement where innovation matters, why not? Marketers strive to keep content fresh, and DataOps cuts out some administrative overhead by empowering users.

ADVENTURE: **Tyranny of Industry Averages**

I'm a cautious proponent of using metrics to improve operations and strategy, but I'm a skeptic when it comes to using benchmarks to compare organizations. My advocacy comes from seeing metrics used to identify areas for improvement. My caution comes from Goodhart's Law, which is often paraphrased as, "When a measure becomes a target, it ceases to be a good measure because people will figure out how to game the system."

My skepticism comes from seeing benchmarks misused to pursue industry averages, as though average is the right goal. Executives who would be offended if you said their kids are average are quite content to proclaim that their organizations are going to cut costs until they reach the industry average—while their competitors are investing in innovations.

Rookie Mistakes

Another reason for skepticism comes from participating in multiple benchmarking programs and learning firsthand how hard it is to accurately and consistently measure anything within organizations, let alone across organizations. Even when the organizations are responding to a questionnaire, I'm skeptical because I've reviewed enough research to know that authors make rookie mistakes. For example, consider this simple question and answers:

What is the age of the software application you maintain?
1 year
2–4 years
5–10 years
>10 years

The main error here is "polytomizing a continuous variable." Rather than using a fill-in-the-blank question to ask for age in years, which would make statistical analysis legitimate, the best that can be done with the available responses above is to count the instances in each age category. But the categories have different widths, which makes any comparisons dubious.

The question itself is also problematic because it is ambiguous. Is age measured from the original go-live date (version 1.0), or the last major rewrite (version 7.0), or just the version the respondent maintains (version 7.3)? Three respondents could honestly give three different answers for the same application.

Bear in mind that this is just one example. Improvised questionnaires often have a spectrum of problems.

We're in Too Deep

Problems can also crop up during data analysis. When the lead researcher on a widely used benchmarking study presented his questionnaire, the equations for calculating an overall best practices score, and the resulting benchmarks, something didn't feel right. A hundred questions boiled down to 10 dimensions combined in an equation that further boiled down to good practices minus bad practices, so higher overall scores were better.

Closer examination revealed that there were no negative overall scores because the implementation of the equations was wrong. It added instead of subtracted! Bad things raised the overall scores instead of depressing them, as intended. Thus, the benchmark was hopelessly confounded. When notified of the error, the researcher just shrugged and continued using that implementation anyway "because companies had already been scored and a correction would just be confusing." In some ways, this was a testament to human ability to imagine patterns in any data, no matter how confounded it might be.

This was not an isolated incident. When a graduate student discovered errors in a famous economics paper because he could not duplicate the findings, the authors replied this way: "It is sobering that an error slipped in, but we do not believe it affects our central message." [Alexander, 2013]

Lessons Learned

Lessons learned include:

- Industry averages don't account for a multitude of things that make organizations different. Figuring out what matters and what doesn't is the hard part.
- Improvised questionnaires often contain novice errors that make the findings questionable. And it can be impossible to fix problems after data has been collected.
- Researchers who should know better can be reluctant to admit mistakes and issue corrections. Too often, the reaction is denial.
- In an environment of Potemkin numbers and alternative facts, however, retractions and corrections don't always set the record straight enough to change minds.

Constraint Management

Can data be the enterprise constraint? Yes. In an enterprise where the principal flow is data as a product or service, something producing data can be the constraint or data itself can be the constraint if it's acquired from external sources. For example, a television or movie business depends on revenue from sales, licenses, and other services, such as broadcast and distribution. Some inventory may be produced internally while the rest comes from outside. In either case, too few programs or movies means the buffer is in the red zone because future revenue is in jeopardy.

In the North American Industry Classification System (NAICS), the Information sector includes these industries where data is the primary product or service:

- Publishing
- Motion Picture and Sound Recording
- Broadcasting

- Telecommunications
- Data Processing
- Web Search

Financial sector includes these industries where data can be a primary or secondary product or service:

- Credit
- Securities
- Insurance
- Funds
- Real Estate

Professional, Scientific, and Technical Services sector includes these industries where data can be a primary or secondary product or service:

- Accounting, Bookkeeping, Payroll
- Research
- Advertising

In enterprises where data is not a product or service but is used for management purposes, data is more likely to be a limitation than a constraint. For instance, delays and errors when submitting shipping data to customs can lead to late shipments of goods or quarantine upon arrival, but data handling itself isn't something that would undergo constraint management.

When data is collected, processed, and disseminated continuously, the Constraint Management production application may apply. On the other hand, when cleansing and migrating data, the Constraint Management approach to project management may be suitable.

To see more about how to apply Constraint Management to data, consider the following profiles from multiple management viewpoints of data. This is just an illustration, not a complete picture.

Staff Functions

Several profiles listed next describe staff functions. Policies set by staff and regulations set by regulators aren't constraints because they can't be used in a Constraint

Management application, but they do impose limits on other systems doing Constraint Management.

Staff functions themselves nevertheless do have local system constraints. They are not, however, in my experience, the enterprise constraint. So, where is the constraint in a staff function? It is probably in a process, project, or skills.

- **Process.** Highly repeatable steps, such as reviews and approvals, can be managed with the Production for Services application.
- **Project.** Non-repeating steps, such as a strategic initiative, can be managed with the Critical Chain Project Management application.
- **Skills.** When a goal is clear but the steps to achieve it are unknown or loosely defined, and the number of people in each skill group is large enough to support a buffer, then skills can be managed with the Replenishment for Services application. In most instances, however, people in staff roles are too few and too specialized for a skill buffer to make sense.

When a few key staff are the bottleneck, the Focusing Steps can be used to improve their productivity by eliminating non-essential activities and offloading selected activities to non-bottleneck staff. Removing distractions and non-value-added work to improve productivity is universally laudable, but the leverage is especially high when a key staff role requires extraordinary imagination and inventiveness—that is, "thinking outside the box."

Profile: Chief Data Officer

System	Data Governance
Goal	Compliance with internal policies and external regulations
Flow	Regulations → strategy → plans and execution → governance

The CDO oversees data as an asset. But it is a staff function rather than a line function.

All staff functions provide guidance and impose limits on line functions (to be described next). For instance, the CDO can set a policy that confidential data belongs behind multiple layers of firewalls rather than just one.

Strategy includes surveying data requirements, planning enterprise data, then executing the plan. For instance, the strategy could be to do Systems of Engagement on Cloud Computing.

Governance includes setting policies, defining roles, creating procedures, observing regulations, and ensuring compliance. For instance, governance could establish authentication procedures for access to confidential data.

Profile: Chief Privacy Officer

System	Privacy Governance
Goal	Compliance with internal policies and external regulations
Flow	Regulations → strategy → plans and execution → governance

The Chief Privacy Officer (CPO) is responsible for privacy. But it is another staff function rather than line function.

The CPO works with enterprise and site security teams on establishing and operating safeguards, per external regulations and internal policies. The CPO also deals with privacy cases.

Profile: Chief Science Officer

System	Systems of Insight
Goal	Research reports, intellectual property, and prototypes
Flow	Strategy → plans and execution → reviews

The CSO is responsible for data scientists who acquire and analyze data for their research. Like the other chief roles described earlier, CSO is yet another staff function.

If the research is managed as projects, however, then the Critical Chain is the constraint. In other cases, lack of research data can certainly be a limitation.

Profile: Business Process Owner

System	Business Process
Goal	Business performance
Flow	Business needs → requirements → acceptance testing

The Business Process Owner (BPO) is responsible for a business process, such as Order to Cash or Procure to Pay, plus the systems that enable that process. Like the chief roles described earlier, BPO can be a staff function, or it can be included in line management for the business process.

A business process may be served by various system types. Systems of Record hold data about entities and transactions. Systems of Engagement attract and retain customers. Systems of Innovation improve a process. For example, online banking through a smartphone app is a System of Engagement that provides customer-friendly access to Systems of Record. Data analytics can be a System of Innovation that tells the business which customers are profitable and which are not.

Profile: Data Analyst

System	All types of information systems
Goal	Business-technical innovation
Flow	Business needs → requirements → acceptance testing

Data Analysts (DAs) create data models to serve multiple applications, including specifications for encryption and compression. DAs may also perform ETL. Business-technical innovation means understanding business requirements well enough to access existing data models for familiar requirements or build new data models for novel requirements.

DAs generally perform tasks during development projects, so project management applies. DA tasks tend to come early in the schedule.

Profile: Design Manager

System	Systems of Engagement
Goal	User Experience
Flow	Use cases → functions → results → mock-ups

The Design Manager is responsible for Designers who design the User Experience. They may take guidance from the Chief Content Officer. UX design answers questions like these:

- What are the use cases?
- What should be the look and feel?

- Which functions will be available to each user type?
- How will those functions be invoked?
- What will the results look like?

Designers perform tasks during development projects, so project management applies. Design tasks tend to come early in the schedule.

Profile: Database Administrator

System	All types of information systems
Goal	Technical performance
Flow	Ideas → experiments → prototypes → specifications

Database Administrators (DBAs) implement data requirements. Thus, DBAs create databases, maintain their security, and tune performance (for instance, by creating indexes).

DBAs perform tasks during development projects, so project management applies. Those tasks tend to start midway in the schedule unless they involve new database technology, which requires planning up front. But DBAs also deal with data-related production problems after development is done.

Profile: Development Manager

System	Development
Goal	Satisfy business requirements with technology
Flow	Requirements and defects → code and tests → development environment
Application	**Critical Chain** (or alternate method)
Constraint	Critical Chain
Buffer	Project (time) buffer
Limits	Sufficient skilled software engineers

Critical Chain, the Constraint Management project method, produces system releases periodically. Alternate project methods produce system updates continuously.

For creating, reading, updating, and deleting data, developers may write data services and expose them via Application Programming Interfaces (APIs). This enables other developers to access data without having to know where or how the data is stored.

Profile: Operations Manager

System	Operations
Goal	Service-Level Agreement (SLA)
Flow	Data → production → output
Application	**Production for Services** (Adjusts capacity based on service levels)
Constraint	Hardware and software capacity
Buffer	Sprint capacity and skills
Limits	Sufficient skilled operators

Operators allocate storage and run production jobs that process data and populate databases. Operators also perform backup and restore.

Profile: Data Producers

System	Data Production
Goal	Revenue from data products and services
Flow	Vision → production → data as a product or service
Application	**Critical Chain** (or alternate method)
Constraint	Critical Chain
Buffer	Project (time) buffer
Limits	Sufficient skills

All of the previous profiles describe roles dealing predominantly with data for internal use. In contrast, this Data Producers profile is a catch-all for the multitude of people who produce data as a product or service.

- **Information sector.** Producers, writers, directors, performers, artists, editors, distributors, publishers, telecommunications, etc.
- **Financial sector.** Credit reporting agencies, financial rating agencies, market researchers, market reporting services, real estate brokers, etc.
- **Professional, Scientific, and Technical Services sector.** Accountants, lawyers, consultants, scientists, architects, engineers, advertisers, etc.

When data production can be managed as a project, Critical Chain applies. Otherwise, Constraint Management's Production for Services application could apply.

System of Systems

A system of systems means management decisions and worker actions in one system can have repercussions elsewhere [Chessell, 2013]. Figure 8-1 illustrates how these profiles interact:

- Staff functions (CDO, CPO, CSO, BPO) set policies.

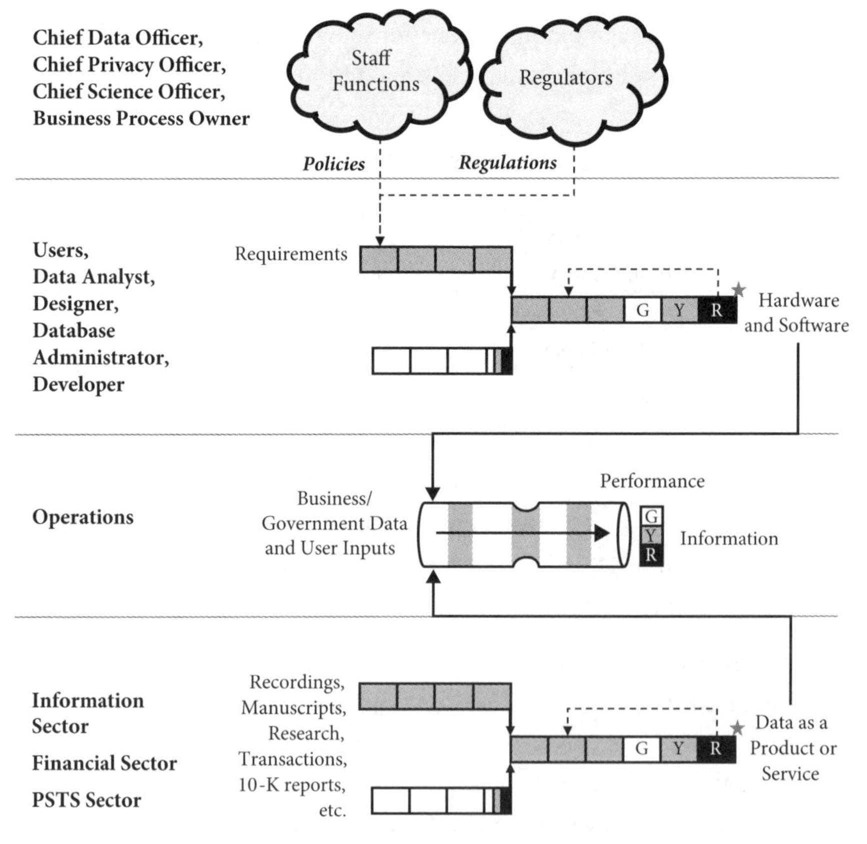

Figure 8-1 Data Constraint Management.

- Regulators set regulations.
- Users set requirements.
- Development projects staffed by Analysts, DBAs, Designers, and Developers create deliverables consisting of hardware and software.
- Operations installs and runs the hardware and software, which turns data and user inputs into information. The enterprise uses that information for its own business processes.
- Data Producers create data as a product or service, but those products and services also require internal data processing to collect revenue, pay bills, schedule work, etc.

The cloud symbols vaguely represent staff functions and external regulators because the specifics of those functions are not shown in detail. Those symbols are not meant to imply anything about Cloud Computing.

Although the entire figure depicts how data is handled, internal data flows happen within the pipe near the bottom representing production. Diagrams of data flows within and between enterprise applications are as intricate as a roadway map. That level of detail is excluded from the figure because internal data flows are unique and typically not where business process constraints occur.

Conclusion

Computing has led to fundamental shifts in business and government. As technology for data handling has improved, some trends have made management more complicated.

First, business and government have become more instrumented. Consequently, managers are awash in data, but whether it's the right data is debatable. Some organizations have so much data that what to do with it is a CXO challenge. Said one customer to its Cloud Computing provider: "You have our data, so why aren't you analyzing it for us?"

Second, despite the plethora of data, managers cannot know everything about their organizations, because their organizations are a lot more complex and operations are a lot less visible than they used to be. Rather than relying on intuition and experience nowadays, organizations have shifted toward data-driven, fact-based decision-making using real-time dashboards. However, rather than filtering, data analytics can overwhelm managers with irrelevant or unreliable information.

Finally, Constraint Management recognizes that even complex systems can be managed at their constraints, so visibility there is critical. On the other hand, visibility elsewhere may not matter, and can be detrimental because it invites attempts to optimize everything.

It's said that data is like closet space: You never have enough. Data is also like closet space because it often harbors clutter.

CHAPTER 9

KNOWLEDGE

Knowledge is the theoretical or practical understanding of a subject. Historically, human intelligence has been the only way to produce, comprehend, and apply knowledge. But that's changing. Technologies have progressed to the point where, in some domains, it's hard to tell whether natural or artificial intelligence is at work.

Although the subjects are related, this Knowledge chapter is separate from the Data chapter because the use cases and technologies are different. Analytics uses algorithms to turn structured data—such as names and numbers—into information. On the other hand, knowledge-oriented technologies use heuristics to turn unstructured data—such as text, audio, images, or video—into knowledge.

While analytics technology is heavy on calculations, knowledge-oriented technologies are heavy on patterns, learning, and reasoning. Algorithms find optimal solutions. Heuristics find satisfactory solutions when optimal solutions are impossible.

If a database is used to store and retrieve data, it's tempting to assume that a knowledge base is used to store and retrieve knowledge. In common usage, however, a knowledge base is a library of documents or other reference material meant for human consumption. That's not what this chapter is about. This chapter is about technologies that can ingest and apply the content of knowledge bases, and many other sources, as well as, or better than, humans can.

This Knowledge chapter breaks with the organization of the Hardware, Software, and Data chapters. It doesn't conclude with a diagram and description of a System of Systems. That point is well established now. This chapter instead focuses on the unique aspects of knowledge.

Knowledge Work

"Knowledge work" typically refers to tasks requiring human intelligence. But the boundaries are shifting.

- In the 1800s, the **Industrial Revolution** replaced human and animal power with machine power.
- In the 1900s, the **Information Revolution** replaced human processing of data into information with computer processing.
- In the 2000s, the **Knowledge Revolution** is augmenting human intelligence with artificial intelligence.

For example, some financial news reports are already computer-generated because humans aren't fast enough to publish it within minutes of an event in the financial markets. For situations like this, the Turing Test was asserted by a pioneer in the computing field: If you can't tell whether you're communicating with a human or a computer, the computer is intelligent.

The difference between traditional data analytics and knowledge-oriented technologies is that the latter can behave in ways beyond their original programming. There is no generally accepted taxonomy of those technologies, but Machine Learning, Artificial Intelligence, and Cognitive Computing are currently prominent subjects.

Machine Learning

Machine Learning technologies generate discoveries from complex data. ML does this by detecting underlying patterns in data sets.

For example, ML can examine subassemblies as they progress through an assembly line and flag those with missing or malformed components. ML can recognize types of farm produce and sort by quality rather than size. ML can recognize human faces and choose to admit or deny access to restricted areas. ML can even predict some voter behavior.

The results are achieved by presenting large enough samples that ML discovers patterns which could elude traditional analytics. Thus, the machine learns what the patterns are instead of being pre-programmed with patterns to look for.

Artificial Intelligence

In simplest terms, Artificial Intelligence is any nonbiological intellect. Today, we associate AI with the performance of tasks that usually require human intelligence, such as speech recognition, image recognition, or language translation. Eventually, however, AI may perform tasks no human can perform because the AI will operate at speeds beyond human capabilities or in environments hostile to life itself. "The question of whether computers can think is like the question of whether submarines can swim." [Edsger W. Dijkstra]

State of the art today is Special Intelligence, which means doing something well only in one specific context. Playing games like Checkers, Chess, and Go was initially used to drive Research and Development. Gradually, however, practical applications have emerged, such as autonomous vehicles.

At the leading edge today is General Intelligence, which means doing lots of things in different contexts. Again, playing a game requiring general knowledge (*Jeopardy!*) has driven R&D. But practical applications are emerging, such as personal-assistant devices that listen for spoken queries and commands, then perform a diverse set of tasks such as dimming lights, adjusting temperature, closing doors, and reporting weather.

Cognitive Computing

Cognitive Computing encompasses reasoning about a domain. For example, CC can explain why a specific treatment might be more appropriate for certain symptoms. Reasons behind a recommendation make it more compelling than black-box AI/ML. But the reasons also provide a means for human experts to assess whether the recommendation is flawed.

Likewise, when CC engages in public policy debate, that requires reasoning. But it highlights differences between humans and computers. We humans understand questions easily but recalling millions of pertinent facts is hard. Conversely, CC recalls millions of facts easily, but understanding questions is hard, particularly when those questions include nebulous concepts. For instance, love of one's mother and one's spouse are different, but neither are even remotely related to a score of love in tennis.

Fairness

Natural intelligence suffers hundreds of biases. For example, we give more weight to recent experience. We overestimate our ability to resist temptation. We have a preference for status quo. And human misunderstanding of probabilities is legendary.

Knowledge-oriented technologies may inherit biases when they learn from humans. For instance, one learned to swear. Another spouted abusive epithets. Yet another invented its own language. A couple engaged in feuds by repeatedly taking actions to correct each other.

Thus, a fairness rating scale is in development. The goal is to influence which knowledge-oriented technologies are trustworthy.

ADVENTURE: Tiptoeing on the Frontier

My introduction to Artificial Intelligence came during graduate school. By today's standards, state of the art was primitive because computer hardware was slow and expensive. For instance, shape recognition was done with logic about edges and vertices rather than machine learning on images. And speech recognition failed famously when "how to recognize speech" was misunderstood as "how to wreck a nice beach."

Game On

One assignment was to write code, an AI player, that would compete in a tournament. Pairs of programs would face off virtually, and the winners would progress to the next round until only one program remained undefeated.

The learning objective was to get students to think about strategies to complete patterns before their opponents. There was no time limit during the tournament, so the AI players could examine many alternatives before selecting their next move. Once initiated, however, the programs played autonomously, without further input from their authors.

The game board was just an unbounded number line of integers, straightforward enough to represent in a computer. Players would take turns making a move by claiming one number on the number line. Winning the round and advancing to the next round was accomplished by claiming a series of six integers at equal intervals.

Thus, each player's code had to examine both offensive and defensive moves. Offensive moves added to the player's own series, while defensive moves blocked the other player's series. A good move did both simultaneously.

A clever move added a number to multiple series. For instance, if a player already had 0, 6, 9, 12, and 24, claiming 36 added to their series based on intervals of 3, 6, and 12. Thus, common multiples was a popular strategy. With enough series nearing completion, an opponent couldn't block them all.

Strategy also had to prioritize offensive and defensive moves. If a player was two moves away from a win, while their opponent was one move away, blocking their opponent meant the player would survive that move even if it meant it was no closer to winning the round. This was the survival-first strategy.

Genuine Stupidity

When the result of the tournament was announced, the winner had completed no patterns in any round. How can that be? The winning strategy was effectively "forced errors." What?

When the winner was about to lose, it would block and then claim a new number outside the range of numbers already played based on the widest partial series it already had. As this happened repeatedly, the size of the board grew exponentially. Thus, the winning player won by not losing.

When the board expanded, opposing players would try to analyze an expanding universe of potential moves and countermoves, then run out of memory in the process. Thus, the winner won, not by clever strategy that completed a pattern, but by forcing all the other players into default.

When asked if this strategy was intentional, the author admitted that it was a desperation move when his player was on the verge of losing. He didn't plan for it to win by default and was as surprised by the outcome as the rest of us were.

Lessons Learned

Lessons learned include:

- Artificial Intelligence may be no match for genuine stupidity. In an unrestricted problem space, edge cases are effectively infinite.
- In languages used for AI, there may be no inherent distinction between data and programs, so it's not unusual for programs to write other programs.
- Functional programming (commonly used for AI) is a different world from procedural and object-oriented programming (commonly used for business, government, and science). Programmers immersed in one world may be unable to comprehend programs in another world.

Data Cleansing

As noted in the Data chapter, data cleansing is often a severely underestimated activity in analytics projects. The same can be said for knowledge projects, although the severity of the problem and the remediation methods may be different.

Considerable data for analytics comes from transactions or from sensors, and it is mostly structured data: product identifiers, numeric quantities, classification codes, time stamps. Erroneous data can be discarded, corrected, or tolerated. The procedures are tried and true.

Considerable data for some knowledge projects comes from devices, and it is mostly unstructured data: audio, images, videos. However, other data for knowledge projects comes from Subject Matter Experts who are the source of even more unstructured data in the form of lengthy narratives, social guidelines, company policies, rules of thumb, and insights about special cases. Procedures for cleansing that data tend to be improvisational.

A passive approach to tapping SME knowledge is to ingest existing documents. An active approach is to interview SMEs or give them tools to record their knowledge. Both approaches are illustrated in the following Adventure section.

ADVENTURE: Trouble with Trouble Tickets

When I was a chief technology officer, one of my business units provided services in response to trouble tickets. Later, when I worked in strategy, we invested in technologies to automate trouble ticket services.

The objective was to ingest trouble tickets, detect patterns, and automate responses to common problems. Uncommon problems would still require action by human agents. If the first- and second-level support agents couldn't resolve a problem, it was escalated to a subject matter expert.

Anecdotal evidence suggested that the 80-20 rule would apply: Solutions for 20% of the problem types could resolve 80% of the tickets. That's terrific leverage if you can get it, so nobody doubted that the objective was a worthy pursuit.

Cleaning Day

The biggest challenge came not from bad data but from data that never was. As we ingested trouble tickets, we saw that most of the cause fields said, "Broken," resolution fields said, "Got it working," and status fields too often said simply, "Case closed." Thus, the majority of trouble tickets contained no useful knowledge whatsoever.

Agents handling trouble tickets were admittedly under pressure to close tickets as quickly as possible, and they assumed that no one would read the trouble tickets later because it wasn't routine. However, that prevented our knowledge project from harvesting knowledge from those tickets. Like sailors tacking to a different course before running aground, we had to pivot the project.

Taking Another Tack

The new approach was to educate agents and SMEs about populating trouble tickets. To get their buy-in, they had to be educated about how the data would be used. In situations like this, everybody wants to know "What's in it for me?" WIIFM thus became a mantra of project and technical leaders, and the project began accumulating new knowledge rather than harvesting existing knowledge.

Lessons Learned

Lessons learned include:

- Before committing to a knowledge project, take the time to conduct due diligence on sources. If the data is bad or missing, that's a major risk factor for it to become a troubled project.
- Data ingestion and cleansing can be as much as 80% of a knowledge project, but without it, the knowledge part will fail. If that effort is not built into the project plan and socialized with the project sponsors and participants, the project may be quite late—and therefore in danger of cancellation.

Digital Robots

What does a robot look like? In science fiction, robots are humanoid. In factories, robots may have just one enormous arm with a grappler, welder, or sprayer on the end. In warehouses, robots may look like storage bins on wheels. In planetary exploration, robots are six-wheeled vehicles with various actuators, sensors, and solar panels. In offices, robots may look like mobile mail rooms. In hotels, robots may look like rogue serving carts. In data centers, robots may look like vending machines as they shuffle tapes in and out of storage bins.

On the Internet, however, bots are practically invisible because they are software scripts that automate routine tasks. Similarly, in operations centers, Robotic Process Automation is entirely software—but not all processes are routine. Thus, these digital robots run on knowledge.

Knowledge comes in two flavors:

- **Institutional knowledge** is documented.
- **Tribal knowledge** is undocumented.

Documentation has been the bane of the Information field for decades because it's expensive to produce, yet seldom referenced. RTFM muttered under the breath stands for "Read the frigging manual."

That said, documentation may be obscured by jargon, hidden in plain sight among thousands of other documents, or devalued by being even slightly out of date. Back when I was learning about database management, I wasted a day debugging a program where the compiler documentation said a function would return just one character from a string. It actually returned two characters—one visible and one not—because the compiler had been ported from a computer with a different hardware architecture.

Tribal knowledge resides in people's heads. It's gathered through experience, and mostly shared orally or on white board diagrams. Subject matter experts may not be conscious of what they know until a situation arises to trigger that knowledge.

This distinction between institutional and tribal knowledge becomes apparent when doing Robotic Process Automation. Writing bots for routine tasks is straightforward because the procedures are either documented or readily recalled. In contrast, handling nonroutine situations is difficult because relevant documentation is often scattered or nonexistent.

ADVENTURE: A Torrent of Messages

In an enterprise computing center, thousands of messages appear on the operator's console every day. Most of those messages are innocuous information messages that scroll off the screen automatically. Others are action-required messages, and they stick on screen until an operator clears them.

It gets complicated when some messages that seem harmless are actually weak signals of an impending problem. Those messages are easy to miss when they occur over a span of hours, days, or weeks.

Because outages disrupt the business and may require service providers to pay penalties for missing a Service-Level Agreement, resolving outages quickly is a top priority. The next priority is making sure those outages don't recur by doing root cause analysis and taking preventive actions.

Riding into the Sunset

When I worked in strategy, I was asked to tackle knowledge erosion. Subject matter experts with deep tribal knowledge about operations were resigning and retiring faster than they were being replaced. And transferring work from one geography to another effectively pushed the reset button on expertise. Expert operations problem-solving was gradually becoming a lost art.

Growing subject matter experts internally can take years because they need to know the enterprise's software applications and business processes, in addition to its computing environment. Hiring SMEs from outside means they bring tribal knowledge from elsewhere, but only a subset of that knowledge will be pertinent at their new site. Indeed, some of their prior knowledge can be downright misleading.

Knowledge walking out the door is an example of a non-constraint on a trajectory to become the new operations constraint. Cognitive Computing offers a way to capture tribal knowledge, but it must be triggered somehow. Unfortunately, signals of impending problems are buried in the blizzard of messages popping up on the operator's console.

Faint Signals

The initial scenario we chose for proof of concept was based on log messages from an actual outage caused by a storage problem. Our objective was to demonstrate a combination of technologies that would have detected and fixed the problem in time to avoid the outage.

Grasping this problem requires some explanation, however. Leaving aside the details of how storage space is allocated, when more files are created than expected, it's possible to run out of storage space. That's what happened in our scenario. The developers were not paying attention to this possibility. As their code created more files than expected, the log messages looked normal. Eventually, however, the system ran out of space.

Connecting the dots on problems whose symptoms appear infrequently is difficult for operators because (1) it requires extraordinary attentiveness, and (2) a given operator may not be on duty when every occurrence appears. Therefore, the first technology we deployed was Anomaly Detection. That tool looked not just at isolated log messages, but at patterns over a specified time horizon, which we set at several weeks because the code that caused the problem ran weekly.

When an anomaly was detected relative to a normal pattern, the tool would raise an alarm, which we fed into the second technology we deployed: Cognitive Computing. The Anomaly Detection tool essentially asked the Cognitive Computing tool, "What is the solution to {log message}?" When operators saw the answer, they could pose further questions in natural language.

Initially, operators or SMEs would take the appropriate action. But in the long run, those actions that could be automated would be taken by Robotic Process Automation, an existing technology.

Lost Tribe

Ingesting documents into the Cognitive Computing tool was relatively straightforward, but curating question-and-answer pairs took considerable time and care. Q&A authors needed tribal knowledge about applications, job schedules, and the computing environment to craft sensible questions and useful answers.

Quality control was exercised by having an editor review every Q&A pair before it was committed. But if the authors or reviewers were not SMEs, the resulting level of quality control wasn't always enough. Duplicate, erroneous, confusing, or conflicting Q&A pairs could have destroyed confidence in the tool. Thus, more of the Q&A curation and reviewing responsibility fell to SMEs. Unfortunately, as some SMEs were headed out the door, the workload on the remaining SMEs increased.

The proof of concept did achieve its objective. We showed that even faint signals can be picked up with Anomaly Detection. We also showed that tribal knowledge can be captured and activated in a Cognitive Computing tool. Both those technologies were effective precursors to Robotic Process Automation.

The next objective was to reach critical mass by covering a larger set of log messages. But it was still a race between knowledge erosion and knowledge retention.

Lessons Learned

Lessons learned include:

- Curating tribal knowledge for basic competence (level 1 support) can take a year.
- Curating tribal knowledge for harder problems (level 2 support) takes more than a year.
- Curating tribal knowledge for the hardest problems (SME-level support) may never end because the problems change as fast or faster than the knowledge.
- Anomaly Detection, Cognitive Computing, and Robotic Process Automation are meant to prevent problems. However, if enterprises don't count the times anomalies were detected and resolved before problems occurred, it may be misperceived as wasted investment.

Hype Cycle

Every emerging technology in the Information field moves through the Hype Cycle: Innovation, Expectations, Disillusionment, Enlightenment, Productivity [Gartner, 2019]. Some technologies are celebrated as success stories during the Productivity stage, while others fade into obscurity during the Disillusionment stage. For example, Data Analytics is solidly in the Productivity stage, while Expert Systems technology never recovered from its Disillusionment stage.

Machine Learning and Cognitive Computing are currently riding grand expectations. Like tides in the oceans, however, academic and venture funding for Artificial Intelligence research comes and goes. The terms "AI Spring" and "AI Winter" are commonly used.

One characteristic of all commercially viable AI technology is it must be trained with enormous numbers of actual cases and then find success in the real world. Making the leap from research lab to practical application is not easy.

Cold Start

The cold start problem happens with every emerging technology. Hiring people with the right skills and experience in a nascent area is always hard because the job market is hot.

Retaining them is hard too because getting started without infrastructure, tools, methods, and colleagues is extra work. Just as some people have the temperament for starting up new ventures, some people have the temperament for starting up an emerging technology. But not all do.

Pseudo-AI

Predictions about technology overestimate what can be achieved in the short run while underestimating what can be achieved in the long run. Hence, some start-ups engage in "fake it 'til you make it" by employing people to perform tasks behind the scenes while implying that technology is doing the work.

Although ethically questionable, the motivation is simple: For some tasks, it's much easier to get people to act like robots than to get robots to act like people. Here are some examples:

- Generating calendar entries from confirmation emails
- Transcribing expense receipts from images
- Conversion of voicemails into text messages

Technological Singularity

Despite these obstacles, worries about a Technological Singularity do appear [Kurzweil, 2006]. In general, a singularity is a point beyond which predictions are impossible because a property is infinite. For instance, the centers of black holes are singularities because they are a huge mass in an infinitely small space. Models that describe most of our universe break down in a singularity such as this.

The Technological Singularity is conjecture that super AI will trigger runaway technological innovation which will lead to profound and uncontrollable changes to human civilization. Whether this will be beneficial or calamitous is a matter of debate, because whether it will occur, when it will occur, or what the outcome would be is all speculation.

One position says that if AI does only what people teach it, it may exhibit undesirable behaviors, such as swearing and deception, but there will be no Technological Singularity. An opposing position says that if AI does more than what people can teach it, there will be no brakes on a Technological Singularity, and it will amplify the worst human tendencies.

Lost in the argument is the possibility that there may be conflict between opposing AI. There is no reason to assume that AI will be a single beneficial, benign, or malevolent entity. Like humans, AI may be diverse rather than uniform. In that context, emergent behavior is impossible to predict.

Such worries seem like remote possibilities, however, to folks struggling with a cold start on an AI Winter's day. Nevertheless, the debate suggests that in all matters technical, we should be careful what we unleash, because even though technology often fails to meet short-term expectations, it often exceeds long-term predictions.

Constraint Management

Constraint Management for knowledge varies by scenario. As always, it depends on where the flow and the constraint are. Can knowledge be the enterprise constraint? Of course it can, if knowledge is what the enterprise sells, and it can't get enough. Red, yellow, and green zones can be established to encourage and track knowledge accumulation, of course, but there are more-likely possibilities.

- Scenario #1 is a knowledge technology vendor. The product is specialty hardware, knowledge-oriented software, or a Cloud Computing implementation. If it's a start-up, the enterprise constraint is probably the research and development project or the sales pipeline. If it's a going concern, the enterprise constraint could be (1) in the market because customers aren't buying, (2) in sales because sellers aren't selling, (3) in the production process for the hardware, (4) in the hardware/software distribution process, or (5) in Cloud operations. The Constraint Management applications for all those alternatives are well-known and described in previous chapters. Nothing special is required.

- Scenario #2 is a knowledge consultant. These services could include strategy and plans for knowledge implementation, installation and configuration of hardware and software, staff training, or knowledge implementation, which includes document ingestion, data cleansing, model training, Q&A curation, etc. If knowledge consulting is the business, then the Constraint Management applications for services apply.
- Scenario #3 is an internal knowledge project in an enterprise not in the knowledge business but using knowledge to manage its core mission. An internal knowledge project probably uses technology from a vendor, which may be provided via Cloud Computing. If so, then technology implementation is probably already done. As described throughout this chapter, however, capturing and curating knowledge is demanding work. But if the enterprise's strategy is to compete on knowledge, that work can be justified as a critical success factor. For example, knowledge-oriented technologies can address manufacturing defects, financial decisions, medical treatments, and government services.
- Scenario #4 occurs when lack of knowledge is not the enterprise constraint, but it could impede a local process. As the adventures with trouble tickets and knowledge erosion illustrated, not having knowledge about key processes can limit what they produce. And the slips can be so gradual that they escape notice until a crisis looms.

Conclusion

Arthur C. Clarke said, "Any sufficiently advanced technology is indistinguishable from magic." That describes knowledge technology: When it works, it's magical. Getting it to work is no small feat, however. Like the wizard in *The Wizard of Oz*, appearances can be deceiving.

Every new technology has the potential to change how we work and live. Autonomous vehicles are not an imminent threat to knowledge work, but they are a threat to skilled work. “Driver” is a common occupation for people without a college education, yet it may be overtaken by autonomous vehicles.

We can only speculate on how far knowledge technology will encroach on knowledge work. But it’s advisable to recognize that as knowledge technology takes on the trailing edge of knowledge work, people will keep expanding the leading edge of knowledge work.

In the meantime, Constraint Management for knowledge means applying principles honed elsewhere to knowledge technologies. Gathering and curating knowledge are tasks amenable to Constraint Management’s applications for engineering, production, and distribution of products and services.

CHAPTER 10

NETWORKS

Networks as we know them today are relatively recent technical innovations. Mobile (cell) phones are replacing land lines. Digital files are replacing fax machines. Virtual Private Networks are replacing leased lines. Fiber optic cables are replacing copper cables and microwave antennas. Packet switching is replacing circuit switching. Internet Service Providers have replaced analog modems. Streaming television is jockeying for position with terrestrial broadcast, satellite broadcast, and cable TV.

Network providers routinely manage constraints from their Network Control Centers because it's their core business. At the operations level, they manage capacity during busy periods, as well as during disasters. At the strategic level, they engineer geographic coverage and capacity to anticipate business and residential growth.

Central questions of this chapter are:

- Can a network be the constraint if you're not in the network business?
- If so, how should a network constraint be managed?
- If not a constraint, can networks be a limitation?

Networks serve multiple purposes and come in various forms. The following sections describe them according to the traffic they carry (voice, data, video networks) or the function they serve (specialty networks). Then this chapter discusses various network topics that could have a bearing on Constraint Management, such as security, disaster recovery, content delivery, and network policies.

Voice Networks

Telephone or telecommunications networks are engineered to carry voices:

- **Plain Old Telephone Service** (POTS) uses analog signals to carry voices over switched circuits dedicated for the duration of each call.
- **Public Switched Telephone Network** (PSTN) is made up of central office switches and trunk switches.
- **Private Branch Exchanges** (PBXs) connect phones within businesses.
- **Mobile phones** digitize voices and carry them wirelessly over the carrier's cell network.
- **Voice Over Internet Protocol** (VOIP) digitizes voices and carries them as a set of packets that can follow different routes over the Internet during each call.

Analog modems made it possible to send data and images (faxes) over voice networks. However, speed, quality, and reliability were lower than modern networks engineered to carry non-voice traffic.

Data Networks

Data networks connect computers and smartphone data plans, as well as devices, in the Internet of Things:

- **Leased lines** are private circuits provided by telcos.
- **Local Area Networks** (LANs) use cables to connect devices over short distances.
- **Wireless LANs** use WiFi to connect devices without cables.
- **Wide Area Networks** (WANs) connect devices over distances too long for LANs.
- The **Internet** is a worldwide public network of networks using the same protocol.
- **Intranets** use Internet protocols for private networks.
- **Virtual Private Networks** (VPNs) are private networks using authentication and encryption to protect traffic carried over the Internet.

Although originally engineered to carry data, the Internet now carries other traffic, too. Unfortunately, the original engineering did not anticipate nefarious uses, so security is a perennial problem. On the other hand, the Internet Protocol was engineered to survive a nuclear war, so there's that.

Video Networks

Video networks have been dominated by entertainment uses, but they can carry information, too:

- **Closed Circuit Television** (CCTV) is not for broadcasting. For example, it's used to connect security cameras to central recording and monitoring stations. Originally analog, CCTV devices are increasingly digital and use Internet Protocol.
- **Terrestrial Broadcast**, **Satellite Broadcast**, **Cable TV**, and **Streaming TV** broadcast entertainment, news, and education.

Video signals tend to consume more bandwidth than voice or data, so compression serves both providers and consumers well. Videos can be encoded at different bit rates depending on what's going on in a scene, with dialog at lower rates than action scenes. Similarly, with more video being viewed on smartphones, encoding schemes can be tailored to the small screens of mobile devices.

Specialty Networks

Specialty networks serve a specific function, and although they serve the public interest, they are not public networks:

- **Global Positioning System** (GPS) is a constellation of satellites in semi-synchronous polar orbits broadcasting worldwide for positioning, navigation, and timekeeping.
- **Remote sensing satellites** in geostationary orbits carry an array of sensors to monitor weather, fires, and search and rescue.
- **Communication satellites** in geosynchronous orbits receive and retransmit voice, data, and video.

One unavoidable characteristic of satellite communications is the latency due to distance. For one-way transmissions, this typically is not an issue. But transmission to and from a geosynchronous orbit takes about half a second per round trip. Round-trip delay on a submarine cable is about one tenth as much.

Security

Networks are an obvious way to subvert physical security without detection. But there are many ways to combat that vulnerability:

- **Firewalls** allow only certain traffic to pass to or from a network.
- **Authentication** determines that you are who you claim to be, and you have permission to do what you're trying to do.
- **Logging** and **monitoring** create records of activities.
- **Encryption** scrambles data, thereby making it useless if breached.

Of course, malware attempts to defeat all those precautions. SQL injections and buffer overflows exploit vulnerabilities created by careless programming. When the technical barriers are insurmountable, cyber thieves use social engineering to entice people to breach security.

Disasters

Natural disasters affect networks, of course. For instance, an earthquake in the Luzon Strait of the China Sea triggered an undersea landslide that severed data cables between Southeast Asia and the U.S.

Human-made disasters are also a threat. Cable cuts by backhoe operators are almost a cliché. But sometimes the disasters are of the "you couldn't make this up" variety: Someone hammering a "buried cable" signpost into the ground managed to cut the cable it was meant to protect.

Disaster recovery is a technical service for business networks and computing operations. In addition to providing duplicate hardware and software and networks, the service uses backup data to restore computing operations. Workers also need workspace, office equipment, and phones, so a full recovery service is more than just computers, data, and networks.

ADVENTURE: World Gone Crazy

Disasters remind us how fragile our lives, residences, and businesses are. If networks go down, the ripple effects can be local, regional, national, or even worldwide.

Ripple Effects

When fire destroyed the switch at a telephone company central office, it knocked out all the phones in a cluster of suburbs, including my own, plus communications between major airports and regional air traffic control. Even basic activities then became much harder:

- With no way to coordinate flights, air traffic was disrupted internationally.
- With no way to call for emergency services, police cars, fire trucks, and ambulances were parked at major intersections. If the service someone needed wasn't there, they could summon it by radio.
- With no way to phone home, there was no way to know which train your spouse, children, or friends would be on. If they needed a ride from the station, the driver had to wait and hope.
- With no way to phone work, there was no way for schools to alert parents about a sick or injured child. In urgent cases, police relayed messages.
- With no data networks, companies outside the affected area opened their buildings to other businesses whose own networks were knocked out.

It ordinarily took 12 months to install a new switch, but the project to clear the debris, repair the building, and install a new switch was completed in about 90 days because it was an urgent priority. All other demands for materials, equipment, and workers were secondary. In Constraint Management terms, this is called "subordinating to the constraint."

When an F-4 tornado leveled a path 1.5 miles (2.4 km) wide and 2.5 miles (4.0 km) long through a city, destroying our family home, the telephone company followed standard procedure and shut down inbound trunks. In theory, someone inside the disaster area should have been able to place one outbound call to a family member or work colleague who could reassure the rest of the family or company. In practice, it took several days for survivors to communicate outside the disaster area, and non-residents were not allowed in, so the period of "no news" was nerve-wracking.

When a hurricane struck a coastal city where my extended family lived, refugees fleeing to the nearest inland city swamped its communication networks, as well as city services, retail, hotels, and restaurants. A few years later, a hundred-year flood drove refugees even further inland.

When bridge maintenance breached a tunnel under the river containing fiber optic cables, basements in the entire business district of a large city were flooded. This knocked out power and communications, which meant buildings were evacuated, residents departed, and business ceased.

When a bug in the GPS network caused clocks in its satellites to lose track of time, other systems dependent on precise timekeeping failed. Cell phone towers lost connection. Emergency communications were disrupted. Automated teller machines went offline. Telescopes halted. And, of course, navigation devices failed. Because GPS is a worldwide network, the ripple effects were worldwide.

When a technician attempted to block a few phone numbers suspected of malicious activity, the network management software interpreted blanks as a wildcard (match to any phone number). Consequently, the biggest telephone outage in U.S. history saw 111 million phone calls blocked.

Disaster Recovery

For disaster recovery, data centers are designed to replicate data and to fail-over. The primary and recovery centers are far enough apart so that they won't be knocked out by the same disaster, yet close enough for data to be replicated in a timely manner. Although recovery sites are engineered to have at least two separate network connections and two separate power sources, part of failing-over includes running on batteries or generators while the power is out.

In large enterprises, the volume of active data is large enough that keeping it all online may not be affordable. Thus, business recovery sites have loading docks and elevators big enough to accept a truckload of tapes containing software and data.

Recovery sites have the same computing hardware as primary sites—or an emulation capability. Though the recovery sites may not run every client's full workload simultaneously, they have more than enough capacity to run the essentials. Thus, those recovery sites are able to replicate primary data networks. This is accomplished with a Software Defined Network. (More about that later.)

Security is a concern. Primary sites may look like a rundown urban warehouse on the outside while running a modern lights-out facility inside. Recovery sites are sometimes disguised on maps: The "satellite view" looks like forest when the site actually has roads, buildings, fuel tanks, etc.

Not My Department

When I was consulting with a city on information technology matters, we did Strengths, Weaknesses, Opportunities, and Threats (SWOT) analysis. When asked about Supervisory Control and Data Acquisition devices, the city's information technology manager said, "SCADA devices aren't connected to the city's secure network, so not a threat." Our reply was, "Yes, but those devices are connected to something, so they deserve a security audit to determine whether the water and sewer systems are at risk from attackers outside the secure network."

SCADA devices were not the city constraint, but if hacked and disrupted, the systems they controlled could quickly become an urgent priority for firefighting and public health. Thus, this was an example of a non-constraint with the potential to become a severe bottleneck on city services.

Lessons Learned

Lessons learned include:

- Natural and human-made disasters can knock out a network. Irregular operations then take over until regular operations can be restored. And a network that wasn't the constraint can become the new constraint.
- Disaster recovery should address the constraint first.
- Organizational boundaries affect how managers see threats that could alter constraints.

Content Delivery Networks

Content Delivery Networks store data and video close to where it will be consumed. This improves performance because the "last mile" transmission distances are shorter and fewer users are accessing content at once. It also improves availability by eliminating a single point of failure that centralized delivery creates.

Software Defined Networks

Software Defined Networks enable enterprises to respond to changing requirements by reconfiguring the network from a console instead of individual switches. The central controller then sends packet-handling rules to switches. For example, suspicious traffic can be dropped or rerouted, and a network can be emulated at a disaster recovery site.

Network Policies

Network policies are rules that authorize access and govern activities over a network. Such policies aren't just documented, they are active rules configured into the network hardware and software.

Some network policies establish access zones based on levels of trust. For example, network firewalls are often configured to grant employees access to a zone with confidential information while limiting business partners and customers to a trusted zone and restricting public access to generic information. User ID, and sometimes a key fob or fingerprint, determines which zones an individual can access.

Some public policies cannot be met with network policies, however. Regulation of cross-border data flows forces companies to put data centers in every country, which can be a prohibitive capital investment or operating expense.

ADVENTURE: **Policy Madness**

When I was CTO in a group doing research and development on Software as a Service, we ran headlong into a conflict between existing network policy and new requirements. What started as an issue with network policy became an issue with company policy, which was to restrict access vertically (by trust level) rather than horizontally (by customer).

Against Company Policy

The existing network policy had been created for Traditional Computing centers, but the new requirements were for Cloud Computing centers. The conflict arose because the public access zone was too restrictive (wouldn't allow access to any servers), while the customer zone was too lenient (would allow access to too many servers).

Alternatives were proposed, debated, and rejected:

- Create a new cloud zone in the existing network? Too complicated.
- Create a separate cloud network? Too expensive.
- Restrict the customer zone? Too dangerous.

Buying the Way Out

Ultimately, my company acquired another company. Problem solved. Their network was already configured for multi-tenancy, which horizontally segmented the network so that customers had access only to their own cloud servers. And that cloud network was already operational.

Lessons Learned

Lessons learned include:

- Network policy was an operations constraint because it restricted access. We could not grow the business without multi-tenancy.
- Company policy was a strategic constraint because it inhibited the revenue opportunity.
- Inertia was a limiting factor because existing policy shaped management thinking.

Braess' Paradox

Braess' Paradox is an observation about network capacity and bottlenecks. Common sense says adding a road to a road network increases overall capacity and therefore should reduce transit times. The paradox says transit times actually increase when too many people choose the route that's best for themselves and that overloads those routes.

Does this paradox affect packets flowing through a network? Fortunately, no, if the routing rules aim to maximize flow across the entire network. However, it can happen if the routing rules maximize flow only to the nearest neighbor and the number of nodes in the network increases. Given the rapid growth of Internet traffic, this is something that network providers are watching.

Constraint Management

Can a network be the enterprise constraint? Yes, if the enterprise is in the network business. That company's revenue can be constrained by the reach of its network or by the content it delivers. Reaching customers in the next county means building out the network. Getting customers to buy premium content means producing better programs.

Many networks are engineering to handle peak demand, however, so the network itself isn't necessarily the constraint even in a network business. When networks are engineered for peak demand, they have slack (sprint capacity) the rest of the time. Thus, the enterprise constraint may be in the market. Also, because building networks and producing programs are capital intensive activities, finance is always a limit if not the constraint.

If an enterprise isn't in the network business, the network may nevertheless be a strategic constraint if company policies work against new business opportunities. However, if the enterprise has its own Network Operations Center, that NOC could use Constraint Management applications for projects, processes, or distribution.

Conclusion

Digitization and globalization are trends increasing business and government use of networks. Social networking and videography are forces behind growing consumer demand for communication networks. That said, if you're not in the networking business or running a network for your organization, a communications network is something that you probably don't think about much—until it's unavailable.

As noted in the chapter introduction, network providers routinely manage their constraint, though not necessarily with Constraint Management applications per se. Here are answers to the central questions of this chapter:

- Can a network be a constraint if you're not in the network business? A network probably isn't the enterprise constraint, but it can be a local system constraint.
- How could a network be managed if it's a local system constraint? With the Constraint Management applications for projects, processes, or distribution.
- How should a network be managed if it's a strategic constraint? If company policy is directed at the enterprise goal, then it will adapt to changing requirements and drive sensible network policy accordingly.

If not a constraint, can networks be a limitation? Yes, like personal data plans that limit how much data a person or family can use per month, enterprises can and do manage their usage of voice, data, and video.

CHAPTER 11

ARCHITECTURE

In common usage, "architecture" refers to the design and construction of buildings that are at least functional but also often have aesthetic appeal. In the Information field, "architecture" refers to the design and implementation of technology that fulfills requirements, which may be functional and aesthetic.

Despite this apparent similarity in meaning, there's practically nothing in common between building architecture and technology architecture. What makes buildings beautiful and software beautiful are entirely different.

An architect in one domain would be lost in the other. Building architecture must obey both the laws of physics and local regulations, while technology architecture has no such restrictions, except in hardware. But even there the similarity is tenuous. How much load a beam can support safely is in no way analogous to how many calculations a processor can perform per second.

In building architecture, the key difference between an architect and an engineer is that the architect focuses more on the artistry and design of the building, while the engineer focuses more on the technical and structural side. In technology architecture, the same difference exists, although sometimes one person fills both roles.

Architecture can depict the current state (As Is view) or a future state (To Be view). Plans describe the route and timeline between them. Projects execute the plans.

Why Technical Architecture Matters

Technical Architecture is meant to satisfy requirements, but sometimes requirements are bounded by architecture. That is, satisfying a specific requirement can be risky and expensive if it conflicts with the prevailing architecture. On the other hand, architecture is often meant to reduce complexity and ensure interoperability. Common requirements thus can be less expensive and less risky than if there were no architecture.

Technical Architecture is relevant to enterprises within the Information field because it generally addresses tradeoffs that affect the marketability and supportability of hardware products, software products, and technical services. For instance, low cost and high performance tend to be incompatible, although compromises are possible. Add in requirements for strong security, high reliability, easy maintenance, backward compatibility, or special functions, and sustaining a workable architecture is challenging.

Technical Architecture is also relevant to enterprises outside the Information field because it affects usability and value derived from their investment in hardware, software, data, networks, and services. In fact, architecture can dictate the location of the constraint and limitations. Therefore, to make the constraint intentional rather than accidental, everyone pursuing Constraint Management should be aware of architecture. For example, if the architecture of systems for sales over the web will not scale readily, holiday season sales are likely to be disastrous. And if the architecture will not support failover and rapid recovery from outages, sales any time of year may be disappointing.

Technical Architecture

Technical Architecture is applied to various scopes because the audiences are different and so are the architects' skills. There are no universally accepted definitions. Here are definitions used in this chapter:

- **Hardware Architecture.** The organization, functionality, and implementation of computer systems, digital storage, or communication devices
- **Information Technology Architecture.** The organization, functionality, and implementation of IT, which includes computers, system software, middleware, storage, networks, and other infrastructure for security, power, and cooling
- **Data/Information/Knowledge Architecture.** The organization, management, and processing of structured data into information and unstructured data into knowledge
- **Application Architecture.** The organization, functionality, and implementation of application software, along with its interaction with databases and middleware

- **Enterprise Architecture.** A holistic view of strategy, business processes, software applications, and computing infrastructure

Hardware Architecture

Despite vast differences in performance and cost, Personal Computing and Enterprise Computing today are both based on Hardware Architectures collectively called Classical Computing. It is further divided into von Neumann Architecture, which has one memory for programs and data, and Harvard Architecture, which has separate memory for programs and data. Furthermore, some processors are designed for Complex Instruction Set Computing, but most are now Reduced Instruction Set Computing. CISC devices can perform multiple steps with one instruction, while RISC devices use simple instructions.

Researchers are working on a radically different Hardware Architecture called Quantum Computing. And if the hardware is radically different, the software will be too. Quantum Computers will do many calculations in parallel that are done today in series, if they are done at all. For instance, protein folding is computationally intense, so it may be greatly accelerated by Quantum Computing. That will aid in development of vaccines and medications.

Although the applications of Quantum Computing are mostly yet to be discovered, such computers will be used in addition to Classical Computing. Just as today's computers use specialized processors for input/output, floating point math, image analysis, and graphics, Quantum Computers will most likely be paired with Classical Computers.

IT Architecture

IT Architecture covers the structure, implementation, and operation of Information Technology, which includes computers, system software, middleware, storage, networks, and other infrastructure for power, cooling, and security. For instance, IT Architecture might say that servers and storage will be installed at two locations; one will be the primary and the other will be a hot standby, and when the primary fails, the standby will pick up processing without interruption. Alternatively, the architecture might say that each site will perform its own processing, but if either fails, the high-priority processing ordinarily handled by the failed site will be picked up by the active site once it is determined that the failed site will not resume service soon.

The first architecture in that example maximizes availability, while the second maximizes productivity. Choosing an architecture always involves tradeoffs of some kind, and those tradeoffs often span Functional, Developmental, and Operational Requirements.

At enterprise scale, Information Technology is inherently complex. IT Architecture therefore strives to compartmentalize complexity so that failures are isolated and multiple teams can each work concurrently on their own subset without colliding with work being done elsewhere. For instance, developers do their work in development environments, and operators do their work in the production environment. Policies and procedures govern how work flows between those environments in an orderly manner. Even in a DevOps environment, where the same people do development and operations, policies and procedures prevent disorder.

IT Architecture looks different to Cloud Computing providers and Cloud tenants for several reasons. First, Cloud providers design their IT to serve multiple tenants simultaneously on shared infrastructure while keeping the tenants' software and data separate. Second, providers design their IT to scale up and down on demand as those tenants grow and adapt their software and data. Third, providers may build their IT to provide hardware servers for high performance, as well as virtual servers for economy and flexibility. The list of differences goes on.

CIOs are rightly concerned about getting locked into a single Cloud vendor—a "walled garden"—for cost and reliability reasons. However, Cloud portability is hard because Cloud platforms differ. The only way to make Cloud-based applications portable is to build them using only the lowest common Cloud capabilities. And even then, there will still be some differences. Developers, on the other hand, want to use the newest capabilities, which may not be common across Cloud providers. Time will tell whether Cloud providers converge rather than continue to differentiate their capabilities.

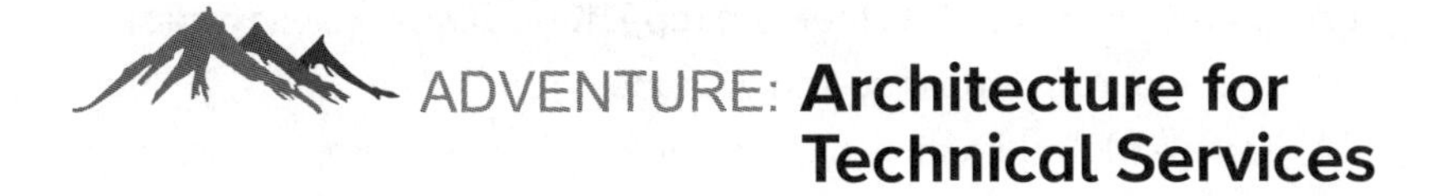

ADVENTURE: Architecture for Technical Services

When I was the senior technical executive in an organization that delivered labor-based technology services, I represented my service line in the division architecture board. You might think that a labor-based services business would have no need for Technical Architecture, but the technology used by people delivering or receiving labor-based services can indeed comply with an architecture.

Technical Services

My service line delivered the following types of service for our own products, as well as those from other manufacturers:

- **Self-service.** Customers could search frequently asked questions, search a question-and-answer database, follow a diagnostic template, or open trouble tickets.
- **Help desk.** In addition to performing the same tasks as self-service, agents could search product databases for additional technical details or escalate the ticket to a subject matter expert.
- **Deskside support.** Technicians would diagnose and repair some problems at the customer's office.
- **Field service.** Technicians would diagnose and repair problems at the customer's computing center.
- **Depot service.** Customers would ship their products to a depot for repair or exchange.

Ten other service lines designed, built, and operated Information Technology infrastructure on behalf of customers.

In a labor-based service business, Technical Architecture contributes to repeatability—along with contributions from methodology, policies, procedures, and training. Customer satisfaction is higher, and services are more profitable, when they are repeatable. Otherwise, every services engagement relies too much on trial and error.

Technical Architecture affected two entirely separate parts of my service line. First, the technology used to capture, retain, and retrieve knowledge was obsolete. Second, the technology to manage contracts and the scope of services entitled by those contracts was unique, which differentiated us from the competition—at a cost.

Lessons Learned

Lessons learned include:

- Technical Architecture applies to labor-based services when it standardizes the technology used to deliver those services, such as knowledgebases and ticketing systems.
- Labor-based services can create a Technical Architecture for clients, then such services can implement and operate Information Technology and application software according to that architecture.

Data/Information/Knowledge Architecture

While IT Architecture is mostly about technology, Data/Information/Knowledge Architecture is mostly about data and information or knowledge derived from it. Thus, Data/Information/Knowledge Architecture is closer to an enterprise view of information and how it supports the enterprise's goal.

For example, Data/Information/Knowledge Architecture defines:

- **Master data.** Facts about an entity, such as a product name or a customer address
- **Transaction data.** Facts about a change to entities, such as a product sale to a customer
- **Reference data.** Facts used in comparisons or calculations, such as sales tax or value-added tax rates
- **Information.** Figures derived from structured data, such as a sales forecast

- **Knowledge.** Know-how/know-why derived from unstructured data, such as natural language understanding

Note that data, information, or knowledge are not necessarily specific to one software application. Often, they are shared across applications because there is less redundancy and fewer inconsistencies when there is just one shared version.

Data/Information/Knowledge Architecture also defines how those artifacts are stored. The most common method is to organize data into tables where rows represent specific entities or transactions and columns contain facts about those entities or transactions. This type of database is called relational because the tables are designed with specific relations in mind. For instance, a table of sales transactions will have a column for customer IDs so that bills can be generated via the relationship between the transaction and customer tables. The language used to create and maintain such databases is Structured Query Language (SQL).

Other database types may be better for specific use cases, however. These are called non-relational or NoSQL databases. For instance, some databases store data as key-value pairs. Others store data in a network of links or in columns rather than rows. Distributed databases don't store all the data at one location. There are also other database types beyond the scope of this book.

Application Architecture

The Software chapter mentioned these concepts:

- **Algorithms** are step-by-step procedures.
- **Design patterns** consist of data structures as well as algorithms that satisfy common requirements.

There are dozens of design patterns. One example is a wrapper, which translates parameters at the interface between modules that would otherwise be incompatible.

Algorithms and design patterns address Functional Requirements. Application Architecture incorporates and extends them with automation and manual procedures to address Developmental and Operational Requirements, too. For example, Application Architecture applies the following principles:

- **Separation of concerns** means software should be divided into modules that each deal with just one set of information or one function. For instance, order-taking and order-fulfillment should be separate because those are distinct business functions even though they share some of the same data. Combining them increases the likelihood that a change to one function will cause a problem in the other.
- **Loosely coupled** means modules should depend as little as possible on algorithms and data structures within other modules. For instance, a module that displays invoices on a mobile device should not know how another module calculates those invoices.
- **Clear interfaces** means the parameters passed between modules and the expected behavior should be unambiguous. For instance, a module that's supposed to receive customer data and print membership cards should not also send emails to those customers as a side effect.

These principles have evolved to cope with complexity. Most software is inherently complex because the requirements are not trivial to implement. But software can also suffer induced complexity due to technical and management decisions, which is where Application Architecture comes in.

Here are some simple examples (not a complete list) of Application Architectures still found in Legacy Systems:

- **Layered** separates applications into Presentation, Application, Business Logic, and Data Access layers, all of which may be running on the same computer.
- **Pipe-Filter** separates applications into a series of modules that each process data and pass the result to the next module.
- **Client-Server** separates applications into servers and clients, which typically are Personal Computers and/or mobile devices. Data processing and storage are done on servers, while input and output are done on clients.

Nowadays, to provide Software as a Service, Application Architectures distribute applications across servers that may be co-located or widely separated. Architectural decisions involve tradeoffs, however [Brewer, 2012]. For example, if an application is not partitioned, its services can be both available and consistent.

On the other hand, if an application is partitioned, the architecture can prioritize consistency or availability, but not both.

- **Consistency** means all instances of a service yield the same result. For example, the retail price of an item is the same, even if the data is stored in more than one place.
- **Availability** means that a specific service instance can be used. For example, if service #1 responds but service #2 does not, service #2 is not available.
- **Partitioning** means service instances are somehow separated. For example, if a service runs at site #1 and site #2, it's partitioned.

One way around this dilemma is to prioritize availability and accept that the partitions will become eventually consistent once they all resume operation. An alternative is to prioritize consistency and tolerate some unavailability even on partitions that would otherwise be active.

Application Architecture also considers topics like these:

- **Caching** speeds repeated use of the same data because it doesn't have to be retrieved and processed again and again.
- **Load balancing** spreads work across servers and adds or drops servers as needed to handle workload.
- **Inertia** happens when changes are hard. Sound architecture during development and minimization of technical debt during maintenance reduce inertia.
- **Release independence** allows parts of an application to be changed and put into production without affecting the rest. In the extreme case, components are updated continuously, and releases are barely discernable to users.
- **Graceful degradation** means a failure in one service does not trigger a failure in other services. For instance, a "circuit breaker" may queue new transactions for later processing rather than immediate processing on an overloaded server.
- **Integration** between new applications and legacy applications is the rule, not the exception. For example, a government agency developed 2.5 million lines of "glue code" between applications on just one project.

- **Orchestration** is coordination of many disparate parts to create a whole that meets requirements and achieves Service-Level Agreements. Migrating monolithic legacy applications to microservices is increasingly common, but that means processing previously done periodically in batches may need to be orchestrated for continuous processing of individual transactions or inquiries.

ADVENTURE: **Architecture for Software Products**

When I was a software chief technology officer, I managed the research and development program that fed the pipeline of new products. My focus was on shepherding selected Horizon 3 research projects into Horizon 2 product development and then into the Horizon 1 sales process.

The software implemented various scheduling, dispatching, and capacity management functions, and sounded alerts when a monitored process strayed out of bounds. It also displayed Key Performance Indicators, which were helpful in performance improvement.

Operations control centers suffered complexity from three sources. The operations processes themselves were complex because the stakes were so high and the information quality so low. The Sense and Respond technology—sensors, cameras, gauges, valves, actuators, locks, keypads, and gates—was complex. The technology to tie all that together with operational oversight was complex.

Our business strategy was to simplify at least one source of that complexity by offering a pre-integrated technical solution. Our underlying technical strategy was to assemble our solution from middleware with a new application layer that received sensor data from, and sent control signals to, the sensor/control layer. The result was a video wall for sensor feeds and real-time dashboards. It was implemented in industrial, energy, transportation, and governmental organizations.

Technically Correct Is the Best Kind of Correct

One of my roles was chairman of the architecture board that reviewed R&D projects at key milestones. This architecture board oversaw dozens of projects staffed by hundreds of developers across multiple countries.

Board members each had a vote, but it was not majority rule. It was more like "Press the pause button when you see something amiss." That is, if any member had reservations about the solution, they could vote "no go," their reservations would be noted and addressed, and their vote changed later to "go" if satisfied. The objective was to identify and resolve architectural issues early.

Frameworks enabled integration of middleware with the application layer, and the application layer with the external sensors and controls. Although the core of the frameworks remained relatively stable, every new customer implementation might require new adapters to reach the core.

Inertia

During meetings with customers, I saw a wide variety in Technical Architectures even among enterprises in the same industry. The prevailing architecture shaped each customer's thinking and limited its options:

- Our solution was **pre-integrated**, so simplification was its main selling point, but reduced maintenance and continued research were other major benefits. Establishing new operations control centers was easier than retrofitting existing centers.
- Organizations with **homegrown** solutions had already done the heavy lifting, and their solutions looked like ours because the requirements were mostly the same. However, they then also owned the ongoing maintenance and research responsibilities.
- Organizations pursuing **best-of-breed** had semi-custom solutions. Their vendors owned maintenance of the components, but the buyer owned their integration. A common CIO request was, "Can you break apart your pre-integrated solution because I only want the parts that would fill gaps in my current components?"

- Organizations choosing **outsourcing** were in industries where huge capital investments meant construction had to be anticipated in their five-year strategic plan. This influenced their decisions about technical innovations, too, even though the technology would have evolved considerably by the time they were ready to adopt it.
- Most troubling were organizations stuck in "**good enough**." Those managers would say they saw no reason to change even when their capabilities fell far short of state of the art.

For example, operations managers in one sprawling facility always had two security officers on duty in the control center—and over 2,000 security cameras in continuous operation indoors and outdoors. When an alarm was triggered or an incident was reported, those officers would watch replays from the nearest cameras, and in about 20 minutes they would have a preliminary reconstruction of what happened. They were proud of that benchmark. Meanwhile, however, officers on scene would have little or no information to go on—and 20 minutes is a long time to be coping blindly during a crisis.

Our solution included technology to monitor all 2,000 cameras simultaneously and raise alerts in real time as incidents happened. For instance, vehicles or people in no-go areas or congregation near secured areas could trigger alerts. With this proactive approach, security officers in the operations center could see incidents in real time, and security officers on scene could know the type of alert and its location immediately.

Lessons Learned

Lessons learned include:

- Technical Architecture in its various forms is not an operational constraint because it doesn't directly restrict what an organization produces. In the case of a software organization, the Critical Chain of projects or critical skills are the operational constraint.

- Technical Architecture can be a strategic impediment, however, if it inhibits management initiatives to the point that the organization gets stuck indefinitely on "good enough."
- Technical Architecture can be a strategic enabler if it drives compliance when projects and products would otherwise drift into incompatibility.

Enterprise Architecture

Enterprise Architecture is a holistic view of strategy, business processes, software applications, computing infrastructure, and technical services. Thus, it's broad enough to encompass Constraint Management for an entire enterprise. Unfortunately, Enterprise Architecture is a scarce skill. Moreover, few enterprise architects are practicing Constraint Management as described in this book. Thus, organizations without an Enterprise Architecture attuned to Constraint Management may inadvertently create conflicts across software applications and business processes.

Geography matters, yet its effect on architecture is rarely acknowledged. Technology developed or maintained in more than one location will have a different architecture than technology from one location. The more different the locations are, the more different the architecture will be. Even technology architected on different floors or in adjacent buildings will be different in subtle ways. Add a few thousand miles, another department, or another language, and the differences become profound. Conway's Law reminds us that the organization chart is usually visible in the architecture.

Just as space matters, so does time. An architecture created during an earlier era may embody its state-of-the-art technology and best practices that appear quaint, if not obsolete, when compared to current technology and practices. Nevertheless, Legacy Systems from any bygone era are sticky if they still fulfill requirements with less investment than a replacement system would require.

Frameworks

A framework is guiding principles and standard practices for building, deploying, and using an architecture. At the lowest level, a framework guides coding and testing via its design patterns. At the highest level, a framework guides the design and evolution of applications, databases, and business processes.

Programmers familiar with a coding framework can generally build new applications faster because some elements of the framework have already been coded and tested, but there is a learning curve up front. However, when requirements deviate from the framework, extra work is required for customization.

Frameworks are thus meant to achieve two outcomes:

1. **Congruence** means the architectures are consistent. For example, if Application Architectures are congruent, those applications are more likely to work together without wrappers or pipes between them to resolve differences.
2. **Compliance** means a plan or implementation is consistent with an architectural framework. Noncompliance may lead to rejection or modification of a proposed project.

Unfortunately, compliance during development does not ensure compliance during maintenance. "Drift" is the subtle deviation away from an approved architecture. This type of technical debt is expedient in the short run but can have severe consequences in the long run as it becomes progressively harder to make changes that comply with the architecture. The worst-case scenario is "no-touch" code, because everyone is afraid modifications will break it.

Constraint Management

Can a Technical Architecture be an operations constraint? No, because there's no way to use that architecture to control production. Technical Architecture is analogous to factory layout plus specifications for machines and operators. Neither is the operations constraint, but both can determine where the operations constraint is.

Technical Architecture can limit what a technology organization produces and how it produces it. That limit can be beneficial by preventing technologies from becoming incompatible, or it can be detrimental by locking the organization into technologies that don't meet current or future requirements. An architectural cul-de-sac means there's an obvious way in, but once trapped, technologists must reverse course to resume progress in a different direction.

Technology Architecture is mostly concerned with making technology work, which is no small feat. However, its ability to make the organization perform better is diminished if no one has identified the constraint. Constraint Management just isn't in the architects' toolkit. If technicians can't think about it and talk about it, they can't do it.

Technical Architecture, however, is good place to design where an operations constraint should be. If that's not done consciously, it will happen somewhat randomly. Creating a Technical Architecture with the enterprise constraint in mind will steer technicians to build different things and operate them differently. Perhaps more importantly, it will guide executives not to sponsor projects to pursue shiny new things that won't move the needle.

Conclusion

Frank Lloyd Wright said, "A doctor can bury his mistakes, but an architect can only advise his client to plant vines." Fortunately, the fate of problematic Technical Architecture isn't quite so grim.

Computing hardware and software are considerably more malleable on small scales. Indeed, they are generally created in anticipation of frequent changes. Thus, refreshing selected devices and software periodically is routine. Nonetheless, Technical Architecture at a more macro scale can be less amenable to change than its sponsors hope. For most enterprises, replacing thousands of devices and several large software applications at once is not at all routine.

Constraint Management is not a familiar subject in Technical Architecture circles. Neither is Technical Architecture a familiar subject in Constraint Management circles. It's not that technical architects don't discuss constraints. They do. A lot. But their touchstone is requirements, and if those requirements say, "optimize performance," the architects will strive for balanced hardware and software capacity. If that happens to optimize the enterprise constraint, the outcome will be a happy one. But if the solution optimizes a non-constraint, the outcome may achieve project objectives while missing the point entirely on demonstrable and enduring business benefits.

CHAPTER 12

SKILLS

This chapter is about skills. The right ones can exceed the goal. The wrong ones can impede the goal. Furthermore, psychological and environmental factors can suppress or amplify skills.

Technical skills come from many sources, including formal education, classroom training, video training, on-the-job training, conferences, self-learning, and mentoring. Over a career, most people get and give skills through multiple means.

Technical skills can be the enterprise constraint. That's true for technology-consuming enterprises as well as technology-providing enterprises. Even if technical skills aren't the enterprise constraint, they can still be the local system constraint in hardware or software development, advisory consulting, technical services, etc. Thus, skills are an essential ingredient in exceeding the goal.

ADVENTURE: **No Skills**

When I was a mid-career hire into a new organization that offered software and services, the senior partner, a non-technical executive, gave what is possibly the worst pep talk ever. Our newest offerings were struggling to attract customers. He deflected blame onto the R&D team for weak sales by sputtering, "You guys have no skills!" No, this wasn't a reverse-psychology ploy: He was lashing out in frustration rather than confronting the real problem.

Beating the Odds

A dozen technical professionals with over 100 years' collective experience just glanced at each other and smiled. After that senior partner left the firm, they caught rising demand in time to become the market leader, according to industry analyst evaluations. Seven years later, that team had developed software, created methodology, delivered training, conducted consulting engagements, documented case studies, and grown to thousands of practitioners worldwide with revenue of over one billion dollars. Not bad for a team with no skills.

Fortuitous Reboot

There were bumps along the way, however. During an office move to larger quarters, a misunderstanding led to the destruction of media containing intellectual capital, such as estimates, proposals, presentations, databases, and metrics.

Losing the archive led to a do-over, however, that produced better intellectual capital than before. Files were lost, but skills were not. Indeed, skills improved when they weren't tied down by precedent set before the team's efforts gained momentum.

Lessons Learned

Lessons learned include:

- Claiming that technical professionals with years of experience have no skills is foolish. They may not all be the right skills, but no skills and skill gaps are not the same.
- Throwing away intellectual capital and starting over—by plan or by accident—can be a good thing if it sharpens skills.
- Existing intellectual capital is written down, but future intellectual capital is walking around in the heads of technical experts.

Education

Entree into the Information field comes from many directions, including education and training. Science, Technology, Engineering, and Mathematics (STEM) subjects that were once the sole province of colleges are pushing down into high school, middle school, and primary grades. Furthermore, not all jobs in the Information field require a college education. Unfortunately, the cost of college has risen tenfold since I was a professor. Thus, students unprepared for college academically, financially, or motivationally should consider alternatives.

That said, for students who are prepared, education is an extraordinary opportunity. But education and training are different things. Preparing students for the workforce is just a subset—often a miniscule subset—of what education achieves, yet it's the entire purpose of training. After graduation, on-the-job training, coupled with online training, classroom training, and self-study, contribute many of the skills that graduates bring to their jobs.

Here are some things they don't tell you in school:

- Higher education is divided into research and teaching institutions. Students can get a great education at both kinds, but the faculty focus is different. At research institutions, faculty are evaluated and promoted primarily based on their publications, and secondarily on their teaching. At teaching institutions, that emphasis is reversed.
- PhD programs don't teach professors how to teach. Whatever teaching skills they have come from emulating others or are self-taught.
- Higher education has regular faculty, assistants, and adjunct faculty. Regular faculty are full-time academics. Research assistants and teaching assistants are mostly graduate students. Adjunct faculty come in mostly two types: (1) full-time practitioners who also teach part-time, and (2) itinerant full-time academics without the benefits or job security of regular faculty.
- Regular faculty may have little or no work experience outside academia. That's why graduates sometimes see a disconnect between what they learned in school and what they later do at work. For example, an introductory Computer Science course teaches how to overload C++ operators to do arithmetic on complex numbers, yet that course does not teach how to write a "Hello, World" program, the simplest kind.
- Programs that new programmers work on are generally 10 to 100 times larger than the largest program they wrote or studied in school. Simply understanding the code, let alone modifying it and testing it, is a level of difficulty way beyond schoolwork.
- Most of a new programmer's time is spent maintaining software that somebody else wrote, and that code was subsequently modified by many

people. Consequently, it has a mish-mash of styles and a backlog of bugs. Some capabilities seem bolted on as an afterthought. Some parts are so mysterious that it's unclear what their purpose is or whether they are ever executed.

What about the "Everyone should learn to code" idea? Having taught computer programming and observed college students struggling with elementary concepts, I don't think universal coding is achievable. If some students can't handle a basic classroom assignment, they will be hazardous on the job. And then there's this graffiti: "Give someone a program, and you frustrate them for a day. Teach them to program, and you frustrate them for a lifetime."

All right then, what about coding camps for those with interest and talent? The jury is still out. Unfortunately, some coding camps are a scam, taking enrollees' money and teaching little of practical value. On the other hand, some coding camps that are affiliated with potential employers do seem to work, but it remains to be seen what their graduates' long-term employability will be. Coding is a small percentage of total effort on Information projects, and someone who knows only coding is a one-trick pony.

ADVENTURE: **Novice Errors**

My first computer class taught computer programming in assembly language. Back then, you had to prove you knew how CPU registers worked and how move, arithmetic, and comparison statements affected registers before you could be trusted with a higher-level language. Fortunately, things have changed for the better. Instead of focusing on how computers work at the bits and bytes level, the focus today is on fulfilling business requirements.

Like many programmers, my first brush with the real world of computing was as a summer intern. Most of my work was satisfactory, but I still managed to blunder into two incidents.

Halt and Catch Ire

As a novice programmer striving to write good code, whenever I wrote a conditional statement, I included code to handle unexpected conditions. While testing a "can't happen" condition, the following incident occurred. Module A called module B, which correctly detected an error condition, printed a message describing the error, and returned control to A, which I thought would terminate.

Due to a defect in my code, A instead called B again, and the cycle repeated until the system halted when it ran out of space. For non-programmer readers, that's called an "infinite loop" because the program doesn't end until some outside force stops it. What I should have done was terminate the program after printing the message just once. The operator was unhappy with me when hundreds of other jobs waiting for my job to finish were also delayed indefinitely. Oops.

Joneskowski & Smitharelli

While making a minor enhancement to the program that printed the entire organization's phone directory with over 30,000 numbers, I noticed that the program moved spaces into a field before moving each person's surname into that field. Assuming that a move into the surname field would be padded with blanks carried over from the fixed-length sending field, I deleted the instruction to move spaces first. I tested it, of course, and everything looked fine.

Once the program went into production, however, I soon got a call from my manager who asked whether I noticed that the organization suddenly had lots of strange surnames. Much to my embarrassment, the compiler did not behave as I expected. I expected it to move a fixed-length string of characters, when it in fact moved a variable-length string. Thus, shorter surnames over-wrote only the beginning of longer surnames, whose endings peeked out. Users reverted to a week-old directory temporarily, so the result was more embarrassing than damaging, fortunately. What I should have done was resist the urge to optimize a trivial program that was working fine as is. Oops again.

Lessons Learned

Lessons learned include:

- My manager shrugged off my misadventures because interns are expected to make mistakes, and new hires are not expected to have high productivity for a year.
- During many years of insisting that programmers check for conditions that can't happen, such conditions have never failed to happen. What the code would have done without trapping the "can't happen" condition ranges from amusing to disastrous.
- Reliable testing of an optimization requires production-level volumes of data, not just test data.

Interviews

Finding a job that you want and getting hired are rarely easy. Recruiters and hiring managers sometimes get carried away describing necessary qualifications; seeking people with five years' experience on a technology that's only a year old is folly.

Interviews are not a one-time thing. Lifetime employment is mostly gone, so most Information professionals go through job interviews many times during their careers. Of course, having a series of jobs is not the same as having a career. Thus, the interview techniques change as candidates seek senior technical positions, management positions, or executive positions.

For technical positions, hiring managers want to know that candidates can do the job. That's hard to assess in an informal interview. Therefore, they have taken to evaluating technical skills several ways. Naturally, these techniques lead to consternation among candidates:

- **Thinking questions.** Show general knowledge and reasoning abilities. For example, "Why are manhole covers round?"
- **Technical questions.** Demonstrate mastery. For example, "How do you find open ports?"
- **Stress interviews.** Intimidation reveals how candidates handle pressure. For example, "You don't have enough experience, so why should we hire you?"

- **Whiteboarding.** Diagram a data structure and algorithm from memory.
- **Side projects.** Make contributions to open-source projects.
- **Homework.** Build an application from scratch.

Although these techniques are in wide use, candidates question their effectiveness and fairness. For instance, whiteboarding is notorious for having little to do with actual development work. And homework requires a sizeable investment of candidates' own time (32 hours is not unusual), so unless the candidates are under final consideration, that's quite an imposition.

Sometimes, however, things go awry unintentionally. During a hiring binge, one company scheduled interviews on a Saturday but forgot that the building was locked on weekends. Employees could badge in, but interviewees were left standing outside until the interviewers figured out why appointments were being missed.

Companies naturally want to publicize good news to employees. For anyone who's been in a management job for a while, being interviewed should not be a big deal. But if you've barely got your feet on the ground, being interviewed on camera is unsettling.

Catching Waves

When I was hired as director of software engineering, I had been on-board less than a week when my new employer insisted on a video interview for broadcast to all employees. The company was fairly stable but striving to diversify, so the purpose of the interview was to publicize this senior technical position.

In the second case, several years later, I was hired through a hiring freeze as an executive consultant. The business was in transition from Manufacturing to Services, so it was seeking different skills. Again, I had only been on-board a few days when I was asked for another video interview, and the company flew me cross-country to get it. This video was to reassure employees about the company's future.

Unbeknownst to me, in both cases the companies were on the brink of upheavals in their respective markets. Demand for their products and services was changing, so they needed radically different skills, and I was lucky enough to catch that wave twice.

Lessons Learned

Lessons learned include:

- Changing an organization's skill mix is a major undertaking.
- Rehearse your answers to likely questions before an interview.
- Wearing TV make-up for the first time is a strange feeling.

Imposter Syndrome

Even after years of study and relevant experience, imposter syndrome is a common affliction. It happens when someone feels like everyone around them knows a lot more than they do. It's especially common during first jobs, but it can happen at any point in a technical career when someone is thrust onto a new team or into a new domain.

Imposter syndrome is a symptom of increasing complexity in the Information field. There's so much to know about the field and an employer's systems that it can be overwhelming.

Company or team culture contributes to imposter syndrome. For the first year at least, new members are constantly reminded that they aren't quite insiders yet.

Several solutions are possible. One solution is talking with mentors or coaches because they will have a more objective perspective. There is also self-study or training to close specific knowledge gaps, and teammates who've worked through imposter syndrome themselves can offer understanding and encouragement to new members as they integrate into the team. The same goes for support from managers and executives.

Modifying versus Rewriting Code

Many programmers prefer to rewrite code instead of modifying code because it's easier to write code than to read code. (Yes, that's counterintuitive.) Rewriting can be a problem, however, because old code has been debugged and new code hasn't. Both modifications and rewrites require testing, but testing new code takes more time. So whatever time might be gained by rewriting may be lost in testing.

The breakeven point appears to come around 25%. If a quarter or more of the code would have to be modified, it's probably more efficient just to rewrite it—assuming the requirements are well documented. If they are not, someone will have to read the code thoroughly enough to glean all the requirements, which may take longer than just modifying a subset.

This debate over modifying versus rewriting code only makes sense, however, on "average" code. If the code is quite large or extremely complex, rewriting only makes sense if it's funded as a major replacement project. For instance, there are programs in production today with over one million lines of code because they grew incrementally for decades and rewriting was never a priority. Packaged software may include tens of millions of lines of code. At that scale, refactoring becomes a major undertaking, too.

Specialists versus Full Stack Developers

When the Information field was young, a software application might be written in two or three languages, and it would run on one computer. Applications today may be written in a dozen languages, and their services may run on servers spread around the globe.

As languages and servers have proliferated, the field has become specialized:

- **Front-end developers** create user interfaces, such as web pages and portals.
- **Back-end developers** create databases and business logic.
- **Full-stack developers** do both, and in small shops, that may be appropriate.

In large shops, however, specialization may occur by system type:

- **Systems of Record** (transaction processing). Back-end developers
- **Systems of Engagement** (social media). Front-end developers

- **Systems of Insight** (business intelligence). Analytics developers
- **Systems of Innovation** (business processes). Researchers and business process owners

And specialization may occur by role in the methodology:

- Planning
- Project management
- Requirements gathering
- Design thinking
- Development
- Testing
- Operations

Where there is overlap today in previously distinct roles, such as DevOps, it is to gain speed or economy.

ADVENTURE: **Faculty Development**

For 15 years, I taught college and graduate school. In addition, I taught for seven of those years in an advanced faculty development institute that prepared business school faculty to teach Information courses. My day job was with a telecommunications company, so I taught data communications, database management, and R&D methods in the faculty institute.

Something Worth Doing

At the time, top-rated business schools had their pick of newly minted Information PhDs, but second- and third-tier schools could not hire enough qualified faculty because supply was too low. Because it alleviated this critical skill constraint, the academic organization that sponsored the institute said it was the most successful program they ever held, based on enthusiastic feedback from college deans. A cohort of students who entered the Information field could not have done so without this faculty program.

Anything Worth Doing Is Worth Overdoing

In the decade that followed, however, new PhD programs proliferated, and old ones expanded. They grew until they produced more than demand, thus leading to a PhD glut in all fields, not just the Information field. Tenured positions were abolished as incumbents retired and use of part-time adjunct faculty increased, thus decreasing demand for PhDs. For new PhDs seeking a career in academia, it was a perfect storm of increasing supply and decreasing demand.

Lessons Learned

Lessons learned include:

- Skill constraints can happen anywhere, including academia. Overproduction can happen anywhere when Constraint Management isn't applied to the supply chain.
- Even good deeds can have unintended consequences. As PhD supply grew and demand shrank, the skills constraint moved from colleges (internal) to the academic job market (external).

Introverts, Extroverts, and Ambiverts

Introvert and Extrovert are personality types often discussed in the Information field because technical work seems to attract Introverts while managers are more often Extroverts. Here are some stereotypical differences:

- Extroverts are charismatic. Introverts are reserved.
- Extroverts prefer talking. Introverts prefer writing.
- Extroverts are energized by social interaction. Introverts are depleted by it.

The Myers-Briggs Type Indicator is meant to classify individuals as unambiguously Introvert or Extrovert, but it has no rigorous foundation because the authors weren't scientists or psychologists.

Here is what neuroscience says:

- Some people have more dopamine receptors in their brains, which makes them less sensitive to this feel-good chemical. Thus, they need more stimulation to produce it. An "Extrovert party" is the quest for more stimulation.
- Some people crave acetylcholine, the alert-and-relaxed brain chemical, so they need less stimulation to protect it. An "Introvert hangover" is recovery from overstimulation.

Thus, it is a neurochemical difference. However, behavior is also affected by context and age. Some people are more extroverted in private than in public. Many Introverts are not shy, they are just weary from having to maintain a heightened state of awareness. However, they can speak to, or perform for, large audiences. Nevertheless, over their lifetimes, even Extroverts become somewhat more introverted.

If you think psychometrics means "crazy numbers," that's about right when it comes to statistics on this subject. Some widely cited figures, such as "Introverts make up only 25% of the general population," are just guesses. Empirical studies find that, using the conventional definitions, Introverts and Extroverts are each around 50% of the general population.

An old saying goes like this: "There are two kinds of people: those who divide the world into two kinds of people, and those who don't." Coercing people into just two categories doesn't account for people who are legitimately hard to classify.

Indeed, there's a school of thought that says most people are Ambiverts because they exhibit varying degrees of both introversion and extroversion. Here's the Ambivert range:

- Charismatic Ambiverts are 0.25 Introvert + 0.75 Extrovert
- Balanced Ambiverts are 0.50 Introvert + 0.50 Extrovert
- Reserved Ambiverts are 0.75 Introvert + 0.25 Extrovert

Using that definition, Ambiverts are about 68% of the general population, a quantity that happens to be within one standard deviation of the mean in a normal distribution. That leaves only 16% Introverts and 16% Extroverts as extreme cases in the tails.

Compare these general population statistics with a survey of workers in the Information field which found about 55% Introverts, 25% Ambiverts, and 20% Extroverts. Thus, it appears that there really are more Introverts in the Information field—three times as many, in fact [IDG, 2009].

Things get even crazier when management ranks are studied. One study says Introverts in management range from 12% at supervisor level to just 2% at the senior executive level. However, one executive admitted, "Nobody knows I'm an Introvert," as though it must be hidden to survive as an executive [Kahnweiler, 2009]. On the other hand, another study says 25–30% of CEOs are Introverts. That's a huge disparity. Unfortunately, no management studies have quantified Ambiverts, so there's good reason to suspect that the Introvert percentages are inflated.

If Introverts are overrepresented in technology roles but underrepresented in management roles, that's a problem when managers assume that technical workers are like themselves—or should be. The takeaway here is that treating everyone the same isn't the best approach to maximizing what they produce. Some people need more excitement. Others need less. Extroverts who feel the urge to fix Introverts should instead capitalize on their strengths and accept the differences as neurochemistry.

Conservatives, Liberals, and Centrists

Steve Yegge observed the following general differences between software developers:

- **Conservatives** want mostly to limit risk. They view liberals as undisciplined.
- **Liberals** want mostly to enable change. They view conservatives as bureaucrats.
- **Centrists** are moderates. They see both sides.

Although he adopted those labels from politics, there is no known correlation between developers who are politically conservative and those who are professionally conservative. The same goes for politically versus professionally liberal. Where there is a clear parallel, however, is the fervor with which each tribe pursues its agenda.

The following is a sample of some specific differences:

- Conservatives think bugs are evil. Liberals think they are just a nuisance.
- Conservatives think data stores should be rigid. Liberals think they should be flexible.
- Conservatives think interfaces should be rigorous. Liberals think they should be informal.

These differences explain why some system development shops have radically divergent philosophies. It also explains why some projects turn into epic battles between developers with opposing viewpoints. But the big revelation is it explains why some mergers and acquisitions run off the rails immediately and irretrievably, eventually resulting in the loss of some truly talented individuals who don't get along with the prevailing philosophy.

ADVENTURE: **Mentoring**

Throughout my career, I was fortunate to receive excellent instruction, and mentors who steered me into opportunities and away from pitfalls. As giveback, I've been an educator, trainer, mentor, and coach.

Paying It Forward

Educator and trainer may sound like they're the same, but they're not. Here is the distinction:

- **Educator.** Teaches general knowledge
- **Trainer.** Teaches job-specific skills

Likewise, mentor and coach may sound like they're the same, but they're not. Here is that distinction:

- **Mentor.** Has done the same job as a protégé and can therefore advise from experience
- **Coach.** Has not done the same job, so asks relevant questions that lead protégés into discovery

Some benefits come from either role, such as making connections between a protégé and the broader technical community, recommending contemporaneous recordkeeping for performance appraisal and promotions, quantifying results rather than just listing duties, and fostering skills currently in short supply. However, coaching is a viable alternative when mentors are not available.

Women in Tech

Some pioneers of the Information field were women [Thompson, 2019]. Unfortunately, women are underrepresented in the Information field today [Reges, 2018]. The problem begins in middle school because, around that time, many girls decide against a career in Science, Technology, Engineering, and Mathematics (STEM). Although that's not an immediate disqualifier, because it's a reversible decision and there are other routes into the Information field, it certainly is a handicap.

Of course, this isn't just a career-choice issue. It's a hiring issue too, because in some organizations, the hiring process resembles fraternity recruitment. It's also a retention issue because too few organizations make accommodations for family responsibilities. It is a promotion issue as well because there is a glass ceiling in too many organizations. Add it all together, and it is unfortunate that many women decide a technical career just isn't worth the pain.

Men Are Parents, Too

Although the Information field is made up of predominantly men, work-life balance by a two-career couple can be a juggling act. During an especially busy ramp-up of a new offering, the consulting team I was a member of held its staff meetings on Saturday mornings because everyone traveled on weekdays. One Saturday I had no one to care for my infant son at home, so I brought him to the staff meeting with me. Thankfully, he slept through most of it, and my colleagues behaved as though my caring for an infant during a staff meeting was routine.

Lessons Learned

Lessons learned include:

- Organizations that are not enthusiastically hiring women into technology and creating a work environment to keep them are missing out on a genuine opportunity to address a technical skills constraint.
- Mentoring and coaching are great techniques in general, but they are vital to retaining and promoting women in technical fields.

Collaboration versus Concentration

About 80% of all offices now have an open layout. Walled offices gave way to cubicles, which are now being replaced by tables. The usual justification is an open layout facilitates communication and collaboration. But a not-so-hidden management rationale is the smaller footprint reduces the cost of rent and furnishings.

That's unfortunate because research is clear: Open offices destroy the productivity of anyone performing work that requires deep concentration, especially programming. Techies wear headphones in the office, work at night, or work someplace else because working in an open office is like being in an all-day meeting.

Computer scientist and entrepreneur, Paul Graham, noted that this disconnect between managers and makers stems from the effects that interruptions have:

- **Manager's schedule.** Each day is divided into hours, so an interruption costs an hour or less.
- **Maker's schedule.** Major tasks take several days and require total immersion, so repeated interruptions and distractions destroy entire days.

It takes 20 minutes to get into mental flow time, and it can continue for hours if uninterrupted. One of the mysteries of the Information field is why organizations hire top technical talent and then suppress productivity via an open office.

Remote Work

Remote work is not a recent innovation, even though trade press articles celebrate its frequent rediscovery. When coworkers are in another building and customers are on another continent, does it matter where someone works? Instant messaging and video conferencing weaken the argument for colocation.

Research shows that people disciplined enough to work remotely can add the equivalent of a full day to each week by eliminating interruptions and turning commuting time into work time. Yet some people seek out coffee houses because they can't concentrate without some background activity. One size definitely does not fit all.

Wrap-around Days

Work can be remote in time as well as space. Technical work used to be done at night because computer systems were less busy. Nowadays it's still done at night to escape frequent interruptions. Of course, night work also accommodates customers and colleagues in distant time zones.

Workdays that start today and end tomorrow are called "wrap-around days." They aren't inherently unproductive, unless they exceed 12 hours, sometimes mockingly referred to as "half days."

"Symbolic overtime" happens when people stay late at work because their boss or coworkers expect it—but no real work gets done. Some national and company cultures promote this. Fortunately, it's far from universal.

Ever since personal computers and networks enabled remote work, I worked mostly from my office at home. That includes stints as a first-line manager of teams spread across the country, an upline executive over teams on another continent, and a corporate strategist running projects worldwide.

The notion that professionals can only be managed in person is an anachronism from the Industrial Age. The notion that professionals can only collaborate in person is a repudiation of the Information Age.

- How is remote management possible? By measuring what people produce or accomplish, not how many hours they work.

- How is remote collaboration possible? By using the collaboration tools that the Information field has bestowed upon us.

Work–Life Imbalance

Working from home was a huge benefit when my wife traveled for business too, our children were young, and we had no extended family nearby. Juggling family as a two-career couple was exhilarating, exhausting, and unsustainable without the flexibility of occasional work from home.

Exit Poll

During an ill-fated consulting engagement in a distant city, I entered the skyscraper where our offices were located and rode the elevator up. As work progressed, day turned into night, then night into day.

When exiting the building, the security guard accused me of failure to sign in, which was a breach of security. I asked him if he could verify that I signed in yesterday. He did verify that, but said it was no excuse for today's breach. I then asked him if he could verify that I signed out yesterday.

He could not, of course, and said he'd have to notify his supervisor that I had somehow accomplished security breaches on two consecutive days. At that point, I explained that there was no subterfuge, only sadness, behind my wrap-around day.

Lessons Learned

Lessons learned include:

- Remote work is not just a method of sustaining productivity by escaping interruptions, it's also a benefit to families.
- Managers who ban remote work are revealing their own insecurity.
- When I stopped working crazy hours, nobody noticed because I was focused on delivering value while ignoring distractions.

Non-technical Work

Technical workers do non-technical work, too. Also called "shadow work," here are some examples:

- Finding out who owns hardware, software, or a business process
- Locating whoever has a password or door key
- Submitting purchase requests
- Getting approvals
- Writing proposals
- Receiving and sending packages
- Making travel reservations
- Submitting time and expense reports
- Communicating and coordinating with project managers and resource managers

Granted, some of these tasks are more amenable to self-service than they used to be. But if projects are delayed or canceled due to staffing issues, distracting technical workers with non-technical work that could be delegated wastes time on the constraint.

Non-technical meetings drag productivity down. When a large software company reduced its workweek from five to four days, productivity rose 40%, mainly due to fewer and shorter meetings.

Solo versus Team Work

Some technical tasks are solo work. One person can swap out a failed storage unit. One person can answer a call coming into a help desk. One person can write a program. Nevertheless, except in small organizations, many Information deliverables come from teamwork.

Indeed, much technical work is done by teams. Requirements, design, coding, testing, and deployment are different skills. Pair programming is effective in some organizations. Code reviews require at least an author and a reviewer. Separation of development and test responsibilities is common because developers are blind to some of their own mistakes. Separation of operational duties is mandated by security concerns.

Productivity versus Anti-productivity

One of the legends in the Information field is that superstars are 10 times more productive than non-stars. While productivity among technical workers does vary, as it does in any field, the 10X claim is dubious. Research behind this legend was done long ago in simpler environments, with small samples, and the task was just coding.

Coding is a small portion of development projects, and it isn't necessarily the most difficult part. Furthermore, productivity is notoriously hard to measure. What matters is not just the volume of code written, but whether it's the right code and whether it's good code. Knowing when not to write code—and when to reuse code—is a hallmark of technical leaders.

Technical leaders are certainly more common, and arguably more valuable, than superstars. While superstars have higher individual productivity, great technical leaders generate higher team productivity. Yet their positive influence isn't always obvious: One manager remarked that his technical leader wasn't personally more productive, yet the entire team worked a lot better when he was around.

Net negative producers, on the other hand, drag whole teams' productivity down to their level. It's especially tragic when they vastly overestimate their own abilities, a condition known as the Dunning-Kruger effect. While they generally do produce something, they are incompetent or toxic to the point that the rest of the team's productivity suffers markedly. Such persons may be in the wrong jobs or on the wrong teams. Or maybe they're just not suited for technical work. The key question is: Can they be rehabilitated? If not, getting them off the team can raise its productivity significantly.

Stack Ranking

Unfortunately, separating employees has been institutionalized in large organizations via stack ranking. In one variation, all managers rank all employees under one executive. In another variation, employees also rank each other and their manager. The result in either case is each employee is classified on a three- to five-point scale. However, the percent in the top rank is capped, and the percent in the bottom rank is mandated. If the organization hires only great people, some great people inevitably wind up in the lowest rank. And if the organization does not hire great people, there are better solutions than hire and fire.

Those ranked highest get the most bonuses, raises, and promotions. Those ranked in the middle get little or nothing. Those ranked lowest are put on performance improvement plans or lose their jobs outright. This differentiated reward-and-punishment system creates disincentives for cooperation and risk taking. It sends morale spiraling. It also ignores transient causes of low performance, such as divorce or illness. Most damning is the method's inconsistency: I have seen the same individual ranked highest, then lowest, then highest again based on the personality of managers more than the capability of the individual.

Stack ranking starts cutting fat, but eventually cuts to the bone. Executives are sometimes shocked to discover that it has created single points of failure:

1. All employees with a critical skill will retire within five years,
2. Only one person remains who understands a vital Legacy System, or
3. New hires have no skills or interest in older technology.

Legacy Skills

The half-life of knowledge on the leading edge of the Information field is just a few years. That's why lifelong learning is vital. However, some technologies and systems have lifetimes longer than the people who created them. Thus, the field has both leading edge and trailing edge problems. While the leading edge moves fast, the trailing edge moves slowly, if at all.

Legacy skills is actually a misnomer. The skills themselves aren't legacy; the technologies are. Some of the same skills used to build systems today apply to Legacy Systems, but the languages and tools are different.

Employee attrition leads to institutional memory loss. For instance, why does the system process transactions that way? Nobody knows because the person who wrote it is gone and the documentation is lost. So, somebody changes it, and three departments start screaming about accounts that no longer balance, special rules for favored customers that got lost, and how much the fines cost last time this happened.

That is not to say there aren't any rays of sunshine. Security through obscurity happens when systems are so old that hackers don't know how to break into them.

Age, Jobs, and Careers

The alternative to aging is dying young. Both have their disadvantages.

Careers in general reach a plateau around age 40. Granted, some achievements come later: All my patents were issued after 40. In the Information field, however, a plateau can come early because a technical career is like running in a hamster wheel that spins faster and faster. In more than a few organizations, developers had better have reached the senior ranks by age 30, because by age 40 they will be considered elders. Thus, mid-career changes are common.

Some tech workers jump over to the management track, and for reasons outlined earlier, technical managers have valuable experience. But most tech workers keep grinding away at technology because there is always something new and interesting around the corner. That, of course, assumes their employers allow it.

This takes us to the difference between jobs and careers. Employers provide jobs. It's up to employees to manage careers. A job gets a paycheck. A career achieves fulfillment in other ways. Thus, job-hopping is not the same as career advancement.

Skill Sets

One strategy for sustaining a technical career is to hone not one, but a set of skills. Indeed, that benefits both employees and employers.

T-shaped skills combine deep knowledge in one area, such as a technology, with broad knowledge in a related area, such as an industry using that technology. Multiple T's are even better. For instance, knowing multiple technologies, multiple industries, and multiple job roles means an employee's resume checks a lot more boxes.

Constraint Management

Can skills be an operations constraint? Yes, but to be a constraint, skills must be subject to Constraint Management. That means establishing buffers and managing them according to actual demand rather than a forecast. Otherwise, a skills shortage is a limitation, not a constraint.

The Constraint Management approach to resource management assumes that client projects and internal teams need people on demand. Thus, those projects or teams cannot wait several weeks or months for people to be assigned. If they can wait, skills aren't a constraint.

An organization that relies on commodity skills can alleviate a shortage simply by hiring more workers. However, few jobs in the Information field are commodity skills. Thus, higher education, summer internships, rigorous interviews, classroom training, on-the-job training, remote work, and mentoring or coaching are common. Even with such preparation, however, new graduates aren't expected to be fully productive for a year, and even experienced hires aren't productive on day one. On the other hand, having a surplus of high-tech skills is expensive, and correcting an over- or under-supply of high-tech skills is painful.

In the Information field, skills can be clustered into groups. Front-end and back-end developers are separate skill groups. Alternatively, skill groups might be defined as analysts, architects, designers, coders, testers, operators, consultants, or researchers. A global company may have dozens, or even hundreds, of skill groups when location and language figure into groupings.

A skill buffer includes people with the same skills who are not yet assigned to a team or project, but who are available for assignment. When assignments end, people go back into the buffer, pending another assignment. Consulting organizations call this being "on the bench." While not good for extended periods, being on the bench occasionally is normal, and bench time can be put to good use for training or administrative work.

Skill groups come in these types:

- **Commodity skills.** Available on short notice
- **Core skills.** Available with significant lead time
- **Critical skills.** Chronically in short supply

Because the organization can acquire commodity skills as needed, the target buffer level is set to zero. For critical skills, such as Technology Architects, the organization will hire whenever a qualified candidate is identified, and even if a target buffer level is greater than zero, the actual buffer level is typically zero. It's that middle group, the core skills, where the majority of skill groups fall and where Constraint Management is most active.

Supply and demand ebb and flow at different times and different amounts. Forecasts exacerbate the problem because they rarely predict turning points. Thus, staffing to plan eventually leads to over- or under-staffing when turning points are missed. In contrast, Constraint Management ignores forecasts and reacts to actual demand. That way, staffing to buffer catches turning points. Constraint Management calls this application Replenishment because it replenishes resources as they are assigned rather than as a forecast predicts.

Details of how skill buffers are sized and managed are beyond the scope of this book because they are covered in *Reaching the Goal* [Ricketts, 2008]. In a nutshell, however, the target buffer level is the number of people needed for new assignments during the time it takes to acquire them. If the actual buffer level is stable, supply matches demand, and the buffer level is in the green zone, so no action is required. If, however, the buffer level falls into the low red zone, hiring is needed. And if it rises into the high red zone, hiring stops until attrition brings it down.

When I wrote *Reaching the Goal*, I was a consulting partner doing professional development, then quality control, and then R&D for a firm in Professional, Scientific, and Technical Services [Ricketts, 2008]. The trigger for our Constraint Management work had been a severe staffing imbalance. With the skills solution we developed, that capacity problem was corrected in a few months.

Lateral Thinking

Constraint Management has always been about thinking differently. The idea that decisions improve when forecasts are ignored is heretical in many management circles. Yet Constraint Management has demonstrated its value over and over in Manufacturing and Distribution companies.

When my colleagues and I applied Constraint Management thinking to Professional, Scientific, and Technical Services, however, there was a problem. Constraint Management in Manufacturing and Distribution carefully manages inventory, but in services there is no inventory. We eventually realized that even in a services business, there are multiple places to look for constraints, and each is an opportunity to explore buffer management. Skills are just one example.

Will Anyone Care?

After successfully applying Constraint Management principles to Services, I thought others might be interested in what we learned. It took some convincing for the publisher to see potential value in a book about services management when most other books with its imprint were about hardware and software products. Fortunately, the book did find an audience, and it's been rewarding to hear from readers who have found ways to apply Constraint Management to their own service operations.

Lessons Learned

Lessons learned include:

- Constraint Management for Goods creates buffers of tangible objects, such as raw material, finished goods, and parts inventories. Constraint Management for Services creates buffers of intangible objects, such as skills, time, and service requests.
- Writing a book which covers fresh territory is a leap of faith that readers will find it, read it, and gain as much value as the author hopes. Another author once observed that if you take the total number of books sold per year about the Information field and divide it by the number of people employed in that field, the number of books per person per year rounds to zero.

Conclusion

The Information field depends on skills that must be refreshed and extended frequently. Education is, of course, a starting point for many, but Larrabee's Law reminds us that "Half of everything you hear in a classroom is crap; education is figuring out which half is which."

The Information field attracts Introverts, while senior management attracts Extroverts. That's a real conflict, though not insurmountable.

The Information field is rather tribal: Technical Conservatives and Technical Liberals have different notions about how best to develop and operate technology. However, the variety of organizations and their technologies means that these opposing philosophies are each successful in some circumstances.

Unproductive working conditions are sometimes mistaken for skills issues. That includes open offices and prohibition of remote work.

Constraint Management does apply to skills if projects and teams need skills on demand and cannot wait for the usual hiring cycle. Skill groups consist of people with similar skills. Skill buffers consist of people waiting for new assignments. Constraint Management prevents over- and under-staffing by managing those buffers.

CHAPTER 13

METHODOLOGY

A methodology is a set of principles, rules, and tasks for developing, operating, or maintaining a specific set of deliverables and work products. In the Information field, deliverables include hardware, software, data, services, etc.

Work products flow between tasks in a methodology and may or may not be part of a deliverable. For example, a test plan is a work product that defines (1) which types and what quantities of test data should be included among the work products, and (2) what the expected test results are. However, the test plan, test data, and test results are not delivered to customers.

Following a methodology enables teammates to collaborate effectively because everyone knows their role, what they need to accomplish, when it's needed, and what comes next. Otherwise, they may unnecessarily work at cross purposes and fail to satisfy requirements.

Although methodologies apply in every industry, this chapter focuses on the Strategy and Constraint Management implications of methodologies for the Information field. Widely used methodologies in the Information field can be divided into two broad families: Planned versus Agile, also known as Waterfall versus Iterative, or Traditional versus Modern. Within each family are variants.

Why Methodology Matters

Consider this analogy. If a team is small, it can get along with just a shared goal, informal guidelines, and a lot of improvisation. On the other hand, a large team without a methodology is like trying to get an orchestra to play a symphony with no sheet music and no conductor. The result is not likely to be harmonious.

This chapter does a high-level comparison but does not teach specific methodologies. Many books, articles, and videos are available with in-depth tutorials for project sponsors, project managers, users, and developers.

Pioneers and evangelists for any methodology may legitimately claim success that is difficult for mainstream adopters to replicate and impossible for laggards to complete. Thus, publications and presentations often promote one methodology with particular fervor, and perhaps a few recommend blended methodologies that attempt to combine the best features. That said, some methodologies are more appropriate in specific circumstances.

This chapter is about the logic behind Planned Methodologies and guiding philosophies behind Agile Methodologies. The following chapters are about execution of projects and processes following those methodologies.

As previous chapters have described, most jobs in the Information field are hard work if done well. Computer hardware and software are particularly hard because they are among the most complex artifacts created by mankind. Thus, when things start to go a little bit wrong, they can end up catastrophically wrong. Methodology doesn't make enterprise-grade Information work easy. Methodology makes it possible.

ADVENTURE: Methodology Shapes Outcome

My first brush with methodology was informal and ineffective. Way back when I was coordinating production through a Heat Treat department (see the Prologue about the enterprise constraint), some expensive parts were missing. They had passed through Heat Treat but never arrived at the Assembly Area.

Although they were key components in an expensive industrial product, they had no intrinsic value to a potential thief, not even as scrap metal. Thus, they were in all probability somewhere in a sprawling factory with four floors.

Bucket List

While most parts were large enough to transport on pallets or in steel bins roughly three feet square by four feet tall, the missing parts were tiny. Thousands of them fit in the bottom of a round bucket. My target was tiny amongst a sea of pallets and bins.

I traced every possible path from Heat Treat to Assembly. On each path, I spoke to drivers, operators, and foremen on every shift. Heat Treat was the last place anyone recalled seeing the parts bucket. I widened my search to adjacent departments, but still saw no sign of the parts.

What Goes Up, Must Come Down, and Up and Down . . .

What I did not realize, however, was my path deviated slightly from the path the parts took. I sprinted up and down the stairs while parts rode an adjacent industrial elevator. The elevator was ancient: One solid outer half-door rose from the floor while another descended from the ceiling, and a slatted inner gate rose while another descended, and once they met all had to lock in place before the elevator would move. It did, however, lift thousands of pounds of steel parts, plus transport vehicles and drivers, so safety precautions were mandatory and sensible. For someone on foot, however, taking the elevator was slow—and it annoyed the transport drivers because it wasn't a passenger elevator.

Of course, that's where the parts bucket was. Inexplicably, someone had taken the bucket off a transport vehicle and set it aside. The bucket had been riding in a corner of the elevator for three days before we found it.

Lessons Learned

Lessons learned include:

- My search methodology had a gap that I overlooked repeatedly even though my target was hiding in plain sight.
- Years later when I taught methodology to hundreds of consultants and trained the trainers who taught several thousand more, I taught them to "engage their brain" because no practical methodology can cover every scenario.

Domains

As the terms will be used here, methodologies can be used on various domains:

- **Project.** A set of dependent tasks that produce unique deliverables during a finite duration under a project manager unless the team is self-organizing (for example, a project can design new computer hardware or refactor legacy software)
- **Process.** A set of dependent tasks that produce the same deliverables on-going under a production manager and process owner (for example, a process can deliver a series of incremental software releases or technical services, such as hardware repairs)
- **Program.** A collection of related projects or processes under a program manager
- **Portfolio.** The entire set of projects, processes, and programs being planned or executed by an organization

Multi-project portfolios may or may not consist of programs. For instance, the U.S. space program was a series of projects that flew Mercury, Gemini, Apollo, and shuttle missions, which were quite diverse. On the other hand, residential construction can be a set of independent projects that builds houses, condominiums, or apartments which are identical, or nearly so.

Resources can be people or things, such as a punch press or an aircraft hangar. Programs encompass shared goals without necessarily sharing resources. Conversely, multi-projects have shared resources without necessarily sharing goals.

Constraint Management has different applications for processes and projects. See the Production and Project applications in the earlier Constraint Management chapter.

Information Projects

Some project management mavens claim there's nothing special about Information projects: In their view, a deliverable is a deliverable. However, project managers from other industries rarely manage Information projects, unless it's for embedded hardware or software, because the concepts and terminology often have no counterparts in other industries. Thus, embedded Information components may be delivered as projects separate from the products that embed them—until final assembly and integration testing brings the deliverables together.

This suggests that although the methodologies may be the same in theory, the environments, expectations, and best practices for Information projects can be rather different in practice. For example:

1. The work breakdown structure for software development is more like product development (engineering) than manufacturing. IT operation, also known as production, is more like manufacturing in the sense that both are highly repetitive.
2. While other industries do design and build as separate projects, software development generally combines them in a single project.
3. While many projects deliver the same products or services repeatedly, software projects always produce unique deliverables because it would be faster, cheaper, and safer to reuse existing software that already meets requirements.
4. Initial delivery of software as a minimum viable product that will be incrementally enhanced is common practice nowadays. This contrasts with manufactured products that are often unusable in a partially completed state.
5. Software projects are not directly affected by weather, while all projects relying on a physical supply chain or outdoor venue are affected by weather.
6. Skill requirements advance rapidly because the Information field evolves rapidly.
7. Services projects range from well-defined (example: install servers and storage) to ill-defined (example: create a comprehensive Technical Strategy).
8. Information projects range from highly repetitive (software releases) to highly inventive (research and development) to highly innovative (productization) to highly valuable (business model change or business process change).

9. Rapid technical obsolescence means a significant delay can miss the opportunity window.
10. Information projects tend to over-engineer technical solutions rather than simplifying business problems.

Iron Triangle

The Iron Triangle of project management notes that there is an unavoidable trade-off between requirements, resources, and time. As the old saying goes, "You can have it good, cheap, or fast: Pick any two."

Planned and Agile Methodologies approach the Iron Triangle differently:

- **Planned Methodologies.** Requirements are fixed. Resources and time are variable. Thus, large deliverables are completed mostly at the end by a team that varies in size over time depending on skills needed, and the overall duration may be substantially longer than the initial estimate.
- **Agile Methodologies.** Requirements are variable. Resources (and sometimes time) are fixed. Thus, small deliverables are produced at frequent intervals using the same team, but the content of those deliverables will vary substantially, and the overall duration is not just hard to forecast, the project may continue indefinitely.

Of course, nothing is totally fixed. Every methodology expects some flexibility. Yet there are limits:

- **Brooks' Law** recognizes that there is an absolute limit to adding resources to finish sooner. "Nine women working together cannot produce a baby in one month" is the classic catch phrase [Brooks, 1995].
- **Minimum Viable Product** recognizes that there is a practical lower limit to sacrificing some requirements to finish sooner. As requirements are stripped away, products eventually reach the "Why bother?" stage.
- **Death March** syndrome recognizes that there are personal limits to excessive overtime [Yourdon, 2003]. Developers can and do make personal sacrifices, but eventually health and family issues arise.

Choosing a Methodology

Choice of methodology may depend on factors such as:

- **Architecture.** Which takes priority: consistency, availability, or partitioning?
- **Deliverables.** Will delivery be incremental or complete?
- **Requirements.** Are users personal, work group, enterprise, or military?
- **Complexity.** Some deliverables are inherently complicated and require the highest skills.
- **Skills.** Novice and commodity skills are often most plentiful.
- **Collaboration.** The larger the team, the more time is spent communicating.
- **Location.** Some methodologies assume that team members are co-located.
- **Documentation.** Paperwork is a major activity on Planned Methodologies because written communication takes place across phases, but much less so on Iterative Methodologies where oral communication takes place across iterations.

A method that works well on a small project (a "two-pizza team") can fail miserably on a large project with tens to hundreds of team members. The converse is also true. Therefore, this book advocates situation-appropriate choices.

In some organizations, however, methodology is not a choice. One methodology has already been designated as the standard, and it is used regardless of fit for purpose. For a uniform portfolio of projects, that may be sensible; for a diverse portfolio, it can be myopic.

Planned Methodologies

Planned Methodologies strive to fulfill a complete set of requirements. The two main variants of Planned Methodologies discussed in this chapter are:

- **Critical Path.** Tasks are arranged by dependencies and precedence. For example, "Task A and task B can be done in parallel, but both must finish before task C can start." Contingency is added into each task estimate to guard against late completion. Tasks may be aggregated into phases.

- **Critical Chain.** Tasks are arranged by dependencies and precedence, but contingency is gathered into a time buffer that protects a series of tasks against late completion. Thus, aggregation reduces overall duration while increasing the likelihood of on-time project completion.

The Information field also uses some alternate terminology:

- **System Development Life Cycle (SDLC).** Critical Path projects that have phases for planning, analysis, design, development and implementation, testing and deployment, and maintenance.
- **Waterfall.** SDLC with a huge software integration task at the end, which too often generates rework because requirements were missed or implemented incorrectly, thereby making the project late and over budget. This name comes from the arrangement of tasks in the Gantt chart so that the work products flow down from one phase to the next like a waterfall. Thus, the overall flow is relatively linear.

As will be seen later, Agile Methodologies were invented to address problems with Planned Methodologies. However, Agile Methodologies are not without their own problems. Thus, projects where partial completion is unacceptable are still done with Planned Methodologies. For instance, data center construction, disaster recovery, application migration to a new platform, and legacy application renovation are still done with a Planned Methodology, though they may not follow a Waterfall pattern.

For some projects, there is no gain from early completion. For example, an Olympic village is not usable early, but there is huge loss from late completion. However, most projects do benefit from early completion because it saves cost and accelerates benefits. Therefore, projects following a Planned Methodology are frequently under pressure for early completion—and Critical Chain supports this better than Critical Path.

Critical Path

The critical path is the longest series of dependent tasks in a project plan, taking task and resource dependencies into account. It therefore determines overall project duration. Thus, any delay to tasks on the critical path delays the entire project. However, the schedule is just one portion of a project plan. Budget, staffing, deliverables, responsibilities, and assumptions are also major elements.

Key concepts behind Critical Path Method include:

- **Estimates.** Work effort (person hours) and duration (calendar days) used to calculate the project schedule.
- **Contingency.** Time added to cover inaccuracies in base estimates, as well as likely disruptions during execution. A common objective is 80 to 90% confidence that each task can be completed on time.
- **Milestones.** Commitments based on the assumption that if every milestone is met, the entire project will be completed on time. Milestones often mark the end of phases.
- **Starts.** Work on tasks begins as soon as predecessor tasks are complete.
- **Percent Complete.** Task status is tracked as percent complete. Late tasks are less than 100% complete when their planned finish date passes.
- **Expediting.** All late tasks receive additional management attention.
- **Replanning.** The critical path can and often does shift during execution as some tasks fall behind, thereby triggering changes to the schedule.

The intended purpose of milestones is to provide interim goals and monitor progress toward them. Their unintended consequence is, however, to trigger gold plating (doing more than required) when progress seems ahead of schedule or expediting when progress falls behind. Unfortunately, excessive gold plating or expediting do not contribute to effective projects.

Task durations, and corresponding work effort in hours, are often estimated by analogy: Tasks on a proposed project are assumed to deserve the same schedule, budget, and staffing of whichever past project it most resembles. Of course, if the past project was initially underestimated and only succeeded through a heroic effort, the proposed project is likely to be underestimated, too.

A better alternative is estimating by metrics. Unfortunately, a major difference between software engineering and other types of engineering is that software engineering has no generally accepted metrics or equations for converting them into estimates.

Estimates vary depending on who does them and when they are done. Estimating by experience happens when the person who will perform a task makes the estimate using their own personal history. However, if estimates must be made before a proposal can be presented to the client, on services projects that's often well before individuals are assigned. Thus, proposal templates for recurring project types include work breakdown structures and task estimates.

Despite its long-standing use, here are some problems with Critical Path Method:

- **Student Syndrome.** Although tasks may be begun in order to report "on-time start" status, other priorities often intrude until most work is done at the last minute, which squanders contingency embedded in task estimates.
- **Parkinson's Law.** Work fills the time allotted, even when an estimate is increased or decreased. There is an inherent conflict because managers want to compress estimates while workers want to expand them.
- **Early Finish Punishment.** Tasks completed early tend to have future estimates reduced by management, thereby creating disincentive for work to finish early.
- **Late Finish Accumulation.** With few early finishes, late task finishes tend to accrue, thereby making projects later and later.
- **Multitasking.** People are sometimes called upon to do more than one task at a time on different projects, if not the same project. The time taken to switch between tasks slows progress on both tasks.

On Information projects, multitasking hinders progress by taking people out of mental flow time when they are most productive. Multitasking creates the illusion of productivity, and this belief is so common and so strong that it usually requires a demonstration to prove it's false.

Therefore, do the Alpha-Number-Shape Exercise. It only takes a few minutes. Try it yourself, then have your team do it.

1. Find a blank sheet of paper, plus a pen or pencil. Start a timer. In one column, write the alphabet from top to bottom. In a second column, write the integers from 1 to 26, top to bottom. Finally, in a third column, draw these shapes repeatedly: circle, square, triangle. Stop and save the timer.
2. Turn the paper over. Start a new timer. Fill in the same alphabet, numbers, and shapes—but do it row-wise instead of column-wise. For instance, the sequence should be A-1-circle, B-2-square, C-3-triangle, etc. Stop and save the timer.
3. Compare the timers. A typical result is the second timer is twice the first timer.

Thus, the second timer simulates multitasking as our brains switch more frequently between the alpha, number, and shape tasks. If you want to make the exercise more challenging, do it with the alphabet in reverse order, Roman numerals instead of Arabic numerals, and these shapes: parallelogram, pentagon, and hexagon.

Most multitasking within and between projects is management-driven or client-driven, so a solution must start there. As will be seen, Critical Chain discourages multitasking, and Critical Path has since adopted the same practice.

Critical Chain

The critical chain is the longest series of dependent tasks in a project plan, taking task and resource dependencies into account [Leach, 2014]. That should sound familiar because it's the same definition as the critical path. However, the critical path and critical chain are never actually the same because they handle uncertainty differently. Even if the same tasks happen to form both the critical path and critical chain, which isn't guaranteed, the overall planned durations will be different, and so will project execution.

Critical Chain Method comes from Constraint Management [Newbold, 1998]. The Critical Chain is the project constraint, so managing it manages the entire project. A third or less of all tasks on a typical project fall on the critical chain, so the method focuses management attention.

Key concepts behind Critical Chain Method include:

- **Resource Leveling.** No resource can be assigned to more than one task concurrently. Hence, no multitasking, and no fractional resource assignments.
- **Estimates.** Work effort and duration based on pure task time: no waiting, multitasking, or contingency. Each task should have about a .50 probability of on-time completion because it's expected that some will finish early and some late.
- **Project Buffer.** A time buffer that protects the entire project. Typically, it is about 50% percent of the Critical Chain, which is significantly less than the sum of embedded task contingencies on Critical Path.
- **Feeder Buffers.** Time buffers that keep non-critical tasks from delaying the Critical Chain.
- **No Milestones.** The only commitment is that the entire project will finish by the date at the end of the project buffer because, if permitted, milestones would have to have their own time buffers.
- **Just Enough Detail.** Project plans with more than 300 tasks are too detailed.
- **Starts.** Tasks are scheduled to start as late as possible because this minimizes work in process. However, once started, tasks are completed as soon as possible.
- **Relay Race.** This work rule says that once a task has started, work on it and nothing else until it's done, then immediately hand off work products to whoever is doing the next task. Like a relay race, work proceeds briskly without concern for the estimate or the schedule.
- **Work Remaining.** Instead of percent complete, Critical Chain Method tracks work remaining on late tasks. Thus, it focuses on work to be done, not work already done.
- **Buffer Penetration.** The amount of buffer consumed by late tasks. For example, if the latest task on the critical chain was estimated at five days, three days have been worked on it, and the estimate of work remaining is three more days (because the task turned out to be harder than originally assumed), buffer penetration is one day beyond that task's planned finish date.

- **Expediting.** Only tasks causing buffer penetration receive additional management attention.
- **Replanning.** Once designated, the critical chain is not changed, even if a non-critical path causes buffer penetration. However, if scope changes affect the critical chain, then replanning is done.

When estimating, it's important for workers to understand that contingency is being moved out of task estimates and into the project and feeder buffers. Thus, shorter task estimates mostly cut out wait time, not work time. However, past estimates may have been wrong: Some tasks aren't compressible (for instance, machine time), while others may have been severely underestimated (testing especially), so their estimates may grow. A simple rule of thumb says new task estimates should be about 50% smaller, on average, but good judgment is necessary. Projects have huge uncertainty, so striving to be more accurate than the noise isn't effective.

Organizations can achieve 95% on-time delivery within one year of adopting Critical Chain Method. This enables about 30% more projects to be completed with the same resources. Sometimes, however, predictability matters more than speed because customers or business processes depend on timely project completion.

Here are some challenges on Information projects:

- The due date is often tied to a business event or budget cycle, but the start date is "now" or "as soon as possible" (ASAP). That puts the project in a time box that no Planned Methodology is equipped to handle because actual duration is determined by scope and resources more than arbitrary due dates.
- On Information projects, things happen way beyond normal task variation. Key persons may leave the company. Clients may want to increase scope but not schedule. The software design won't work. Requirements are too expensive or technically impractical.
- Information projects include verification, validation, testing, and certification, which can suspend work until approval checkpoints are passed.
- Critical Chain assumes minimal coordination between parallel tasks. To the degree that there are lots of technical dependencies, a gain in one place is hard to exploit.

Although Critical Chain Method isn't as well-known as Critical Path Method, it has reached the tipping point on mindshare. The Project Management Institute has embraced Critical Chain in its Body of Knowledge. Critical Chain has been used on thousands of projects, including mainframe software, application software, hospital management, oil rig manufacturing, and aircraft maintenance. For the latter, the team won the Shingo award for world-class performance.

Like all Planned Methodologies, Critical Chain works best in stable requirements, but Information projects are rarely stable. Ambiguous requirements and mid-project changes in scope are the rule, not the exception. That's why Agile Methodologies were invented.

ADVENTURE: **Creating Methodology**

Commercial methodologies come with documentation, sample work products, sample deliverables, training, consulting, and project management templates, such as time and expense reporting. Custom methodologies are created when specific deliverables are not supported by commercial methodologies.

Should Islands Have Turf Wars?

During my career, I contributed to several methodologies:

- Software metrics gathering, analysis, and performance improvement
- Software reengineering, refactoring, and quality assessment
- Portfolio analysis and technical strategy

These methodologies were used to deliver Professional, Scientific, and Technical Services.

Methodologies for new system development were readily available, but methodologies for legacy portfolios were not. New service offerings had to produce unique deliverables. Therefore, we developed methodology to close the obvious gaps.

Eventually the company methodology keepers came calling and demanded that work on new methodologies cease in favor of using their existing methodologies. Eventually truces were established when we showed how our methodologies closed gaps in the existing methodologies. But the grumbling about divergence from standards never went away entirely. The final solution was to apply a common name across a set of essentially unrelated methodologies, thereby confusing everyone who wasn't paying attention.

Lessons Learned

Lessons learned include:

- Creating a custom methodology from scratch is a lot of work. One methodology had four major phases. It took years to evolve through practice with hundreds of clients.
- Rather than stifling innovation by prescribing how deliverables are to be produced, methodology is a way to harness innovation. There is no inherent conflict. For instance, every methodology spawned software tools and best practices to facilitate execution.
- In a methodology for services, it's impossible to anticipate every task. During execution, some new tasks may be identified, and old ones deemed unnecessary. Thus, services methodologies tend to anticipate scope creep and use formal project change documents to manage scope. Services methodologies generally include client responsibilities, too.
- When a services methodology is driven by countable objects, such as computer programs, petabytes of data, or user stories, it helps to have an estimating tool that scales the project plan accordingly. That way, when scope changes, the plan changes.

Agile Methodologies

The Agile Manifesto laid out a dozen principles of Agile software [Beck, 2001]. They include customer satisfaction as top priority, face-to-face collaboration between business people and developers, development at a constant pace, continuous delivery, and self-organizing teams. Agile practices spring from those principles.

Agile was invented specifically for software development. Planned Methodologies, on the other hand, apply to more than software. Whereas Planned Methodologies strive to fulfill a complete and stable set of requirements on a predictable schedule, Agile copes with uncertain and changing requirements by rapid discovery. Whereas Planned Methodologies complete large batches of deliverables after lengthy schedules, Agile completes small batches of deliverables during rapidly repeated cycles. Thus, integration and testing during each Agile iteration address the late integration problem in the Waterfall Method.

Decades before Agile, Prototyping was a methodology for rapidly building small-scale examples and getting user feedback which fed into the requirements gathering phase of Waterfall Methodology. Prototyping still works with Planned Methodologies today. Where Agile exceeds Prototyping, however, is Agile responds to changing requirements, not just initially uncertain requirements.

Agile Methods range from strong business orientation to strong technical orientation. By some counts, there are 40 Agile Methods, and more are being invented. Thus, there is no consensus on what Agile is. Agilistas insist that Agile is a process, not a methodology, but less fervent Agile practitioners have no qualms referring to Agile projects. So, recognizing that Agile is controversial, here are some prominent examples:

- **Design Thinking.** Identifies customer goals and behavior, then designs and tests solutions repeatedly
- **Lean.** Redesigns business or Information Technology processes to remove waste
- **Extreme Programming (XP).** Uses engineering best practices and tools to build and integrate software
- **Scrum.** Breaks down user stories into discrete units, which are worked until done
- **Kanban.** Prioritizes tasks into a queue, where they are worked until done
- **DevOps.** Gets software and Information Technologies into production often

These Agile Methodologies have the following common characteristics:

- Dedicated teams minimize multitasking between projects.
- Work rules minimize multitasking within projects.
- Work flows in small batches.
- Testing is most often the bottleneck.
- Customers can add, change, delete, or reprioritize features at any time.
- Developers can release software at any time instead of only in major releases.

The Agile Method most often used today is Scrum, but Kanban and DevOps are growing. These are the Agile Methods described in following sections.

Scrum

Scrum is a process with specific roles, rules, and practices. Roles described in *The Scrum Guide* [Schwaber, 2017] are:

- **Product owner.** Represents customers and other stakeholders. Selects user stories to go into each release.
- **Development team.** Does everything from analysis and design to coding and testing and documenting.
- **Scrum master.** Schedules meetings, plans releases, and shields team from distractions.

There are typically three to nine teammates, all with general skills. All are co-located or have easy online collaboration. Note, however, that Scrum does not have a project manager role per se because there is no plan beyond the next release.

Key concepts behind Scrum include:

- **User stories.** Simplified requirements, including user type, user wants, and reasons.
- **Story points.** Estimates of effort needed to develop user stories. Sometimes quantified as T-shirt sizes: S, M, L, XL.
- **Product backlog.** Prioritized collection of user stories.
- **Sprint.** Time-boxed milestones, usually two weeks long.
- **Burndown chart.** Day-by-day measure of work remaining, which bounces around but trends down.

- **Stand-up meeting.** Short meeting, often 15 minutes daily, for communication between team members, mostly about issues.
- **Exit criteria.** User stories are implemented. Testing is done. Software is usable.
- **Release.** Software ready to be shipped covering stories from 2 to 12 sprints.
- **Review.** Meeting to identify areas of improvement.

Just as dates aren't supposed to slip on Planned Methodologies, scope isn't supposed to slip on Scrum when a sprint covers a customer critical situation or commitment. Scope is otherwise flexible on non-critical user stories.

Kanban

Kanban is a method for managing and improving delivery of software and services [Anderson, 2015]. It makes workflow visible by using physical or digital boards to move work between these categories: (1) needs to be done, (2) in process, and (3) completed. It regulates the introduction of new work in process based on completion of previous work. Thus, Kanban manages work in process so that developers, the constraint, are not overloaded. This pull approach contrasts with the push approach of Critical Path Method.

Kanban has roots in Constraint Management [Anderson, 2004], but Kanban is more like Constraint Management for production than for projects. That is, Kanban sees developers as the constraint rather than the schedule as the constraint. Kanban therefore manages work flow. Key concepts include:

- **Customer needs.** Focus on them.
- **Flow.** Work is the progression of value.
- **Commitment point.** Ideas have been proposed and accepted, so work is about to begin.
- **Delivery point.** Work items are complete.
- **Work in process.** Items committed, but not yet completed. Limiting WIP is a core practice.
- **Lead time.** Time between commitment and delivery.
- **Delivery rate.** Number of items delivered during a period.

- **Bottlenecks.** Places where workflow is slowed or blocked.
- **Policies.** Defining a process explicitly.
- **Roles.** There are no required roles, but service request manager and service delivery manager are common. These roles correspond to commitment and delivery points.
- **Visualization.** Visibility aids collaboration, but boards take many forms.
- **Kanban meeting.** Daily stand-up discussion of work items and blocking issues.
- **Self-organization.** Manage the work and let teammates organize themselves.
- **Improvement.** Do better through feedback loops and evolutionary change.

Unlike Scrum, which breaks work into sprints of fixed duration, Kanban prioritizes work, but doesn't prescribe time to delivery. Thus, Kanban is best suited to short, high-velocity projects where a task list is good enough because work breakdown structures and schedules distract from getting work done.

DevOps

DevOps is short for "Development Operations" [Kim, 2014]. As of this writing, there isn't a single authoritative definition because the subject is quite broad [Mueller, 2010].

Two key aspects stand out, however [Kim, 2016]. First, DevOps is Agile for operations. Second, DevOps narrows, if not closes, the gap between development and operations. In some implementations, the same team does both development and operations for the application it owns. In others, operations are substantially automated via software. Site Reliability Engineering is a related concept that says systems should be designing for automated operations from the outset. Automatic scale-out is one example.

DevOps enables continuous delivery [Humble, 2011]. That should sound familiar because frequent delivery, instead of occasional software releases, is a hallmark of Agile development. But continuous delivery requires continuous integration and continuous testing, too. Thus, DevOps is not a good fit for Planned Methodologies when integration and testing are deferred to the end.

One way to foster continuous integration, testing, and delivery is through daily builds. This is the practice of compiling, linking, and possibly testing all software components to ensure that dependencies are satisfied, and no new bugs have been created. One reason that Legacy Systems may be ill-suited to DevOps is their builds can run more than 24 hours.

Hybrid Methodologies

Given the descriptions of Planned and Agile Methodologies earlier, it might seem that they are totally incompatible, but that's not true. Some organizations practice both types, albeit separately. But it's also possible to combine them selectively.

One way for these methodologies to co-exist is to use Planned and Agile Methodologies for different types of systems. However, as explained in the Two-speed Technology adventure, cross-system dependencies make this difficult.

Another way to combine disparate methodology types is to use Critical Chain for development, then use Agile for maintenance of the same software. That may make sense when development is too big for Agile, and maintenance is too urgent for Critical Chain.

Yet another way to combine them is to adapt Critical Chain and Agile so they can be applied simultaneously on the same project [Ujigawa, 2016/Hutanu, 2015]:

- Create a feature list and prioritize it.
- Identify tasks using Critical Chain, but instead of creating absolute time estimates, create relative time estimates, such as T-shirt sizing: S, M, L, XL.
- Rather than executing Agile in time boxes, expect tasks to vary in duration.
- Plan the project based on velocity (story points per day), plus a project buffer.
- Create a Critical Chain plan, plus a scope buffer in case requirements expand.
- Execute the project using Agile for tasks and Critical Chain for buffer management.

Thus, Critical Chain is what's shown at the executive level, while Agile is followed at the developer level.

ADVENTURE: Two-speed Methodology

Two-speed methodology happens when some projects are Agile (churning out minor releases every few weeks) while others are Planned (grinding out major releases a few times a year). Some companies have spent years adopting Agile methods on their Systems of Engagement, only to realize that their Systems of Record cannot do two-week sprints effectively. That's a problem because Systems of Engagement rely on Systems of Record to fulfill data requirements.

At one such company, Systems of Record development teams were directed to slice their work into two-week segments, but compressing their Waterfall Methodology into ultra-short release cycles drove up the defect rate to unacceptable levels because it was impossible to test changes thoroughly. The executives who initially pressed for those complex enterprise applications to be reimplemented on a distributed platform eventually realized that would be unaffordable because dozens of developers were maintaining millions of lines of mission-critical code.

Finding Another Gear

The near-term objective was coordinated releases across Systems of Engagement and Systems of Record with Systems of Engagement setting the pace. Longer term, the objective was to rearchitect the Systems of Record with microservices that not only could be delivered rapidly but could also be executed more frequently than batch jobs.

Enterprise DevOps tools were necessary for the Systems of Record development teams to adopt Agile methods. But in addition to integrating those tools into their computing environments, the source code had to be migrated into those tools, which was a major undertaking.

Agile Ain't Pixie Dust

Although Agile was an appropriate destination, it was an inappropriate route to that destination. Software tool implementation and source code migration were inherently a Waterfall project. Requirements were known and stable. Methodology was tried and true. Tools were capable. Sizing and estimates were fact-based. Experts had the necessary skills, experience, and availability.

Daily scrum meetings, however, created delay, expense, and frustration. Three project managers, each catering to a different management audience, sought coordination of status and issues in daily stand-up meetings which turned out to be unnecessary overhead. Progress was slowed because management repeatedly intervened at the wrong time in the wrong way. The lead consultant had to run interference to minimize disruptions to experts doing the work.

Including gaps due to budget wrangling, architecture board approval, and other developer priorities, this migration project took four times longer than previous projects of the same type and scope following the standard methodology. It doesn't make sense to hire experts and then tell them how to do their jobs.

Lessons Learned

Lessons learned include:

- Coordinated application releases across disparate computing platforms are much harder than coordinated or independent releases on the same platform.
- Starting Agile migration with the largest, oldest, most complex, mission-critical application is not a good strategy. Too heavy. Too slow. Too laborious. It's not a safe route to buy-in, even if the technical and business objectives are achieved.
- Trying to manage a Waterfall project as if it were an Agile project only fools the uninitiated. Nevertheless, this was not a failed project: It did achieve its technical and business objectives; it just took a lot longer to do so than necessary.

Dark Agile

Agile has its Agilistas, but also its skeptics. Indeed, Agile's success has brought turmoil. For example, developers and clients sometimes have different Agile experiences [Schindler, 2013]:

- Developers see Agile as flexible, but clients see it as disorganized.
- Developers see Agile as iterative, but clients see it as unpredictable.
- Developers see Agile as focused, but clients see it as disruptive.
- Developers expect Agile to uncover deadbeat clients early, but clients see it as ambiguous.

Developers sometimes have their own complaints:

- Agile was conceived by programmers, but as practiced today it's dominated by managers, who have different objectives.
- Too many key decisions are made by managers with minimal understanding of technology.
- Managers assume all developers work the same way, so tasks can be assigned to anyone, but Agile only works with the right skills and enough of them.
- Distributed teams and large teams are not agile.
- Heavy processes are the opposite of agility.

Some developer complaints are specifically about Scrum:

- Story points are meaningless, yet tracked, and often used as performance metrics.
- Task management tools waste developer time. Burn-down charts are not Agile.
- Scrum is not flexible: *The Scrum Guide* says it's immutable.
- Scrum is not consistent with *The Agile Manifesto*: Scrum uses time boxes (limited duration), but the manifesto does not require them, and Scrum is heavier on management.
- Scrum has become big business with lots of process and training, sometimes referred to as "the Agile Industrial Complex," but certifications are destructive because the bar is so low.

Therefore, Agilistas have disavowed Dark Agile, also known as Faux Agile. They advocate a return to *The Agile Manifesto* [Fowler, 2018/Holub, 2014].

Technical Services

Professional, Scientific, and Technical Services is a sector in the North American Industry Classification System. It includes organizations that do consulting, research, training, maintenance, repair, help desks, call centers, business operations, IT operations, and more [Ricketts, 2008].

Routine services are driven by simple checklists and performed by individuals or small teams. On the other hand, non-routine services are guided by sizable methodologies and delivered by extended teams because the work is complex.

Here are some ways that services methodologies may differ from hardware and software methodologies:

- Planned and Agile Methodologies generally assume that workers make their own task estimates, but service proposals often require sizing and estimates before individuals are assigned. Therefore, service estimates may be done by estimating tools based on metrics, or by proposal teams based on their combined experience and informed judgment.
- Planned Methodologies usually make no allowance for undiscovered tasks. Instead, service contracts often get client sign-off on changes to scope, resources, dates, and deliverables with project change request documents that alter the contract.
- As a service provider improves its own performance, the project constraint may shift to subcontractors where the provider has less control.
- On service contracts, clients always have responsibilities, and may be co-producers of work products or deliverables, but clients often have windows of availability. For instance, they are seldom available during busy season.
- Planned and Agile Methodologies omit management time, but it's required for billing on service contracts. Management time is elastic because it's determined by project duration.
- Client sophistication affects what the provider shows about service project plans. For instance, if the client expects multiple milestones or

the client will demand on-time completion of the last task rather than the project buffer, it's not wise to show a Critical Chain plan.

- When the client knows less than the service provider yet wants to manage the project, that's a risk factor.

Intellectual property ownership is an issue for services as much as for hardware and software. Service providers need to retain enough ownership to stay in business while enabling clients to use deliverables. This applies to methodology, too.

Constraint Management

In common usage, a constraint is anything that restricts what a system can produce. In Constraint Management, however, a constraint must be something used to manage the system. Lots of things restrict what a system can produce, but they can't all be used to manage that system. For instance, fire regulations limit restaurant patrons, but no establishment is managed solely by limiting patronage. On the other hand, kitchen staff are a constraint that managers routinely use to manage service levels.

With this difference between limits and constraints in mind, methodologies approach constraints differently:

- **Critical Path** refers to the "triple constraint" of scope, schedule, and resources—but it treats them as limits and assumes that local optimization of individual tasks adds up to global optimization of a project, which is not a valid assumption.
- **Critical Chain** manages buffers on the longest path through the schedule, the constraint, which does achieve global optimization of the project schedule. That assumes, however, that skills are not the constraint.
- **Agile** manages scope by dropping low-value features. But managing only the current and next iterations is another kind of local optimization.
- **Scrum** manages schedule by time-boxing iterations.
- **Kanban** manages work in process by regulating task initiation.
- **DevOps** doesn't have a universal constraint, but skills are a leading contender.

Sometimes points on the Iron Triangle are non-negotiable. Some projects have firm due dates and non-negotiable scope, such as regulatory compliance and income tax withholding. Commercial contracts may have penalty or bonus clauses that dictate scope and schedule. However, haggling with clients may let that start date slide while the due date doesn't budge, resulting in further schedule compression.

Yet speed isn't the only thing some clients care about. If early completion will result in delivery of materials or goods that the customer has no space for, on-time completion, not early completion, should be the goal. Same goes for services: It doesn't do the client any good for the technicians doing server and storage installation to show up before the equipment is on site and the building has power and cooling.

Can methodology be the constraint? No, because there's no way to regulate project execution by modulating a methodology. However, choice of methodology can determine where the constraint is located: Planned Methodologies manage schedule, while Agile Methodologies manage scope.

Conclusion

Tribes (see the Skills chapter) are drawn to different methodologies. Technical Conservatives want to limit risk, so they favor Planned Methodologies. Technical Liberals want to enable change, so they prefer Agile. Technical Centrists want it all, so they choose hybrid methodologies.

Choosing a methodology necessitates tradeoffs. Planned Methodologies are roughly 40% analysis and design, 20% coding, and 40% testing. Agile methods are roughly 10% analysis and design, 20% coding, and 70% testing. Thus, the deliverables are different, with Planned methods producing more documentation. Whether that's good or bad depends on your role. Developers often hate writing documentation, but Operations and Tech Support say lack of documentation makes their work considerably harder.

After successfully applying Constraint Management to projects for many years, Clarke Ching observed the following [Ching, 2013/Ching, 2014]:

"Both movements are held back by their founders.

- Agile was created for small teams of above-average staff, and now it's struggling in the world of average folk.
- Critical Chain is limited because its creator has gone, and its competitors' messages are easier to absorb."

There is no such thing as an optimal plan, because every time the plan is changed, it changes how the project is executed. Thus, it doesn't pay to polish a plan to the detriment of executing it. However, there is a huge difference between feasible and infeasible projects, so getting into the feasible zone is vital.

A major misconception about all methodologies is that compressing the schedule does not necessarily yield the same outcome in less time: It creates a different outcome, no matter what the new plan says. Likewise, cutting the budget by substituting lower skills changes the outcome, but changing scope changes not only the outcome, it also changes the route to that outcome. As shown in this chapter, changing methodology can compress schedules more effectively.

Nevertheless, there are frequently external factors that take projects to places their plans couldn't anticipate, which is why execution ultimately matters more than a plan does. That's the subject of the next chapters.

CHAPTER 14

PROJECTS

The previous chapter on Methodologies was mostly about techniques for planning projects and practices for executing projects. This chapter is about what actually happens during execution. Why the emphasis on execution? *The Project Management Body of Knowledge* is 98% planning, 2% execution [PMI, 2013/Woeppel, 2014]. Yet no matter how good the plan and how robust the practices, success ultimately comes down to execution. And there are many ways for projects to go awry.

Success versus Failure

One of the enduring myths of the Information field is that most projects fail. Ponder that claim for a moment. Does it match your experience? Probably not, because a high failure rate is not sustainable. A fail-fast strategy is a cute way of saying experiment until you find something that works. Some start-ups do prosper that way, but the unlucky majority perish. Likewise, for going concerns, a high failure rate is a path to bankruptcy.

Sure, most organizations have a disappointing project now and then. If they don't, they're not taking risks. But a high endemic failure rate is an astounding claim. The most-cited study of software projects triggered anguish about a software crisis when it found:

- 31% were canceled.
- 53% were challenging.
- 16% were successful.

"Challenging" was defined as not meeting the original schedule, cost, and scope objectives, even though the objectives of many projects change while underway. Nevertheless, the conclusion most often drawn is 84% were failures [Glass,

2005]. The numbers have drifted over the years, but the myth has not changed: Most software projects still fail.

Why the original project parameters were considered sacred is never explained. Because they are established before any work has been done and they often reflect aspirations without evidence, they have the highest uncertainty and are therefore the values most likely to depart from what can be accomplished during execution.

Are all canceled projects failures? No, about half of canceled projects are canceled despite being well managed [Boehm, 2000]. Those projects are approved with reasonable business justification, based on sound technology, well managed after launch, and nevertheless terminated before completion because the technology or the business changed. Would an organization be better off if such projects were completed but the deliverables were never used? Not likely. However, even canceled projects can produce learning or artifacts for future projects, so cancellation can mean "partially complete" and does not always mean "worthless" [Emam, 2008].

Are all challenging projects failures? No, the definition is one-sided: Early completion, cost underrun, and over-delivery of scope are ignored. Moreover, a miss on any single measure thrusts projects into the challenging category [Eveleens, 2010]. Also, the definition doesn't allow for reasonable deviations: Some delays, some excess cost, or some altered functionality are to be expected when uncertainty is high and the original objectives are intentionally a stretch. When the definition of "challenging" is altered to allow variance within reasonable ranges, the overall success rate on Agile jumps to 64% and the failure rate drops to 6%. For Waterfall, success jumps to 49%, failure drops to 18% [Ambler, 2014].

Furthermore, there is considerable variation around those rates. In organizations with skilled management, the overall software project success rate is around 90% [PMI, 2017].

Is there room for improvement? Of course. Is there a software project crisis? Nope.

Work Rules

Execution determines project success more than plans do. Therefore, ad hoc project execution fares poorly. It's better to stay out of trouble than to recover from trouble. Toward that end, here are some distinctive work rules.

When to start a new task? Critical Path Method says start tasks per the plan, which schedules tasks to start as soon as possible because that preserves task contingency. In contrast, Critical Chain Method says start tasks as late as possible, taking buffers into account, because this minimizes work in process. Similarly, Kanban says start another high-priority task when a current task is complete because that minimizes work in process. Scrum, however, says start highest priority tasks with the objective of completing them within the next iteration.

How to perform tasks? Critical Path says strive to finish every task on time, like a train schedule, because if every task is on time, the project will be on time. In contrast, Critical Chain says once a task is started, work as fast as possible, like a relay race, because if tasks are completed without delay, the project will be on time—or early. Unfortunately, the Critical Path work rule is based on an erroneous assumption, that local optimization adds up to global optimization. By aggregating contingency into buffers, and ensuring that early task finishes offset late finishes, Critical Chain achieves global optimization directly. Thus, Critical Path projects have about a 0.8 probability of on-time completion, while Critical Chain projects have about a 0.95 probability.

Does multitasking improve productivity? Conventional wisdom says it does, so Critical Path overschedules resources unless resource leveling is enforced, and Scrum tolerates self-imposed multitasking. On the other hand, a simple exercise shows that multitasking reduces productivity, so Critical Chain and Kanban prohibit it. But these positions on multitasking refer to project work. Every project, regardless of methodology, is subject to multitasking between project and non-project work. Mobile phones are a non-project distraction. In addition, "piling on" happens when people think they must attend every meeting due to expertise, authority, or risk. Hence, effective project managers and scrum masters counsel developers against piling on.

Does a "full kit" improve productivity? A full kit means all preceding work products must be done before the tasks that depend on them can start. Thus, full kit replaces mid-stream wait time with up-front wait time. This reduces work in process but may not improve productivity. The extreme form of full kit separates requirements and design from development and testing. That approach works well in Manufacturing and Construction, but the proliferation of Agile Methods demonstrates that it is not practical on most software projects.

Traceability is another desirable project attribute. It means every requirement can be traced to a deliverable. And every deliverable can be traced back to requirements. If there are gaps, the project has under-delivered or over-delivered—sometimes both.

Which is more effective: Push or Pull? Critical Path pushes individual tasks for on-time task completion. Critical Chain pulls work through entire projects to on-time completion. Scrum pushes work into a product plan based on priority. Kanban pulls work into progress based on priority plus resource availability. Overall, both Planned Methods and Agile Methods benefit from pull instead of push.

Status of Planned Projects

Ideally, when project execution begins to falter, that shows up in status reporting early enough to trigger remediation. But some reporting procedures are better at early detection of issues and insights into possible corrections.

By portraying tasks as horizontal bars against a timeline, Gantt Charts are a well-worn way to portray planned schedules, but they can also show status. When the Critical Path shifts during execution, that shows as a modified Gantt Chart. On the other hand, the Critical Chain never shifts during execution, unless the project is replanned to accommodate scope or resource changes.

For work in process, Critical Path subtracts hours spent from the task estimate to infer hours remaining. In contrast, Critical Chain asks directly for hours remaining, which may differ significantly from the task estimate and hours already spent. Tasks causing buffer penetration then shift on the Gantt Chart.

Critical Path reports percent completion of all tasks, not just critical tasks, which sometimes creates the illusion that a project has made more progress than it really has because non-critical tasks cannot make the project late. In contrast, Critical Chain focuses on hours remaining on critical tasks. Consequently, Critical Path projects are notorious for sticking at the 90% complete level for prolonged periods, while Critical Chain projects home in on just the tasks that penetrate the buffer.

When a critical task is late, Critical Path typically asks, "What caused the delay?" Unplanned tasks, rework, and wait time are major causes of late finishes. In contrast, Critical Chain asks, "What is the task waiting for?" The former question cannot be answered objectively, but the latter can—and the necessary action is clear.

When a critical task is early, Critical Chain's relay race work rules exploit that early finish to offset late finishes elsewhere. However, this assumes minimal coordination between tasks. To the degree that there are lots of dependencies, as is often the case in Information projects, early finishes are harder to exploit because the successor task is waiting on more than just the one predecessor that finished early.

Critical Chain has a unique status tracking approach called a Fever Chart. (See Figure 14-1.) The horizontal axis is percent complete of tasks on the Critical Chain. The vertical axis is percent of buffer consumed. The Critical Ratio is consumption divided by completion. Ideally, that ratio is less than one because that says the project is on track for on-time completion.

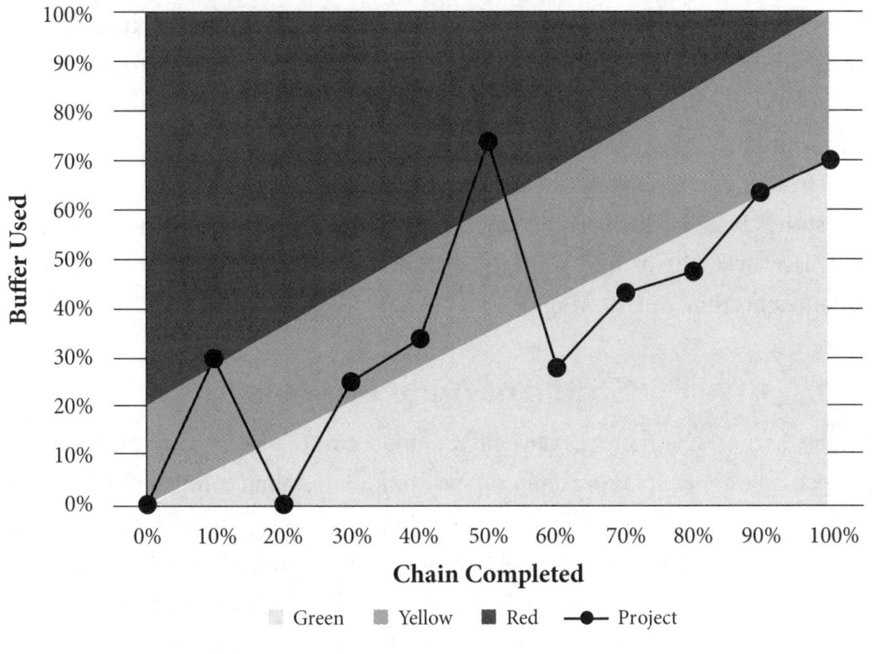

Figure 14-1. Fever Chart.

Because both axes are percentages, a Fever Chart can be used on any size project. As a project is executed, its status will generally move from the lower left corner (mostly incomplete, and little buffer consumed) toward the upper right corner (mostly complete, and possibly much buffer consumed).

Within the Fever Chart are three zones where current status can fall:

- The upper left triangle is the red zone: The buffer is being consumed faster than tasks are being completed.
- The lower right triangle is the green zone: Tasks are being completed faster than the buffer is consumed.
- Between the red and green zones is a yellow zone band: Completion and consumption are roughly balanced.

Thus, for a single project, a Fever Chart shows buffer zones and levels over time. A project in the green zone needs no management attention. One in the yellow zone should be monitored. And one in the red zone needs help.

Critical Path tends to plan and track in detail. For instance, Critical Path may restrict tasks to 40 hours apiece because that facilitates weekly reporting. In contrast, Critical Chain recommends never more than 300 tasks per project even if the task estimates are larger than 40 hours because more detail doesn't help manage the project.

Critical Path strives to meet every milestone. Critical Chain does not use milestones because they require their own buffers, which lengthen the project plan. However, clients may need to be shown how buffer management leads to better results than milestones.

Status of Agile Projects

While it's true that milestones can deliver some subset of project scope on a planned project, milestones are more often at phase boundaries than at finished deliverable drop points. In contrast, Agile Methods recognize the 80–20 rule of thumb, which says 20% of the deliverables probably generate 80% of the value. Therefore, Agile Methods strive to deliver the essential 20% first. Planned Methods have no way to accomplish this, except by planning phases with finished deliverables.

As Scrum executes each sprint, the daily Burndown Chart of work remaining should generally trend down. As Scrum completes each sprint, the task backlog should likewise decline unless new or changed requirements have emerged. For a software business, this task backlog may never reach zero, though it will receive less funding as the product approaches end of life.

Similarly, as Kanban executes, work in process is monitored, and excess work in process is controlled. Kanban measures due date performance. Kanban also monitors the delivery rate as throughput, which is deliverables per period. Kanban measures quality in terms of escaped defects (defects in production). And Kanban removes bottlenecks whenever possible to improve throughput and quality. Thus, Kanban is the Agile Method most aligned with Constraint Management principles [Anderson, 2010].

Spin

Some aspects of status reporting are independent of methodology. For instance, a multiday task is more likely to finish on Friday than another day of the week, and a task is unlikely to finish by 2:00 p.m. unless it is less than a full-day task.

Nobody wants to be the bearer of bad news. A colleague calls this "being economical with the truth." The most common kinds of misinformation about projects are [Glass, 2008]:

- **Misestimation.** Developers pad their estimates. Managers cut those estimates.
- **Misreporting.** Developers overstate progress and understate work remaining. Managers set the tone for truth or fiction.

Eventually, misinformation surrenders to the truth when projects get into trouble. When a project is deeply troubled, and it's a project worth saving, it's time to bring in professionals to get the project back on track.

ADVENTURE: **Troubled Projects**

With apologies to Charles Dickens' *A Tale of Two Cities*, "It was the best of times, it was the worst of times, it was the age of wisdom, it was the age of foolishness" could be the motto of teams that promote best practices and rescue troubled projects. While an executive in a technical services organization that did thousands of Information projects every year, I oversaw a team of top-gun project managers with that mission.

The overwhelming majority of those projects followed best practices and finished successfully. A few, however, needed help, and some were beyond help.

Search and Rescue

The outcome of one troubled project was not what you might expect. A good client had launched a major software development project that was key to their Horizon 2 strategic initiative. They were betting the future of their firm on that project. Although we did other work for that client, we were not involved on that project. As work progressed, however, the executives became uneasy when the project missed its early milestones, and we were brought in to do an independent assessment.

Step #1 was to gather data and plug it into our estimating tool:

- Functional, developmental, and operational requirements
- Skills, tools, and methodology
- Information Technology and business environments
- Progress and issues to date

Step #2 was to generate new estimates across a range of alternative actions.

Step #3 was to present findings, conclusions, and recommendations to the client.

When we told the client that its project had been substantially underestimated, you might think that the executives would reject that finding or punish the managers who had committed to it. But they took a more enlightened approach. The executives confirmed the evidence, added staff and budget, then replanned the project.

After this pivot, the project proceeded to finish on time and within budget, according to the revised plan. And my firm won a larger follow-on project because we had earned the executives' trust by telling them the truth and then helping them turn around their troubled project. This was a case where getting it done right was more valuable than speed and cost because the client needed predictability to align this project with others in its strategic initiative.

Hair on Fire

Unfortunately, most troubled projects do not have such enlightened oversight. When the response to trouble is more pressure, people running around as though their hair is on fire are not focused on the constraint. (More on this in later adventures.)

Troubled projects are predictable because the top reasons they get into trouble are remarkably consistent:

1. **Fuzzy deliverables.** "Are we there yet?"
2. **Scope creep.** "Who moved the finish line?"
3. **Unrealistic expectations.** "If everything goes according to plan . . ."
4. **Underestimated effort.** "Seems pretty simple on paper."
5. **Schedule slips.** "We'll just have Christmas next July."
6. **Understaffed and under-skilled.** "If only we had more and better staff."
7. **Client not fulfilling its responsibilities.** "We'll get around to it eventually."
8. **Low quality.** "We tested once, and everything was fine."
9. **Technical issues.** "What do you mean, the technology doesn't work at scale?"
10. **Wobbly priorities.** "Never mind what we said last week."

Fortunately, for every worst practice there is a corresponding best practice. But favorite practices are sometimes called best practices without evidence that they really are best. It pays to pursue practices with a sustained record of success.

Some best practices are context dependent. For instance, a small, simple project may have no trouble doing multiple builds per day, while a large, complex project may struggle to do daily builds. Or a military project will deserve more inspections and testing than a commercial project to build a new website.

In addition to project rescue, which happens after the fact, the mission of the delivery team was also education, which should happen well before trouble arises. That education should focus on bottlenecks, such as testing, but it can delve deeper, such as conditions leading to defects escaping into production, which is a topic beyond just testing.

Too Agile

Can Agile projects get into trouble? Sure, if they take too many iterations, cost too much, distress the client, or build what the client says they want but that doesn't actually solve the business problem at hand. For instance, clients routinely want better forecasting, but better forecasts are notoriously difficult. Instead, by being demand-driven, Constraint Management of inventory makes forecasts irrelevant. Thus, getting a wayward Agile project back on track may mean working with clients on requirements more than working with developers on delivery.

Lessons Learned

Lessons learned include:

- Any project can get into trouble. All other things being equal, however, shorter projects are generally safer projects.
- Causes of troubled projects are not a mystery. The same ones appear over and over.
- Solutions to a troubled project are not a mystery. Setting the project on fire doesn't help.
- A learning organization endeavors not to make the same mistakes repeatedly.

Scope Creep versus Scope Surge

Scope creep is many small changes in requirements over time. Scope surge is a large change in requirements all at once. Their causes and consequences are different. Scope creep often arises as clients learn about the deliverables. Scope surge more often comes from an exogenous source, such as a rescheduled business deadline or a merger.

Clients sometimes expect small changes to be accommodated without changing the schedule, staffing, or cost. That may be possible for one instance, but even small changes add up. Thus, experienced project managers think beyond the current request.

On an Agile project, small changes get folded into the product backlog. They float toward the top or sink toward the bottom based on priority. Thus, scope creep is business as usual. One aspect that Agile projects must consider, however, is how users will react to interface changes. If the changes require relearning, users may resist changes that the project sponsor requested. When a user interface changes more than once, users may complain about instability.

On a Planned project, even small changes deserve a formal project change request because it reminds clients that scope changes aren't free. When seemingly minor changes accumulate into significant schedule and cost deviations, clients may decide that the scope change isn't worth pursuing. Alternatively, when project managers can anticipate scope creep, perhaps based on history, planning subsequent projects with a scope buffer specifically for missing or changed requirements makes it easier for those projects to accommodate change.

Whereas scope creep is stealthy, scope surge is conspicuous. And scope surge can multiply project size with deceptively simple decisions about architecture and deliverables. On Planned projects, scope change of this magnitude is not containable within the current schedule, staffing, or budget—and probably not within a scope buffer either. If so, the best response may be replanning.

Expediting

When a project falls behind or acquires additional scope, there's more to expediting than just exhorting developers to work harder. Here are some actions project managers or a scrum team can take:

- **Reassignment.** Get someone with higher skill or more availability.
- **Crashing.** Put more people on a team to compress the schedule.
- **Swarming.** Get several people to concentrate on just one critical task.
- **Crunching.** Work more overtime than usual.
- **Fast tracking.** Perform some tasks in parallel instead of in sequence.

Beware that such actions are not sustainable forever, so it's important not to burn out teams with overwork.

ADVENTURE: Scrum Falls into the Gap

Every project depends on skilled resources. Strange as it sounds, individuals may not become a bottleneck until they're gone. This adventure in project execution involved two separate Scrum teams.

Developer Down

Team #1 maintained a mature software product. However, the scrum master was not the resource manager. The developers took direction from the scrum master, but they were direct reports to a department in another division led by the resource manager. The logic behind consolidating skilled developers across many scrum teams under one resource manager was that there would then be one place to do hiring, transfers, and other personnel administration, thereby unburdening the scrum masters. An unintended consequence however, was this tied the hands of scrum masters at critical moments.

One of the developers fell off a ladder at home, injuring his back and preventing him from working. His skills were about average, so getting a replacement should not have been difficult. However, the division where he reported was under a hiring freeze, so the position went unfilled for months. Pleas and escalations went unanswered, and the work backlog grew, but there were no consequences for the resource manager because he was reaching the objective for his job role.

Developer Down and Out

Team #2 maintained a different software product. On that team, one developer was the focal point for trouble tickets, which meant investigating and resolving production issues. He had deep tribal knowledge about the business and the technology. Unfortunately, he died.

Although a new developer was assigned without much delay, he was incapable of handling trouble tickets as well, because he had no experience with the business, the product, or the production environment. Although there was no resource gap according to allowable headcount, there was a gaping hole in terms of actual knowledge and skills.

Lessons Learned

Lessons learned include:

- Constraint Management strives to avoid floating constraints because the buffer protecting the previous constraint no longer meets that objective.
- When it's not possible to fill developer positions with the requisite training and experience, the project will suffer a decline in productivity and perhaps a schedule slip.
- "Yes" has many meanings:
  - **Commitment.** "I'll do it."
  - **Confirmation.** "I hear you. (But I'm not going to do it.)"
  - **Counterfeit.** "Go away. (We never had this conversation.)"

Replanning

Replanning happens when a plan is no longer credible. This can be caused by insufficient progress or by scope expansion.

Hofstadter's Law says, "It always takes longer than you expect, even when you take into account Hofstadter's Law." Gardner's Corollary adds, "Hofstadter was an optimist."

Critical Path projects tend to be replanned frequently because a shift in the Critical Path affects tasks on the old path as well as the new one. The expediting methods listed earlier may be used to compress the new plan. However, performance measures will be misleading if the replan makes the new plan the same as the actual schedule, resources, and cost to date.

Critical Chain projects are replanned if scope expands sufficiently—and this does happen to Information projects—but they are not replanned due to insufficient progress because the Critical Chain does not shift. Once a project is in execution, the longest penetrating chain is what matters, which may not be the Critical Chain. The expediting methods listed earlier may be used to accelerate execution of the plan.

On Agile projects, the scrum master may replan the current or next sprint, but generally not the entire product backlog. When reprioritizing items, some may be deleted because they are no longer needed.

Kanban reprioritizes, but doesn't replan, because there is no plan. To keep the product backlog from growing without bounds and accumulating low-priority items, any work items aged beyond a cutoff date, such as six months, can be deleted.

ADVENTURE: Fantasy Collides with Reality

The adventure with the troubled project described earlier ended well, as most did. Unfortunately, a few troubled projects end badly when there is an unbridgeable chasm between what is committed and what can actually be delivered.

One such project was managed by a company's development organization but used contractors to augment their staff. After design milestones passed without the expected work products, the company replanned the project and demanded that its contractors commit to the new milestones. However, the user steering committee that owned the business process and funded the system asked for a second opinion. In addition, the contractor's project manager, sensing that the dates were too aggressive, yet knowing that the firm would throw resources at a contractual commitment, brought my team in to provide an independent assessment.

We've Seen the End of This Movie

Here's what we saw. The objective was to replace a large, complex Legacy System that had been in production for over a decade. It was a

System of Record that supported many concurrent users across a geographic region. Thousands of business rules were buried in the source code. Healthcare and financial regulations added to challenging performance, reliability, privacy, and security requirements.

Not only was the replacement system meant to be running in production in a matter of months, the Legacy System was to be taken out of production in a big-bang cutover. There was no provision for parallel systems and a staged cutover.

Furthermore, the original one-year plan was predicated on a new methodology, new software architecture on a different hardware platform, a new software tool that the client had no experience with, and a new user interface. The tool vendor claimed that it could generate high-quality code directly from requirements, so developers were engaged in data and process modeling, and little time was allocated in the plan for code generation and testing. Unfortunately, the vendor claims were based on much smaller, much simpler systems. Therefore, the work was proceeding slowly, and the generated code was not performing well even at low testing workloads.

It's Called the Iron Triangle for a Reason

When directed to replan the project, the company's project manager had back-scheduled from the due date. He thus compressed every task in the plan, most to absurdly short durations. Inexplicably, he did not conclude that more resources were needed. Nor did he consider whether the original due date was feasible, let alone the compressed plan. The new system would have to be designed, built, and tested, the data converted, and users retrained in record time. Hoosiers call this "ten pounds of stuff in a five-pound sack."

Rather than assuming the due date was attainable, we used our project estimating tool to take size, complexity, skills, tools, and other factors into account while solving for the schedule, resources, and budget needed. Not only was completion by the committed due date infeasible, there was reason to believe that a project of this scope and complexity could never be completed with the chosen approach and tools. Though the old system might have been replaced by taking a different approach, such as refactoring with a strangler pattern, it was debatable whether the benefits would outweigh the costs. And this alternate approach would not have moved the replacement system to a different platform.

It fell to me to tell the user steering committee what we concluded. Even with the most wildly optimistic estimating assumptions, it was still a multi-year project. After my presentation, crickets were chirping in the conference room. Recriminations came later.

TITSUP

"Total inability to support usual performance" (TITSUP) was how this project ended. Contractors completed the design per the current contract but were not awarded additional work on this project. Frankly, I was relieved because we had seen other troubled projects like this tally huge losses. The due date passed without anything resembling a workable replacement system, and the company's project leadership was summarily disbanded. The last I heard, the Legacy System was still in production.

Lessons Learned

Lessons learned include:

- Understanding and replacing old systems is hard work. It's sometimes harder than building a replacement system from a blank canvas because Legacy Systems accumulate decades of arcane business rules and design optimizations.
- Tool vendor claims should be investigated thoroughly before making commitments. It's not enough to compare checkmarks in marketing brochures or watch prerecorded demonstrations on toy problems.
- Combined learning curves from new methodology, new tool, new architecture, new user interface, and new platform will increase project complexity and duration.
- Back-scheduling projects from arbitrary due dates is malpractice. If project managers don't know where the boundary is between feasible and infeasible, their organization is going to have a bad time.

Nontechnical Executives

As the Information field has evolved, more nontechnical executives have assumed key roles. For instance, the Chief Information Officer is increasingly a nontechnical executive because it gives the business a greater voice. Some technical responsibilities that were once the purview of the CIO have migrated to the Chief Technology Officer role or to technical directors reporting to the CIO. Of course, executives sponsoring Shadow Computing may have no technical expertise themselves.

This evolution happened for several reasons. The main one is that Information Technology has become more powerful out of the box. But the easy business problems have already been solved, so the technology wish list is to solve increasingly hard business problems. Finding people who are both snow-proof on technology and adept at business can be difficult.

Some new business-oriented roles have emerged, too. A business process owner knows how the business works and how the technology supports it even if the owner doesn't know at a deep level how the technology works. A business process outsourcer performs a business process on behalf of the client, thereby enabling the client to focus instead on its core business.

And then there are roles specifically at the interface between business and technology. Scrum masters, for instance, fill that role on Agile projects. Systems analysts fill that role on Planned projects.

Overall, the increase in nontechnical executives has restored some balance, compared to the days when the Information field was dominated by technical executives. That said, when nontechnical motivations overrule technical warnings, the pendulum may have swung too far.

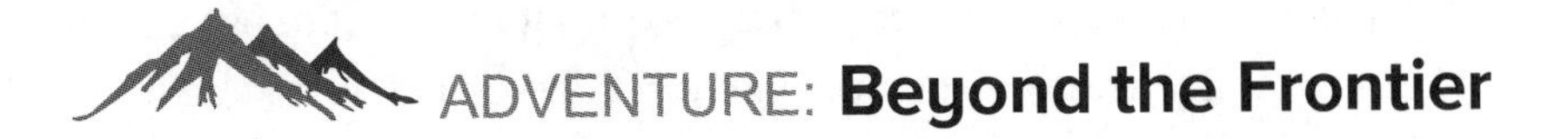

ADVENTURE: **Beyond the Frontier**

The previous adventure was about a troubled project that dashed beyond rescue before warnings were raised. This adventure is about a troubled project that launched on a mission beyond the frontier despite urgent warnings beforehand.

During its kickoff, business process owners and application owners were told about a project to replace multiple Legacy Systems with an enterprise software package. It was not a meeting to gather feedback. Commitments had already been made. The project was launching regardless of any misgivings.

The largest Legacy System—both in terms of technical construction and business revenue—processed thousands of transactions per customer per year. The biggest customers had over 100,000 transactions per year. If every transaction generated paperwork, the volume would be massive. Therefore, the business had evolved not just to paperless transactions, but also to consolidated billing with various surcharges and discounts. As service requests were entered and fulfilled, the service contract and its billing were adjusted automatically. That saved labor for both the customer and the provider, but it meant that the system had become quite large and complex. Because it supported customers in many countries, local customs, laws, and regulations added to the complexity.

Kobayashi Maru

When teams expressed concern about the packaged software's capability, they were told to hold their concerns because the package was already being used successfully by other companies. However, upon further investigation, it turned out that those other companies were in a different industry. Warning signs were evident because a business process by the same name is not necessarily consistent across industries. In this case, it was radically different.

When the package capabilities were compared to the Legacy System, there were hundreds of issues. About a third were useful but nonessential features. Another third were useful and able to be handled by the new package with configuration and custom coding. However, that still left a third that were beyond what could be reasonably handled by the package.

When shown those issues, the project sponsor demanded that the business process owner reengineer the business process to fit the package. This was a stunning directive because the line-of-business executives had not bought into that approach because it would have meant giving up capabilities that generated competitive differentiation.

As the project crawled forward, package performance became a show-stopper. The software hit the wall at one percent of Legacy System capacity. Hence, the vendor would have to implement special processing—and the client would have to fund some portion of the development work. That, of course, was in neither the spirit nor the budget of this project.

As the drama was unfolding on this business process, similar discussions were happening elsewhere. Creating a consolidated data model covering many Legacy Systems was not for the fainthearted. Some same-named data had different meanings. Some different-named data had the same meaning. Lots of data didn't map easily to the software package because it used its own schema to establish relationships. Extract, transform, and load would be tedious when master data was migrated.

Developers couldn't make the package work with the current business model. Business process owners couldn't transform the business model to fit the package. And the project sponsor couldn't afford to keep the Legacy Systems because the project was investing big on the premise of saving big. Thus, it was a classic no-win scenario.

Plan B Is Try Harder on Plan A

Commitments were based on the consultants' study and proposal, but the consultants had not spoken with business process owners or application owners before the kickoff to assess feasibility. The proposal had been justified by claims and assurances from the package vendor. Of course, that's a dangerous way to justify a project.

Because package implementation couldn't stop, the new direction was "minimal integration and limited deployment." In the short run, the Legacy Systems would have bidirectional interfaces to the package. Over time, functionality and data would migrate from old to new, and the Legacy Systems would eventually be retired. However, the project collapsed when it ran out of funding and senior executive patience.

Lessons Learned

Lessons learned include:

- Package implementation may look easy on paper, but requirement mismatches and data migration can make it ferociously hard.
- Reengineering a business process to fit a software package is easier said than done, especially if the Legacy Systems being replaced support a differentiated business model.
- Talk to the business process owners and application owners before committing to a major package implementation project.
- When throwing a warning flag on a project before it is troubled but after commitments have been made, don't expect a career-enhancing reaction.
- "Not my first rodeo" refers to a goat rodeo, which is slang for things going disastrously wrong.

Constraint Management

Critical Chain and Kanban are the only methodologies discussed in this book that practice Constraint Management during project execution, but they do it in altogether different ways:

- **Critical Chain** manages time buffers on the longest path through the plan, the schedule constraint. It takes scope and skills into account, but they aren't considered the constraint.
- **Kanban** manages work buffers, the backlog and work in process, by regulating task initiation based on available skills, the resource constraint. It takes target due date into account when setting priorities, but due dates aren't considered a constraint unless a class of service is defined for them.

Having some buffer penetration is normal for Critical Chain projects. Such projects aren't troubled unless they are in danger of consuming the entire project buffer and thereby finishing late. Fever Charts expose trouble and track its resolution.

Having a product backlog and work in process is normal for Kanban projects. Such projects aren't troubled unless high-priority work in the backlog grows too large for its completion when the client needs it. Various Kanban metrics expose trouble and track its resolution.

If the constraint isn't in the project time or work buffers, the next most likely place is in scarce skills, such as hardware or software architect. But as one of the adventures showed, even a shortage of commodity skills can derail a project.

Obviously, projects have their own local constraints, but can projects be the enterprise constraint? Information projects are generally not the enterprise constraint because they don't control what the enterprise produces. However, if the enterprise sells hardware, software, or services and a Horizon 2 project is high priority, it could become the enterprise constraint. For instance, the R&D project behind a new generation of hardware or software might determine when the new revenue stream begins. Unless the enterprise is a start-up however, it's more likely that the enterprise constraint is in the production of current hardware, software, and services. Or if sales aren't keeping up with production, the constraint could be in the sales function or in the market.

Conclusion

Contrary to conventional wisdom that says most Information projects fail, most are successful when studied evenhandedly. Thus, there is no software project crisis, although there is always room for improvement.

Much of the project management literature is about planning because a bad plan sets up a project for failure. But there are many ways that bad execution can derail even a good plan.

When Information projects fail spectacularly, it's less often due to a bad plan or faulty execution than to unrealistic expectations. Projects targeted beyond the feasible frontier are literally in the danger zone from day one, even though the evidence often emerges later.

Every methodology has work rules and status reporting to keep projects on track. Of course, if rules aren't followed and status is distorted, projects get into trouble. Getting them out of trouble may require outside expertise, especially if the project is trapped in the Iron Triangle with no obvious way out.

Scope creep is a common trigger for Information project trouble because tasks missing from the plan reveal themselves during execution. At that point, expediting may or may not be enough. When it's not, reprioritizing or replanning are necessary. Nevertheless, troubled projects sometimes do pass the point of no return: No amount of reprioritizing, replanning, and expediting can save a fantasy project.

The most widely used project methodologies, Critical Path and Scrum, do not manage constraints explicitly. However, Critical Chain does Constraint Management on Planned projects, while Kanban does Constraint Management on Agile projects.

Here are some aphorisms about projects:

- "The best thing about milestones is that whooshing sound they make as they go flying past." [anon]
- "Agile is a way to get nowhere faster: it's like the product is perpetually in beta test—always surprising, never stable." [anon]
- "Plans are useless, but planning is indispensable." [Dwight Eisenhower]

CHAPTER 15

PROCESSES

Constraint Management has a long history of process improvement. The first Constraint Management solution was for Manufacturing processes, and it's still used today, albeit often in a simplified form.

The Information field likewise has a long history of process improvement. Information has enhanced recordkeeping, decision-making, and commerce in every industry. It has also automated previously manual processes. Naturally, the Information field is on a never-ending quest to improve its own processes.

Projects versus Processes

Recalling how the terms were defined earlier, projects and processes are distinct ways to produce deliverables:

- **Project.** A set of dependent tasks that produce unique deliverables during a finite duration under a project manager unless the team is self-organizing
- **Process.** A set of dependent tasks that produce the same deliverables on-going under a production manager and process owner

Here are a few characteristics of processes:

- **Manual versus automated.** Labor-based (example: consulting) versus technology-based (example: software)
- **Scheduled versus on-demand.** Performed as planned (example: payroll) versus as requested (example: shipping)
- **Queued versus real-time.** Wait (example: toll booth) versus no wait (example: toll transponder)
- **Grouped versus individual.** Batched (example: groceries) versus separate (example: haircut)

- **Inventory versus no inventory.** Tangible (example: bolts) versus intangible (example: cleaning)

As seen in the previous chapter, projects have internal constraints. Projects just can't get enough work done fast enough to suit some clients, even when Constraint Management optimizes the project plan and execution.

Processes, however, follow a different pattern. Processes often start with an internal constraint, but when Constraint Management optimizes the process, an external constraint emerges because processes can then deliver more products and services than the market will buy.

Why? Projects do not have economy of scale because they are unique, but processes do because they are repetitive. Projects are mostly manual, but processes can be semi- to fully automated. Projects may not make physical inventory, but processes often do, which serves as a buffer between production and the market.

This chapter covers processes in the following somewhat indistinct categories:

- **Manufacturing.** Produces products for external customers
- **Service.** Performs tasks for external clients
- **Business.** Performs tasks for internal clients but interfaces with external customers
- **Information.** Delivers high-tech hardware, software, data, and services

Manufacturing Processes

Manufacturing can produce products via projects or processes. For example, a large, complex product like an oil rig is often manufactured via a project because it involves a unique design and a complicated set of interlocking tasks to build it. However, manufacturing the drilling pipes that an oil rig will eventually use is done via a process because the same relatively simple, serial tasks are repeated on every pipe.

Constraint Management solutions for projects, Critical Chain and Kanban, were described in the previous chapter. The Constraint Management solution for production processes was described earlier, in the Constraint Management chapter, and is outlined again next:

- **Original operations.** The constraint is internal: It's the machine or person with the least productive capacity. Jobs are released into the shop when the buffer ahead of the constraint is in its red zone because maximum productivity happens when the constraint, and only the constraint, is fully productive.
- **Simplified operations.** The constraint is external because the market will not buy everything the manufacturer can produce. Jobs are released into the shop with enough lead time to finish by the due dates on orders. Finished goods is a buffer between manufacturing and the market. No person or machine is 100% utilized, including the former internal constraint, but Constraint Management nevertheless prevents excess work in process.

As described in the Prologue, some manufacturing processes operate without any awareness of an internal constraint. The quest for high utilization, in the mistaken belief that it maximizes productivity, actually leads manufacturing into a more chaotic condition than if it pursued Constraint Management or another method that minimizes work in process.

The original solution is distinctive because it relies on unbalanced capacity to regulate the entire process. This contrasts with Lean Manufacturing, which uses balanced capacity. Both minimize work in process, but Lean applies best to stable processes because it is more susceptible to disruption. (More about Lean in a later chapter.)

The simplified solution is valuable because manufacturers compete on speed as well as price and quality. Minimizing work in process reduces cost, but it also enables speed because the factory isn't clogged with unnecessary work in process.

Both the original and the simplified solutions rely on unidirectional buffers. Green zone (high buffer levels) says no action is required. Yellow zone (moderate buffer levels) says get ready for action. Red zone (low buffer levels) says replenish the buffer.

High capital investment and the logistics of getting more machines makes manufacturing capacity relatively fixed and oftentimes adjustable only in large increments. To the degree that services do not rely on capital investment, service capacity is more flexible and adjustable in small increments.

Service Processes

Just as parts move through tasks in a manufacturing process, service requests move through steps in a service process. Just as a manufacturing process can have an internal constraint, so can a service process. But both are increasingly likely to have external constraints, which means they must respond to changes in demand. This is accomplished with a skills buffer for labor-based services or a capacity buffer for technology-based services.

Service optimization depends on adaptability more than predictability. Hence, the Constraint Management solution for service processes relies on two applications. One monitors service delivery, the other adjusts capacity as demand rises and falls:

- **Delivery monitor.** Service requests are released as received or with enough lead time to meet a due date. When service levels fall below a threshold, a signal to increase capacity is generated. When service levels rise above a threshold, a signal to decrease capacity is generated. Work is pulled through the process to meet target service levels, but the constraint is external because capacity adjusts to demand.
- **Capacity management.** Buffer management enables flexible capacity. For instance, if servers automatically scale in and out based on demand for online game playing, the available server capacity is a capacity buffer. Likewise, if available labor hours are increased and decreased based on demand, spare labor capacity is a skills buffer.

Core skills are available from the job market, but not instantaneously, so they are buffered. The skills buffer is sized to cover what would otherwise be a shortage during periods of increased demand. People without a current assignment are said to be "on the bench," though there are always useful things they can do, such as give or get training. As explained in the Skills chapter, commodity skills are readily available on short notice, so there is no need to buffer them. In contrast, critical skills are chronically in short supply because they are not readily hireable, so buffering them is not possible.

Service capacity can also be adjusted on a more granular level. For example, during peak periods in a call center, as wait time rises, more agents can be brought in to handle calls—or in-bound calls can be diverted to a callback queue, which smooths demand. During slack periods, when there are fewer calls to answer, some agents can be assigned other tasks or sent home.

Services to businesses often have Service-Level Agreements written into the contract. Services to consumers more often have unwritten service-level expectations.

Some services, such as consulting, use no physical inventory at all; other services, such as repair, rely on parts inventory. Warranty replacement services rely on reconditioned products inventory. The latter services must therefore manage their parts and products inventories at the same time as they manage their service capacity. The Constraint Management solution for distribution manages such inventories. However, physical inventory takes up space and may require shelter and environmental controls. A backlog of service requests does not.

In contrast to manufacturing buffers that are unidirectional, capacity buffers are bidirectional. In other words, buffer management happens both when the buffer level falls into the low red zone (too little capacity) and when it rises into the high red zone (too much capacity). The green zone (adequate capacity) says no action is required.

ADVENTURE: **Predictability versus Adaptability**

My initial foray into Constraint Management led to the realization that solutions originally devised for Manufacturing wouldn't work in Services without some adaptation. For example, the Manufacturing solution uses unidirectional buffers because the workflow is one-way, but Services needs bidirectional buffers because their resource flow is circular even though their workflow is one way.

The Allure of Forecasting

Conventional wisdom says adjusting service capacity in anticipation of future demand requires a forecast. Forecasts are always subject to assumptions about cycles. That gets complicated when there are cycles within cycles: daily, weekly, monthly, etc. Although busy hours of the day, busy days of the week, and month-end cycles are relatively predictable, forecasting the business cycle (recession and expansion) is so difficult that even economists don't know we're in a different phase until we've already been there for months. Thus, demand forecasts never seem to be quite accurate enough, and sometimes they are catastrophically wrong. My own business unit had mistakenly forecast demand to rise long after it turned down during a recession, and this led to excess capacity.

Capacity on Demand

Constraint Management provides an alternative: Adjust capacity on demand rather than to a forecast. Skill buffers can be sized to cover demand during lead time for hiring and training, typically weeks or months. Asset buffers can also be sized to cover demand during lead time, although with automation, the lead time to adjust capacity may be just a few moments.

Faith in forecasts is deeply rooted, however. Capacity management on demand isn't taught in most management programs as an alternative to forecasting. We found logical arguments and math examples to be insufficiently persuasive. Simulations, however, enabled managers to compare forecasts to capacity on demand and draw their own conclusions, much like they do during the multitasking exercise. For managers who have been using forecasts for years, simulations can shake them out of their rut.

Lessons Learned

Lessons learned include:

- Skill and asset buffers in a services context act like shock absorbers, preventing demand variations from disrupting service levels.
- Sizing buffers is easier than making forecasts, which decrease in accuracy the further into the future they reach. Service capacity then reacts to demand instead of anticipating it.

- Simulations are an underappreciated technique for demonstrating alternative methods to managers and getting their buy-in.

Business Processes

As the terms are used here, service processes are performed for external clients, while business processes are performed for internal clients. In practice, however, service processes and business processes may be identical. Thus, solutions for one often work for the other, and that's true for Constraint Management.

Business processes are sometimes separated into these classes:

- **Front-office.** Customer-facing (example: direct sales and online sales)
- **Back-office.** Out of sight (example: order fulfillment and shipping)

Constraint Management works for both.

When organizations outsource business processes, such as payroll, to service providers, they typically have two objectives. First, they want to focus on their core business by off-loading non-core business processes. Second, they expect the service provider to perform the business process better than they do it themselves. This can be accomplished in many ways, including some addressed by Constraint Management:

1. Process automation
2. Manual procedure standardization
3. Paperwork elimination
4. Economies of scale
5. Deep expertise
6. Additional languages
7. Labor arbitrage
8. Follow the sun (service in all time zones)
9. Forecasting
10. Capacity management

Manufacturing and Service processes have trended toward automation whenever technology-based processes are less expensive and more reliable than labor-based processes. That same trend affects business processes. For example, processing of travel expense transactions was automated decades ago. However, travelers were expected to enter their own expenses into the system, which saved on some clerical positions but distracted revenue-generating workers with non-billable work. Current state of the art, however, is for travel expenses to flow automatically from travel service providers to enterprises with traveling employees. Then, only exceptions to travel policy attract attention.

ADVENTURE: **Bumpy Ride for Road Warriors**

Consultants experience a lot of travel services just getting to and from client sites. This adventure began like any other business trip: leave work, catch ground transportation to the airport, then a flight to a distant city, arriving well after dark for a drive to the hotel and a few hours' sleep before a breakfast meeting kicks off the new day.

Missing the Cutoff

While en route, however, bad weather delayed flight arrivals. Once outside the terminal, road construction required the shuttle bus to take a detour onto partially flooded roads. The bus was full. The time was late. The ride was long. And things were about to get worse.

Instead of picking up my rental car by 10:00 p.m. as planned, a couple of shuttle buses of fellow travelers arrived with me at the rental car lot at precisely 2:10 a.m. Seeing 50 people in line at the rental counter was not a good sign. When I reached the counter myself, the rental agents wearily explained that all unclaimed contracts had been automatically purged by the computer at 2:00 a.m. They were therefore writing new contracts by hand.

Empty Offers

As a frequent traveler, on another trip I got a text message from the rental car company asking if I would like a different car than the type I had reserved. When I clicked on the link immediately, the app said no other cars were available. Why didn't the system figure that out before it offered me a choice that it couldn't honor?

Similarly, I got a postcard from a company offering better Internet service, which I rely on when working from home. When I called, however, there was an awkward pause before the agent sheepishly admitted that they don't serve my address. Why didn't the system figure that out before offering me a service it couldn't provide?

Lessons Learned

Lessons learned include:

- Consider manual overrides for automated processes that could inconvenience customers and overwork employees. Otherwise, don't wonder why social media contains so much vitriol.
- Customers get justifiably irritated with inane processes, so until processes have been tested with real customers, they're not ready to release. Exposing them to a few unwitting test subjects doesn't count.

Information Processes

The Information field has enabled semi- and full automation of countless Manufacturing, Service, and Business processes. This section addresses processes specific to the Information field itself and how they can benefit from Constraint Management:

- Development and Maintenance
- Information Technology Operations
- Technical Support
- Disaster Recovery

You may wonder why Development and Maintenance are on the list, inasmuch as the previous chapters covered them as projects—not processes. Well, there are two reasons.

First, some Agile enthusiasts refer to Agile as a process and object to the term, Agile project. On the other hand, Agilistas recoil at the idea that Agile follows a defined process at all. They reserve the term, Agile, for principles and practices, not process. So, there you go. To some developers, development is a process, and therefore amenable to Constraint Management. On the other hand, if development is neither a project nor a process, management probably isn't something those developers care about.

Second, product development is more than software development [Jones, 2000]. That's why product manager and project manager are separate roles. The product manager decides what software capabilities to develop during Horizon 2 and 3 so that the product will be marketable during Horizon 1. The product manager also prioritizes maintenance during Horizon 1 to sustain the product's life. Product management activities are on-going, so it's appropriate to view them as a process, not a project, even though product managers do contribute to projects.

Information Technology Operations does projects to install and configure hardware, software, and networks in the Information infrastructure. But it also continuously executes processes that run the infrastructure. Where is the constraint? Most likely, it's in the infrastructure itself, so upgrading hardware, software, or network periodically elevates that infrastructure constraint.

Sometimes, however, the IT Operations constraint is in work rules and incentives. At one data center, the time to mount magnetic tapes so data could be written to them or read from them was long and getting longer. Therefore, the operations manager set a service level target and tied bonuses to reaching it. With their bonus pay tied to performance, operators wore out their shoes while running to mount tapes quickly.

Infrastructure as a Service is the external service version of what would otherwise be an internal process. For a service provider to offer IaaS, it must have a reasonable amount of automation, so capacity can be adjusted on demand. Robotic Process Automation can make capacity adjustments in IT infrastructure without human awareness or intervention. For instance, RPA can expand storage and initiate fail-over from one Availability Zone to another.

Technical Support installs, maintains, and repairs hardware, software, and networks. In other words, Tech Support is who Operations calls when things break or who users call when things don't work. Tech Support has processes for on-site service, depot service, help desk, and self-help. Each of those processes can have a different constraint, and all are amenable to Constraint Management.

Despite the best efforts to avoid trouble, some disasters are unavoidable. Disaster Recovery services help organizations restore operations, oftentimes from another site. But DR sites must have workspace for people, not just computers. DR sites also must have the same computer environments as the client uses in its own sites, and there must be a way to get client data to the new sites. The most recent data may be online somewhere, depending on what was destroyed and whether applications are designed for availability. However, backup copies may not be online, so a process for restoring primary copies must be executed. This may entail trucking backup tapes from archival storage to the DR site. Because disaster recovery processes are episodic rather than continuous, and conducted under stressful circumstances, it may not be possible to pin down a single bottleneck. It could be in infrastructure or skills or suppliers of essential commodities like diesel fuel for generators. However, when restoring the primary site, Critical Chain applies to those projects.

ADVENTURE: **X-ray for Software**

Executive consultants are often invited into client engagements because something is already off the rails, or close to it. Maybe the client doesn't know where to start. More often, the corporate antibodies have already defeated an in-house solution.

We had done Process Improvement and Technical Strategy engagements many times, so we had a proven methodology, tools, and skills. We were accustomed to clients being defensive, if not combative, when we were first introduced. This engagement was different.

Assimilation

The client's business grew mostly by acquisition, and they were experts at it. After close of business on the Friday when an acquisition became official, the Information team would swoop into every acquired site, replace every piece of hardware, install company standard software, train employees on the client's business processes, and convert acquired data to company standard data. By the time the business opened on Monday, the acquired company's IT had been completely assimilated. Resistance was futile.

The client's enterprise constraint was external, in the market, which is why their business strategy was to grow by acquisition more than organic growth. But the CIO's local constraint was somewhere in IT processes. The IT team had plenty of technology and skilled staff. But the IT team couldn't improve performance with the prevailing processes for Development, Maintenance, and Operations. Those processes regulated the commitments that the CIO could make to the business. That said, this was not a troubled IT shop. Quite the opposite: They were forward-thinking and innovative. They were doing DevOps years before the term was coined. However, their efforts to establish consistent processes across sites had fallen short.

Process Assessment

The client had two computer centers, ostensibly following consistent processes, yet frequently tripping over inconsistencies. One site ran high-performance batch systems for back-office business processes. The other ran high-reliability online systems for front-office business processes. A change to hardware, software, data, or networks at one site typically required a corresponding change at the other site due to data sharing and scheduling dependencies. The sites were hundreds of miles apart, so only executives and managers routinely visited both sites.

During the Process Assessment phase, we conducted process modeling sessions at both sites. Knowledgeable IT staff joined us in a conference room where we had covered one wall with paper. As they explained their Development, Maintenance, and Operations processes, we began drawing diagrams. As they caught on to our symbols, we handed the pen over and talked them through fleshing out the diagrams. The discussions were lively.

Sometimes they admitted that the way they really did things departed from the documented process. Sometimes they disagreed on the actual process: Steps might be in a different sequence or the steps might be altogether different. Where there was disagreement, we used a squiggly halo around those steps to highlight them visually. We also used bold colors to identify critical steps: ones where problems frequently arose, where go/no-go decisions were made, or where there were leaps of faith due to unknowns.

With the As Is process documented on one wall, we adjourned. The next day, we collaborated on what the To Be process should be. Documenting it on an adjacent wall meant it was easy to refer between the diagrams. On steps where problems frequently arose, the To Be process inserted upstream steps that could avert the problems. On the decision-making steps, the To Be process reconsidered what steps were needed to better inform the decision-makers. On the leap-of-faith steps, the To Be process considered how to fill in the missing knowledge.

Sometime later, we brought the teams together. Site #1 diagrams were on one wall. Site #2 diagrams were on another. We then worked with the teams to compare both To Be diagrams and hash out the differences. Eventually, we got to a single To Be diagram, then we redrew all the diagrams using a diagramming tool for legibility.

When IT executives could walk around and see the As Is and To Be processes, they bought into the revised processes immediately. This kicked off the Process Improvement phase.

Process Improvement

The constraint governing Development, Maintenance, and Operations turned out to be Impact Analysis. Before building new applications, before changing old applications, and before altering the IT infrastructure, everyone needed to know the impact of their work. Unfortunately, complex dependencies are hard to document thoroughly and accurately by hand. For example, inserting one new column into a database table means all the programs referencing that data should be modified, recompiled, and tested. Missing just one can be a serious omission, and it was not uncommon to find hundreds of artifacts affected by even a simple change.

To automate Impact Analysis, an Application Understanding tool first needed to catalog all the artifacts and their relationships. Then developers and operators could query the tool to ask, "What's affected if I make this change to that artifact?" Application Understanding is therefore like an X-ray for software: It reveals artifacts and relationships that would otherwise be invisible.

With Application Understanding preceding Impact Analysis, Change Management was the next hurdle. Change Management asks, "Did everything that was impacted actually get changed and tested?" If not, rinse and repeat.

May We Have Another?

To measure process improvement, we helped the client implement standard estimating templates and defect tracking. During the 12 months following completion of this engagement, the client saw its quality improve and cycle time drop every month, with overall performance improving over 40%.

IT staff and managers were jubilant. Not only were they getting kudos from the CIO and from the business for their improved responsiveness and reliability, their personal lives were better too because they were putting in less overtime and they got fewer call-outs in the middle of the night to fix something broken.

Then the unprecedented happened: The IT staff and managers appealed to the CIO for another process improvement engagement. Bottom-up impetus for change is rare. For executive consultants, it doesn't get any better than this.

Lessons Learned

Lessons learned include:

- An organization can be a master at business process standardization, but that doesn't necessarily translate into IT process improvement.
- Tools for Application Understanding, Impact Analysis, Change Management, and Estimating/Metrics can be key to IT process improvement.

- Measuring process improvement against a baseline enables continuous improvement.

Constraint Management

Can an Information process be the enterprise constraint? Probably not. Any process can contain an internal constraint, and that's a common occurrence, but an entire process would generally not be the enterprise constraint because that process most likely contains some non-constrained elements. Perfectly balanced processes are rare, although that's what Lean strives for and comes close to at times in Manufacturing.

IT processes are not as repetitive as Manufacturing processes. An IT constraint easily floats even when capacity is unbalanced because it isn't practical to unbalance an IT process enough to maintain a completely stable constraint. The best that can be done is to remain vigilant about the constraint and recognize when it has shifted.

The conventional approach to process improvement is to improve anything and everything on the assumption that local optimizations add up, but that just overloads the constraint and adds confusion to non-constraints. Therefore, the Constraint Management approach is to improve only those few things that will move the needle on overall performance.

Most cycle time in processes is wait time. In extreme cases, processing time may be as little as 1%. For instance, if it takes a minimum of two weeks to get approval to install something new into the IT infrastructure because the review board only meets twice a month, having the board meet weekly would cut wait time in half. But replacing the approval process with a set of simple and clear guidelines that take zero time for decision yet leave an audit trail would eliminate wait time altogether.

Documentation is another time-consuming aspect of many processes. For example, computer program comments are controversial. Technical Conservatives say comments are guideposts for whoever maintains the code. Technical Liberals say comments more often mislead, so they are inclined to ignore or delete comments. However, in some fields, document automation has unintended consequences. Electronic health record systems with pick lists made it easier for healthcare professionals to document their work, but the volume of such documentation has risen to painful levels [Fry, 2019]. Healthcare documentation today can take as much or more time than face time with patients.

When improving a labor-based process, it's better to improve the process before adding people so that they won't be joining a suboptimal or chaotic process. In some cases, automation will mean the process can be done by fewer people, not more.

Constraint Management warns about inertia after a process improvement, but inertia before process improvement is a more immediate concern. The Not-Invented-Here Syndrome has thwarted more than a few process improvement efforts. That's why Constraint Management has a buy-in process. (See the Constraint Management chapter.)

Finally, a process contribution that the Information field can make to the enterprise is to assist in moving a constraint to a more strategic location because naturally occurring constraints are sometimes in inconvenient places. For instance, if IT can partly or fully automate a previously manual step, the constraint can be shifted to another step in the process that better accommodates scheduling and work-in-process inventory.

Conclusion

The Constraint Management solution for processes can be applied in Manufacturing, of course, but also in Service processes, Business processes, and Information Technology processes. In every instance of an internal constraint, the solution works by throttling work in process in order to maximize productivity on the constraint. In the more common case of an external constraint, the solution modulates processing capacity according to demand because that makes the process less cluttered and reduces unnecessary investment.

Recognizing the broad applicability of the solution is useful because Manufacturers are becoming service providers, too, since the profit margin on Services can be higher than on products. For some manufacturers, Maintenance, Repair, and Operation (MRO) processes are the new cash cow.

CHAPTER 16

PORTFOLIO

Previous chapters covered Information deliverables: hardware, software, data, knowledge, and networks. They also covered Information techniques and resources that produce deliverables: architecture, skills, methodology, projects, and processes. With that foundation in place, this chapter begins to put the pieces together on the way to a discussion of strategy.

Here are some pertinent definitions:

- **Project.** A set of dependent tasks that produce unique deliverables
- **Multi-project.** A collection of coordinated projects
- **Life Cycle.** The stages that systems or products move through
- **Portfolio.** The entire set of projects and programs being planned or executed

Portfolio Management views all systems and products at moments in time. Life Cycle Management views one system or product over time. Thus, Portfolio Management is cross-sectional, while Life Cycle Management is longitudinal.

Life Cycle Management

Product Life Cycle Management typically includes stages like these leading to commercial products:

- **Research.** Prototypes are built, and studies done
- **Development.** The product is built
- **Announcement.** Customers order the product
- **General Availability.** The product is shipped
- **Migration.** Movement to successor software or hardware
- **End of Life.** The product can no longer be ordered
- **End of Support.** The product is no longer supported

The System Development Life Cycle similarly applies to Information Systems for internal use:

- Planning
- Analysis
- Design
- Development and implementation
- Testing and deployment
- Production
- Maintenance
- Renovation
- Migration
- Sunset

Agile, of course, blurs the lines between phases which are separate in the traditional SDLC. With Agile, development never ends unless change requests stop, funding is cut off, or the team is disbanded.

Whatever stages of the life cycle an organization chooses, they can be used to classify assets, describe overall portfolio maturity, and answer questions like these:

- Is the product R&D pipeline robust enough?
- Is maintenance of current systems crowding out development of new systems?
- Which products or systems should be sunset?

Portfolio Management

A typical Information portfolio is composed of products and systems at different points in their life cycle. Portfolio Management considers whether the mix is right, and if not, what can be done to make it right.

Portfolio Management can be performed stepwise:

1. Inventory the **current portfolio**.
2. Understand **current capabilities**.
3. Set **priorities**.
4. Describe the **future portfolio**.

5. Establish **future capabilities**.
6. Decide how to get from current to future portfolio.

The future portfolio may depend on a change in tools as well as skills.

Inventorying the current portfolio may not sound difficult, but a surprising number of organizations cannot answer these deceptively simple questions:

- How much **hardware** does the organization have and where is it operated?
- How much **software** does the organization have and where is it installed?
- How much **data/knowledge** does the organization have and where is it stored?
- How many **networks** does the organization have and what do they connect?

A thorough inventory often reveals a much larger software portfolio than expected if application size is measured and classified by programming language. When the inventory includes application programming interfaces (APIs) to software both inside and outside the organization, critical external dependencies may be revealed. When an inventory includes software produced by users rather than developers, such as spreadsheets and PC databases, deficiencies in Central Information may become apparent. When an inventory includes Shadow Information, the IT managed by business units may rival if not exceed the centrally managed IT.

Understanding current capabilities answers questions like these:

- How many **skill groups** does the organization have and how are they deployed?
- How many **tools** does the organization have and what are their purposes?
- How many **projects** does the organization have and what are their missions?

Because most people have multiple skills, a capabilities assessment is much more than just a headcount. A thorough assessment classifies skills by group, and within group by level: novice through expert. Skill and tool deficiencies may impose limits on which projects are feasible.

Setting priorities ultimately determines which projects to approve. Information portfolio rebalancing is analogous to financial portfolio rebalancing: Both reallocate resources to things that matter most to achieving a goal. However, Information priorities must span strategic horizons, and Technical Strategy should align with Enterprise Strategy. Because there are always more opportunities than resources, Portfolio Management is as much about what not to do as it is what to do.

The distinction between Planned Projects and Agile Projects carries over into Portfolio Management:

- **Planned Portfolio.** Annual plan based on forecasts and control
- **Agile Portfolio.** Rolling requests based on discovery and enablement

A Planned Portfolio starts with detailed problem and opportunity statements, then thorough analysis and planning. It encourages the top projects to fit within the annual budget cycle even if scope, complexity, and effort estimates say more time is needed. But pushing projects has unintended consequences: Projects may launch before the team is ready, and a year later deliverables may be released prematurely.

An Agile Portfolio means less management, not no management [Thomas, 2008]. An Agile Portfolio starts with high-level problem and opportunity statements. A rolling three-month horizon releases Agile projects as resources become available. Then project scope, resources, and schedule are adjusted as the project proceeds. Thus, pulling from the backlog enables projects with the resources and time they need.

The difference between Planned and Agile Portfolios can be summarized as a few big bets versus many small bets, with associated differences in risk. Of course, it's not just possible, it's likely that portfolios in large organizations are a mix of Planned and Agile projects. As explained by the "Two-Speed Technology" adventure in the Methodology chapter, the dependencies must be managed somehow.

ADVENTURE: Portfolio Escapades

During my career, I drove or contributed to portfolio initiatives in several organizations. None were even remotely similar. All revealed some ugly truths.

Playing Close to the Vest

The first initiative included a basic portfolio inventory of internal systems:

- How many systems do we have?
- What technology platforms are they on?
- Where are they in their life cycle?
- How many customers and employees use them?
- How big are the systems?
- What languages are they written in?
- How many people are doing development? Maintenance? Operations?
- What tools are used?
- What methods are used?

The CIO simply wanted to know what was in the portfolio, where the pain points were, and what untapped opportunities existed to better serve the business. Unfortunately, as an outsider managing career employees, his directors were adept at outmaneuvering him (see Goodhart's Law). They dragged their feet on metrics by challenging the definitions. Some data they submitted was suspicious. His executive assistant flogged the directors for accurate data, but they viewed him as a junior executive wannabe, and that just added to their antics. Thus, the process improvement initiative was thwarted until it abruptly turned into "How can we outsource the CIO organization?" Directors who thought they were outsmarting the CIO found that they had no seat when the music stopped.

Reading the Fine Print

The second initiative included an inventory of products, plus the management system. The product inventory was automated because developers used source code control and standard sizing metrics, so many issues with the previous organization were not issues here. However, the management system inventory revealed thousands of internal contracts between managers each year because each department had its own budget.

Every internal contract was evidence of a conflict because if there were no conflict, there would have been no reason to create an enforceable agreement. Both sides would already have been aligned. This method of enforcing agreements had evolved because some managers conveniently forgot or reneged on their commitments and because new managers did not feel obligated by agreements made by their predecessors.

Executive pleas to stop the overhead that went into this practice went unheeded because there was no alternate method to enforce agreements. Executives would not adjudicate a dispute without documentation, so their own policy contradicted their pleas for management by trust.

Turning Around the Questions

The third initiative included an inventory of internal systems, plus the development and maintenance processes. The initiative began as many do, with this dialog:

1. What metrics do we already have? Quite a few, because they're easy to measure.
2. What conclusions can we draw from them? Our process is weak but not broken.
3. What should we change to improve our processes? The metrics don't tell us.

At my urging, however, the team turned those questions around:

1. What is our objective? To improve development/maintenance processes.
2. What questions must we answer? Where the processes could be improved.
3. What must we measure to answer those questions? Design and defect metrics.

The new approach had two advantages. First, it revealed where there were holes in the current metrics. For example, many defects could be traced to design issues, yet the company had no design metrics and no defect metrics. Second, it identified elements of the current data that could be ignored because they weren't relevant. For example, the company was tracking some program metrics that had been discredited by research showing no correlation between those theoretical metrics and actual performance of development and maintenance teams. Dropping them caused no harm, while eliminating some wasted time.

After implementing selected metrics, the company was able to measure improvement in what it delivered to users: fewer defects. Just as important, it could measure software designs and thereby focus developer attention on simplifying designs that were unnecessarily complex, and therefore effectively untestable.

Lessons Learned

Lessons learned include:

- Portfolio metrics are a fertile ground for gamesmanship.
- Matrix organizations evolve management procedures to enforce agreements.
- Portfolio metrics can and should focus on things that move the needle.

Prioritization

Inputs to prioritization include:

- **Business value.** Revenue generation, cost saving, customer retention, new markets
- **Strategic positioning.** Market share, barrier to entry, Horizon coverage
- **Mandates.** Laws, regulations, licenses, professional guidelines

Projects differ in how much is known before the project versus how much is discovered during the project. IT infrastructure projects tend toward the high-certainty end of that continuum, while software projects are on the high-discovery end.

Feasibility

Projects with high priority may nevertheless be infeasible, or at least ill-advised. Feasibility considerations include:

- **Financial.** Can we afford it?
- **Technical.** Can we build it?
- **Operational.** Can we run it?
- **Marketable.** Will customers buy it?
- **Usable.** Will users use it?
- **Supportable.** Can we deliver technical support?
- **Strategic.** Will it achieve the goal?

What Constraint Management calls a conflict, developers call a double bind. For example, a common complaint, especially on Agile projects, is "We've never done a project with these requirements or this technology or this team before, so we have no good way to estimate it. Nevertheless, once we make an estimate, we'll be held to it."

As a former developer, this response is understandable, but as an executive, it's unacceptable. Whether an estimate is three days, three weeks, three months, or three years makes a huge difference. If developers and managers can't estimate at least a rough order of magnitude of effort, how are executives supposed to justify benefits, assign staff, and approve work? Neither T-shirt sizing nor estimating by analogy are expected to yield high accuracy. A better approach is to approve investigation that produces a rough three-point estimate as its deliverable: best case, most likely, and worst case. Then decisions can be based on data rather than hunches.

Missing tasks and underestimated tasks can make scope so large that a Planned project would be judged economically infeasible if that scope were known at the outset. Likewise, an Agile project proposal may be based on a minimum viable product too big to hit the launch window. In both cases, the sunk cost fallacy may lead decision-makers to continue such projects anyway.

Sensitivity analysis may be helpful in such cases. By varying the project parameters and redoing the calculations, decision-makers can get a feel for how much reality can depart from assumptions before a proposal becomes infeasible. For instance, if the proposed project finishes six months late and 25% over budget, is it still worthwhile?

Another way to evaluate feasibility is to compare a proposed project to completed projects, both successful and not. When a proposed project is outside the envelope of successful projects, the risk is significantly higher. Sometimes the risk is acceptable. After all, that's how organizations stretch and grow.

It is the projects beyond the frontier of past successes that warrant scrutiny. The objective is to eliminate Fantasy Information projects. Beware, however, because some project advocates become irate and manipulative when told that their Fantasy project isn't feasible.

Industry benchmarks are better than nothing, but not much better, because they lump together organizations with vastly different requirements, capabilities, goals, and risk tolerance. A better approach is to gather an organization's own data. It doesn't have to be comprehensive or highly accurate because the frontier is usually obvious, as in "The largest project we've ever successfully completed was X, and this proposal is 2X."

For projects beyond the frontier, special handling is in order. For example, it's prudent to assign the most experienced project managers and best-matched technical skills. Also, if the project is for a customer, as in technical services, make sure the client understands the risk and is either willing to shoulder some of that risk or pay a premium for success.

ADVENTURE: **Turning It Up to 11**

When I was an executive consultant in an organization delivering business services with technology, we accumulated a database of engagement data and used it to evaluate requests for proposal (RFPs). For clients wishing to outsource a business process, there was an on-boarding phase that introduced technology, followed by a process improvement phase that sharpened skills, followed by a steady state phase. Thus, engagements followed a learning curve, which meant performance improved over time until it reached a plateau. The performance data formed a visible frontier that was easy to see on a chart.

This frontier meant we knew how much improvement we could commit to and how fast we could get there. Both were helpful in avoiding the winner's curse, wherein a service provider wins an engagement, but the promised results are unattainable, costs exceed revenue, and the client is dissatisfied. It's also helpful in educating clients before they experience buyer's remorse, wherein they choose the low bidder and are disappointed when that vendor can't deliver.

No Man's Land

When we plotted our competitor's bid, it fell well outside our frontier. They were proposing a leap in performance during on-boarding that we knew was unattainable, so we showed our frontier to the client and let them draw their own conclusion.

Because our competitors had no comparable data, the client assessed their bid as riskier, and we won the engagement even though we were not the low bidder. Our own technology plan and process improvement plan continued to improve, but knowing where the frontier was at any point enabled our business to grow without the missteps that sometimes plague evolving service offerings.

Lessons Learned

Lessons learned include:

- The achievable frontier is an existence proof, not a stretch goal.
- Without portfolio data, it's impossible to know whether a proposal is out of bounds.

Justification

Projects are justified based on their return on investment. A common assumption is that executing the project affects only the investment, not the return. Hence, the focus tends to be on avoiding delays because that adds cost. Yet delay in project completion at least delays the return, if not reduces it. Project sponsors do not always attempt to quantify this effect. Yet the increase in revenue from early completion can be greater than the cost saving.

Constraint Management asks, "How will this project affect the constraint?" Will it subordinate non-constraints so they no longer overwhelm the constraint with excess work in process? Will it elevate the constraint so it can produce more than before? Will it move the constraint somewhere better?

When project proposals are evaluated rather than mandated, strategic decisions compare business benefits before the project to benefits anticipated after the project (see the appendix for examples). If the project won't move the needle on business benefits, it doesn't matter whether it's affordable or technically feasible.

Some potential systems projects stay in the backlog because their business value is not enough to rise above the acceptance threshold. These disqualified proposals will accumulate if unchecked, so it's better to let them time out by establishing a policy that proposed projects can stay in the backlog only for one annual Planned Portfolio cycle or six months on a rolling Agile Portfolio cycle.

Initiation

How projects are initiated affects how they are executed:

- **Milestones.** Due dates drive the project
- **Darwin.** Only the strong will survive

- **Death March.** Excessive overtime
- **Freezing.** No new projects until old project logjam is cleared
- **Throttling.** Projects start only when critical resources are available

A project driven by milestones starts when the plan says it should start, even if the resources aren't all available. This may mean the project is behind schedule immediately after launch, but the assumption is resources will catch up in time to make the first milestone. Billing by the hour and progress billing at milestones create incentives to get projects launched on time, despite not having full staff.

The Darwin approach launches more projects than there are resources for, in the dubious belief that the most important projects will attract the necessary resources and the rest will die or be killed along the way. This approach also assumes that some projects will be terminated before completion when they run out of funding. However, an unintended consequence is that all projects go slower when they compete for resources, because they are forced into bad multitasking.

Even more painful are Death March projects that demand personal sacrifice, yet are doomed from the outset by unrealistic dates, understaffing, or impossible deliverables. Such projects may be approved as a stretch goal, then stagger on for years before delivering nothing of value.

Freezing project initiation is an effective technique for clearing a project logjam created by the Milestones and Darwin approaches, but it won't stop a runaway Death March. For that, project termination is the remedy. Industrial projects can be frozen because the physical work in process just sits there until the project resumes. But Information projects are harder to freeze because the intangible work in process becomes outdated rapidly.

Once a project logjam has been cleared, throttling is an effective technique for launching just those projects that can be completed with available resources. Selective release is based on availability of each project's critical resources. Because Information projects don't freeze well, it's more practical to choke their release.

Termination

Sometimes it is necessary to cancel projects [Boehm, 2000]. If the initial justification was sound, cancellation is not a sign of failure. It's a rational business or technical decision. When freezing projects merely delays the inevitable because partial technical solutions don't age well, it's better to cancel them outright and free those resources to do something more productive.

Post-mortem analysis can reveal valuable lessons if conducted as an act of discovery rather than punishment. Discovering something later that would have made such a project better is the kind of lesson to feed forward into future projects.

Consider this old story about the factory worker who was worried his employment was about to be terminated for ruining a batch of parts with an honest mistake. According to legend, his boss said, "I can't afford to fire someone who I just spent 10,000 dollars educating. You are now the person least likely to make that mistake again."

Governance

Governance includes setting policies and making decisions about an Information portfolio. Thus, governance is the process for evaluating, approving, initiating, freezing, choking, monitoring, rescuing, and terminating projects. While project management monitors individual projects or a set of dependent projects, governance monitors all active projects.

Many people have roles in Information governance:

- CIOs
- CTOs
- CFOs
- Research managers
- Product managers
- Project managers
- Operations managers
- User groups for commercial products
- User steering committees for internal systems

Collaboration leads to balanced evaluations. Dominance by people with one viewpoint can skew governance. For instance, emphasis on financials can skew toward cost saving instead of revenue generation. Emphasis on new markets can skew toward business risk taking. Emphasis on Legacy System replacement can skew toward technical risk taking.

Using conventional reporting, senior executives may not know projects are going to finish late until shortly before the due date passes. Fortunately, scrum projects show progress on Burndown Charts, while Critical Chain projects show progress on Fever Charts, and both give early warning. (Kanban projects don't do scrums, so there are no Burndown Charts.) In addition to showing the progress of individual projects on their Fever Charts, Critical Chain also plots the entire portfolio of active projects as points on a roll-up Fever Chart. Reporting this chart bi-weekly alerts executives to projects at risk much earlier, which then enables recovery plans.

Projects of different size and duration can display on the same Fever Chart because both axes are percentages, not size or duration. Red zone projects deserve executive attention. Green zone projects can be managed by the project manager. If someone has responsibility on more than one project, which is not recommended but does happen, the roll-up Fever Chart indicates which projects should get priority: those deepest in the red zone.

Multi-project Management

Multi-project management coordinates the execution of dependent projects. A project may depend on work products or deliverables from predecessor projects, such as a project to deploy software that depends on installation of new hardware and hiring of additional developers. Or multiple projects may depend on a non-sharable resource, such as a hangar bay that cannot hold more than one aircraft at a time. (See Figure 16-1.)

Multi-project management methods include:

- **Super project.** Hierarchical relationship with subprojects
- **Pipeline projects.** Staggered relationship based on a non-shareable resource or integration task
- **Flexible capacity.** No relationship, except shared resource dependency
- **Kanban boards.** Hierarchical decomposition of projects into tasks across multiple teams
- **Scaled Agile Framework (SAFe).** No relationship, except technical dependencies

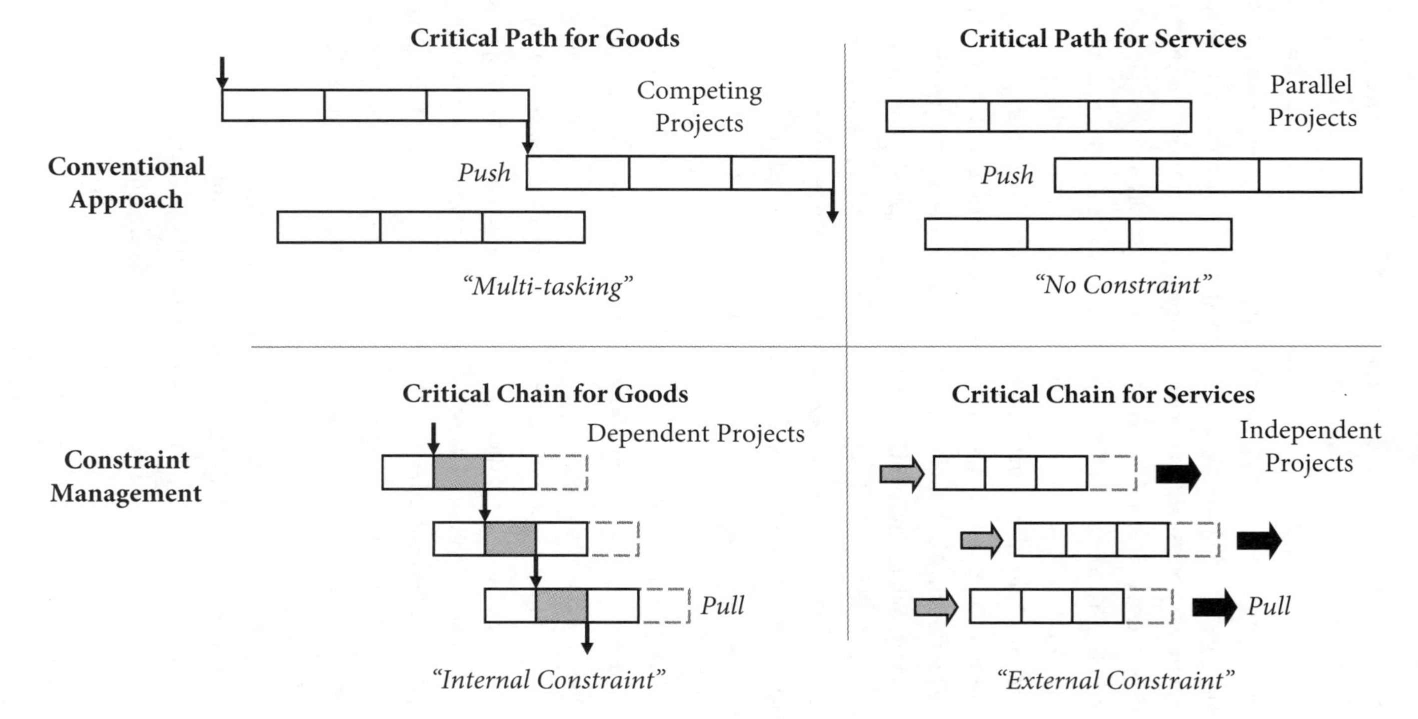

Figure 16-1 Multi-projects.

Super projects are the traditional Critical Path Method for multiple projects. Each subproject is planned separately, then a super project plan is created that fits the subprojects together based on when predecessors need to deliver their work products or deliverables to successors. Thus, phases of the subprojects may be represented on the super project as a single block on the Gantt chart, and milestones mark the dependency points. This reduces detail on the super project plan. It also identifies the super critical path—the set of subproject phases with no slack.

When executing a super project, delays in predecessor subprojects ripple onto successor subprojects. Contingency at the super project level can mitigate this, but the same caveats about local optimization and squandering of contingency apply to a super project. In other words, a super project is susceptible to all of the shortcomings of Critical Path. Nevertheless, a super project can work well with multiple subprojects that each produce unique deliverables and that depend strongly on each other, such as building a ship's hull, its propulsion, fuel tanks, ballast tanks, crew quarters, cargo holds, passenger accommodations, navigation, etc.

Pipeline projects are the traditional Critical Chain approach to multiple projects when there is a non-shareable resource or integration task. Each project is planned separately using relative dates rather than calendar dates. For instance, instead of planning the launch on July 1, each plan starts on day one. The projects are sequenced by priority, and an approximate start date for each can be computed.

To execute a pipeline of projects, the actual start of each project is adjusted based on the availability of the non-shareable resource or the schedule for the integration task, then the relative dates are converted into corresponding calendar dates. For instance, after one aircraft rolls out of the maintenance hangar bay, another should roll in without delay. Or after one project completes its integration task, the next project's integration begins. The non-shareable resource or integration task is thus the cross-project constraint. When plotted on a calendar, the project pipeline forms a stair step pattern.

Pipelining works well when the projects have the same or similar objectives and methods, such as routine maintenance on complex equipment or manufacturing of a complex product. The non-shareable resource can be a physical facility, special-purpose equipment, or persons with critical skills. On software development projects, for instance, architects are often the resource constraint, so design tasks are where projects would be linked to form a pipeline.

Flexible capacity is a nontraditional Critical Chain approach to multiple projects when start dates or completion dates are determined by clients and the service provider must adjust capacity accordingly. Each project is planned independently, and staffed from groups with commodity, core, or critical skills. Because critical skills are always in short supply, they are a resource constraint across projects. Thus, while striving to accommodate client dates, the provider must also strive to assign skilled individuals without forcing them into bad multitasking across projects.

For a given scope, clients can specify the start date or finish date, but not both because that violates the Iron Triangle. If scope is negotiable, then both start and finish dates may be stipulated if the plans include an adequate project buffer. However, regardless of date commitments, the provider must adjust its capacity by monitoring its skill buffers and replenishing them as needed.

Kanban scales up for large projects by putting a hierarchy of requirements on the Kanban board, but it follows the same principles used on small projects. Tasks are still pulled from the backlog into work in process based on their priority. Even when there are more tasks and more teams to perform them, the work still fits on one large Kanban board.

Scaled Agile Framework (SAFe) is an Agile approach to multiple projects. All the teams must be Agile, but they can use Scrum, Kanban, or any other Agile method. And projects unsuited to Agile can be managed with another Portfolio Management method.

While Agile teams typically think ahead at most two or three scrums, SAFe looks across multiple teams and over a longer horizon of six or seven scrums in order to coordinate deliverables at enterprise scale. Cross-team dependencies are identified at the program level (collection of teams). Overall investment and direction are provided at the portfolio level (collection of programs).

SAFe is controversial because it establishes some layers of oversight and coordination that resemble the Waterfall method. To Agilistas, the separation of developers and product owners from portfolio-level decisions doesn't feel agile. However, the SAFe additions are lightweight compared to full-blown Waterfall portfolios.

ADVENTURE: Paradigm Shifts

The Constraint Management field experienced a paradigm shift decades ago when Manufacturing and Retailing implemented solutions and discovered that their constraint moved from internal to external rather rapidly. Instead of production being unable to keep up with demand, sales could not keep up with production.

Because Constraint Management originated in industries where capital equipment is expensive to change, the standard solutions assume that once Constraint Management unleashes the capacity hidden in chaos, capacity is relatively fixed henceforth. Therefore, reducing production capacity until the constraint is again internal isn't a strategy often advocated. The recommended strategy is to increase sales, which is a different problem than adjusting production capacity.

Conversely, flexible capacity is common in labor-based services businesses, which is the problem we were trying to solve because it enables multi-project management with shared resource dependency. The traditional approach to service capacity management is "hire to forecast," but forecasts are notoriously inaccurate and reaction always feels too late. We invented a non-traditional approach using Constraint Management that is called "hire to buffer." Buffer management meant we slowed or stopped hiring when capacity exceeded engagement needs and restarted or accelerated hiring when capacity fell short of engagement needs. This flexible capacity on demand meant we knew how to smooth the peaks and valleys in our resource management. It worked beautifully in simulations.

Executive Champion

Fortunately, we had an executive champion able to convince resource managers to question their own long-standing paradigm. That accomplished step #1 of the buy-in process: Agree on the problem (see the Constraint Management chapter).

When managers have been practicing one way for years, it rocks their world when they are offered a better alternative. That accomplished step #2 of the buy-in process: Agree on the direction of a solution.

One Shot at Persuasion

Step #3 was where things became difficult: Agree that the solution solves the problem. We only had one shot at persuading the managers that we were onto a better way. We knew that talking about it or showing slides about it wouldn't be persuasive. So, we reconfigured the simulator to allow manual decisions using the old paradigm alongside simulated decisions using the new paradigm.

When resource managers could see for themselves that capacity on demand via buffer management led to fewer under- or over-capacity conditions, they were open to trying it for real. To our delight, it worked as well in practice as in simulation. The remaining buy-in steps were easy once we turned this corner.

Lessons Learned

Lessons learned include:

- Paradigm shifts are an antidote to conventional wisdom if the new paradigm is unquestionably better.
- It helps to have an executive champion get the buy-in process started, but follow-through is critical.
- Sometimes you only get one shot at persuasion, and the Constraint Management buy-in process was created for that situation.

Project Management Office

Huge projects, complex multi-projects, and diverse Information portfolios deserve a Project Management Office (PMO). The PMO is where managers and developers set priorities, make plans, freeze/release projects, observe status, expedite laggards, and raise issues.

When projects are managed with Critical Path, the PMO does mostly planning and replanning. When projects are managed with Critical Chain, the PMO does mostly execution enablement. When projects are Agile, if there is a PMO, it may be virtual instead of physical.

System of Systems

As noted in previous chapters, an enterprise is a system of systems, and the location of a local constraint depends on the observer's viewpoint. Local optimization is therefore a hazard. Portfolio Management is where short-term local system conflicts are recognized and resolved. Strategy is where long-term local system conflicts are resolved. (More on that later.)

To the Chief Financial Officer, the local constraint is probably cash. If a troubled project misses its due date and exceeds its budget, expected profit can turn to loss. Or a strategic initiative to drive new revenue may miss its window of opportunity. Risk management across projects pools such risk, but the more troubled projects there are, the more the budget buffer (cash and credit) will be depleted.

To the Chief Technology Officer, the local constraint is probably skills. If new projects cannot be fully staffed because some current projects are troubled, the skill buffer (skills on board and in the hiring pipeline) can be depleted. Skills management across projects pools the risk, but critical skills are always in short supply.

To the Chief Information Officer, the local constraint could be many places, including funding for equipment, staffing for new projects, or maintenance and enhancement of Legacy Systems. While enhancements may be discretionary, considerable maintenance is not because it fixes defects that prevent correct operation and accurate results.

Thus, the CTO and CIO may want to hire more, while the CFO says there's not enough funding for an additional headcount. Several solutions to this conflict are possible. New projects could be frozen. Top technical talent could be redeployed to rescue troubled projects. Selected Agile projects could be switched from development mode to maintenance mode to redeploy some staff to higher-priority tasks.

Most importantly, however, all investments should be reassessed for their impact on the enterprise constraint because improvements to local systems should be in service of the enterprise constraint. For instance, if the enterprise has a sales constraint, improving the sales process should be top priority, even if it's not the local constraint for the CFO, CTO, or CIO. If an enhanced sales system could elevate the sales constraint, that project ought to leap to top priority, alongside mandatory projects.

Constraint Management

Can Portfolio Management be the enterprise constraint? No, because Portfolio Management doesn't directly control what the enterprise produces. The enterprise constraint is most likely external, in the market, or in an internal process within a factory, a warehouse, a service depot, or a sales office.

Although Portfolio Management isn't the enterprise constraint, it does contend with conflicting local system constraints. Portfolio Management can influence where the enterprise constraint and local constraints are located, possibly helping to move them to more strategic locations. These conflicts are generally evident in the Information budget, which can be a local constraint on the financial system. If there is a financial buffer, it is in funding available for contingencies.

Despite the power of Constraint Management, Portfolio Management is unlikely to find constraints unless its practitioners are looking for them. That's unfortunate because Portfolio Management gets easier when it is focused on constraints. Non-constraints can take care of themselves for the most part unless they are in danger of becoming the new constraint.

Because few if any organizations have unlimited resources, Portfolio Management makes tradeoffs based on priorities. However, aspirational proposals beyond the feasible frontier should receive extra scrutiny, and if accepted, receive adequate resources and appropriate oversight.

Information portfolios of mixed projects are harder to deliver because their objectives and resources are diverse, yet mixed projects are the rule, not the exception for large enterprises. Planned and Agile projects are based on different principles. Modern and Legacy Systems are based on different platforms. Yet Legacy Systems cannot be easily discarded or replaced when they enable critical business processes, and they tend to have stringent development and operational requirements.

Conclusion

A typical Information portfolio is composed of products and systems at different points in their life cycles. Portfolio Management considers whether the mix is right, and if not, what can be done to make it right. That may result in launching new projects, freezing old projects, and sunsetting older products and systems. However, the trailing edge of technology tends to move more slowly than the leading edge because the business case for change gets progressively harder to make and so does finding people with the necessary skills.

Planned portfolios use time buffers to synchronize multiple projects, while Agile portfolios use work buffers. That fundamental difference means Planned and Agile portfolios generally move at different speeds. In the time that one company releases software (every nine weeks), another company makes 300,000 changes [Power, 2014]. That's an apples-to-oranges comparison, but it does illustrate that one approach does not fit all.

The upper bound on size and complexity of Agile projects is debatable. Agile advocates claim high scalability, but scalability is relative. The largest Planned projects are staffed with hundreds of developers from dozens of skill groups. Moreover, large defense projects are often Planned because government contracts require specific forms of status reporting, and failure to achieve milestones can result in penalties of over 100 million dollars. On the other hand, most commercial projects are not near that scale.

One subject where Planned and Agile may agree entirely is on controlling work in process because that reduces confusion, maximizes productivity, and minimizes defects. Critical Chain, Scrum, Kanban, and other Agile methods actively limit WIP. Waterfall and Critical Path projects do so only passively, if at all.

Finally, exogenous factors often take portfolios to places that plans and priorities couldn't anticipate. A collapsing market, supply chain disruption, a deep recession, or the emergence of new competitors are a few examples. While it's true that execution ultimately matters more than plans, Portfolio Management raises the response up to a level where it affects multiple projects, if not an entire portfolio. While steering some projects in a different technical or business direction, Portfolio Management may simultaneously freeze other projects, throttle projects not yet launched, and expedite projects deepest in the red zone.

CHAPTER 17

SERVICES

Previous chapters covered products. This chapter covers services. Products and services are covered separately because services are dissimilar to the product businesses where Constraint Management has seen its widest adoption. Perishable agricultural products require different handling, storage, and transportation than durable manufactured products. Likewise, services require different methods than manufactured products because there is no way to put an inventory of finished services in a warehouse.

Of course, covering products and services separately for expository purposes is not meant to imply that they are always separate in practice. While some enterprises offer only products and others offer only services, plenty of enterprises deliver both products and services because products need maintenance and repair and because the profit margin on such services can exceed the margin on the products they support. This coupling of products and services is especially prominent in the Information field.

Altogether, services account for nearly 80% of employment in developed economies, so it's a worthy topic. But first, consider how we got here. (See Figure 17-1.) During the early Industrial Revolution in the U.S. (1850 to 1900), employment in the Agriculture sector dropped from about 85% to 30% as Manufacturing employment rose from 5% to 50%. Subsequently, during the Information Revolution in the U.S. (1950 to present), Agricultural employment further dropped to about 1%, Manufacturing employment declined to 20%, and Services employment rose to nearly 80%. Therefore, the Information Age could also be called the Services Age.

Productivity of the Agriculture and Manufacturing sectors increased markedly in the U.S., thereby enabling this massive shift in employment. (See Figure 17-2.) Between 1950 and 2015, real Agriculture output rose by 150% [USDA, 2019]. Similarly, between 1980 and 2015, real Manufacturing output rose by 150% [Muro, 2016]. Thus, Manufacturing productivity grew twice as fast as Agriculture

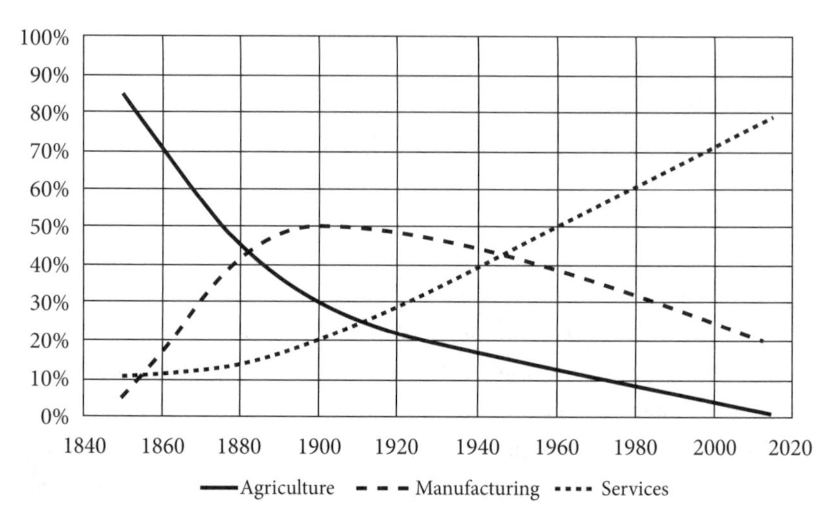

Figure 17-1 Employment by sector.

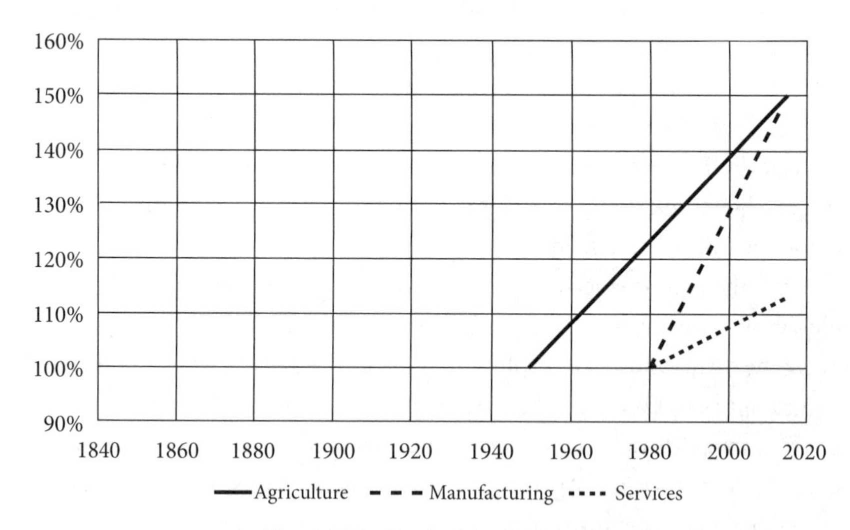

Figure 17-2 Productivity by sector.

productivity [Sposi, 2014]. Yet both the Agriculture and Manufacturing sectors markedly improved their productivity while simultaneously reducing their workforces. How is that possible? Automation and information. Digital Agriculture enables new farming and ranching methods, while Digital Manufacturing enables new design and production methods.

Are there Digital Services? Yes, earlier chapters covered various Information Technologies delivered as a service instead of as a product. This chapter, however, is about labor-based services. Because such services are less amenable to automation and information, Services sector productivity has grown at half the rate of Agriculture productivity and one quarter the rate of Manufacturing productivity. This disparity has vexed economists, politicians, investors, and executives.

Professional, Scientific, and Technical Services

Professional, Scientific, and Technical Services is one of several service sectors in the North American Industry Classification System. Its distinguishing feature is the fact that services are not just labor-based, they are based on expertise. That is, delivery of services is almost entirely dependent on scarce skills [Ricketts, 2008/ Ricketts, 2010].

Compared to the Healthcare sector, which relies on collaborating inputs from high-tech machines and materials, equipment and materials are not as important in PSTS industries. That's ironic because some industries in the PSTS sector do R&D, design, consulting, and repair of Information Technology.

The PSTS sector also includes the Accounting, Architecture, Law, and Advertising industries. Of more interest here, however, are industries that can be referred to collectively as Technical Services.

Technical Services do conventional projects for hardware development and maintenance, software development and maintenance, scientific research, and management advice on technology management. Technical Services also do unconventional projects to bring automation and information to business processes.

Service Organization

Here are some typical Technical Services:

- **Technical Strategy.** Internal studies, industry benchmarks, technology roadmaps
- **Scientific Research.** Lab studies, clinical studies
- **Research and Development.** New products and services, field testing
- **Technical Consulting.** Architecture, design, coding, testing, implementation

- **Operations.** Running hardware, software, and networks
- **Business Process.** Outsourced procurement, payroll, human resources, etc.
- **Technical Support.** Self-help, help desk, maintenance and repair

Executives in Professional, Scientific, and Technical Services are called partners when the organizations are legal partnerships. However, the title is so ingrained that executives in PSTS corporations are still called partners even though it's not accurate.

PSTS organizations typically follow a leveraged model:

1. Many entry-level consultants, scientists, and technicians
2. Fewer senior-level consultants, scientists, and technicians
3. Even fewer managers
4. Much fewer partners
5. Only a handful of senior partners

In other words, the organization is a hierarchy with few levels and a broad base.

Span-of-control issues limit the width of the hierarchy. Adding layers enables wider hierarchies, but then span-of-control issues emerge vertically rather than just horizontally. That is, too many management layers mean that whatever's happening at the lower levels is at best garbled, and at worst, unseen at the upper levels.

In theory, a matrix organization is an alternative to a hierarchical organization. However, in practice, a matrix organization may turn a two-dimensional hierarchy into a three-dimensional hierarchy. For instance, consultants, scientists, and technicians may have two managers: a resource manager and a project manager. In a multi-project environment, however, three dimensions aren't sufficient because one worker with multiple skills may have multiple project managers and multiple resource managers.

Add in industry specialties and geographic locations, and the organization becomes n-dimensional. Consequently, conflicts arise often. Globally integrated enterprises are intended to reduce duplication by having one universal process, but local laws, regulations, and customs mean one universal process still has exceptions.

While Scientific Research has laboratories, and Technical Support has parts inventories, Technical Services organizations are otherwise different enough from product-based businesses that Constraint Management applications work differently. There is more focus on skills and methodology instead of production and distribution.

ADVENTURE: **Presentations 101**

Technical Services necessitate many presentations, from conferences to pre-sales to proposals to engagement deliverables. Most of my presentations were uneventful. A few, however, went astray.

No Show

At one conference, I flew in a few hours before my presentation time only to discover that my session had been scheduled twice and I had already missed the first instance. In many years of making presentations, this was the only time I was scheduled to present the same session twice at one conference.

Apparently, my session attracted enough pre-enrollment that the organizers scheduled it twice on separate days so attendees with a daily pass had more than one opportunity. But they neglected to notify me. Oops.

Puppet Show

On a different occasion, I was scheduled to arrive the day before my presentation, but bad weather canceled my original flight and forced me to arrive the day of my presentation. I couldn't be reached in flight, so my colleagues were anxious about trying to cover my presentation if I didn't arrive in time.

They had a copy, but the subject was technical: portfolio partitioning based on data affinities. In a nutshell, it covered ways to divide many software programs into testable subsets while minimizing dependencies across subsets to reduce rework.

With an audience of about 500 people assembling in the hotel ballroom, there was an audible sigh of relief from my colleagues as I stepped off the escalator minutes before my scheduled session. As they wiped away beads of sweat, they kidded me that they were going to put on a puppet show about portfolio partitioning if I had been a no-show again.

Dinner Show

Several months later, chance provided my colleagues with a measure of revenge for their high anxiety. Another conference was being held where it was only local travel for me, but out-of-town travel for my colleagues.

As I arrived for my morning panel discussion, light snow was falling. By lunchtime, the snow was accumulating. By afternoon, the storm was intense enough that flights were being canceled, and unbeknownst to me, my colleagues were grounded in a distant city.

Unfortunately, one of those colleagues was the featured dinner speaker. I was drafted to give his dinner presentation instead—but I didn't have a copy. As an added complication, he had connectivity issues at the airport where he was stranded, and I had my own connectivity issues at the conference venue. Consequently, faxing his presentation was the last-minute solution. As the conference host introduced me, I was handed the first three pages. Hence, I began making the presentation with no idea how it ended.

Every minute or so, someone offstage would hand me another page fresh off the fax machine. Although I knew the topic well from a technical perspective, the featured dinner speaker was in marketing, so I literally had no idea what was coming on future pages. I hoped it wasn't a puppet show.

Lessons Learned

Lessons learned include:

- Confirm the schedule.
- Build in a time buffer.
- Extemporize as needed.

Service Methodology

Some Technical Services are delivered as projects, others as processes. (See previous chapters on those topics, plus methodology.) Short-term service engagements are often managed as projects, while multi-year service contracts are more often managed as processes. For instance, hardware or software installation and configuration can be done as a project, then operations, maintenance, and repairs can be done as a process. Those might be phases in a single contract or separate contracts.

Why is Technical Service methodology needed if everyone's an expert? Turning everyone loose makes it harder for team members to collaborate, and it results in rediscovery of proven techniques, so it's inefficient. Furthermore, allowing everyone to do things their way creates inconsistent deliverables, so it encourages quality issues, too. Methodologies therefore standardize the repeatable parts of service engagements, while expertise addresses the non-repeatable parts—the problems nobody has solved before and the solutions that are hard for clients to comprehend because they are novel.

Methodology is often supplemented with templates. Templates are models for routine documents, such as estimates and proposals, as well as for deliverables, such as reports and presentations. If software is a deliverable, a template can provide quick starts on everything from architecture to design, coding, testing, implementation, operation, and user guides.

Technical Service deliverables typically include:

- Hardware installation, configuration, maintenance, and repair
- Packaged software installation, configuration, and upgrades
- Custom software development, maintenance, and enhancement
- Technical strategy reports and presentations
- Technical consulting reports and documentation
- Research studies
- Product designs
- Training materials

Unlike hardware and software, some Technical Service deliverables are not testable, at least during the engagement itself. For example, if the deliverable is a Technical Strategy, the intended benefits of that strategy may not be observable for a year or more. Therefore, the proposed strategy may be compared to Technical Strategies in other organizations with similar issues. If it worked elsewhere, clients are more inclined to accept the proposed strategy. Clients may not want to be a first-mover unless the proposed strategy offers sustainable competitive advantage via its boldness and uniqueness.

Service Disruptions

Businesses and governments have regular operations. That's what performance improvement initiatives generally address. However, business and government entities affected by natural disasters or human-made disasters also have irregular operations.

Irregular operations are proactive if given enough lead time and confidence in predictions. Irregular operations are reactive when the disaster arrives before preparations. Sandbags can be filled before rain falls, but a levee can't be built as a dam is breached.

Weather is the great disrupter. Travel and Transportation are directly affected, of course. Planes must maintain extra separation in congested airspace when visibility is below a threshold, nor can they fly when temperatures are above a threshold because wings have insufficient lift. However, it's hard to identify an industry that isn't affected by weather at least indirectly. Even industries with indoor production are susceptible to supply chain disruptions. Service industries such as sports and theaters cancel events when the fans and audiences can't travel even if the venues are undamaged.

Weather predictions are notoriously difficult because high-resolution and long-range forecasts require sensors, storage, computation, and expert interpretation. Where I live, a blizzard left deep snow behind, and a month later, the same area had a heat wave with temperatures over 90°F (32°C). Nobody predicted either event until they were imminent—and their close timing was inconceivable, let alone predictable.

The Constraint Management solution for manufacturing sustains regular operations by making production less susceptible to occasional disruption. A late material delivery or rush order won't throw production into chaos if appropriate buffers protect the constraint. As compelling as that logic is for product businesses, however, it doesn't always work as well for services businesses if there is no way to establish and manage a capacity buffer.

For regular service operations, Sense and Respond works better than Predict and Prepare. Some enterprises are so severely affected by disruptions, however, that forecasting matters. In those cases, predicting the start, severity, and duration of irregular operations is vital. For instance, one unexpected hailstorm at a major airport did over 100 million dollars in damage to just one airline's planes, equipment, and facilities.

The triggering event doesn't have to be extraordinary. Some disruption types are common, so it's easy to predict that some will occur every year. But precisely when and where they will occur is a truly hard prediction problem, and getting that prediction early enough to plan and prepare is even harder.

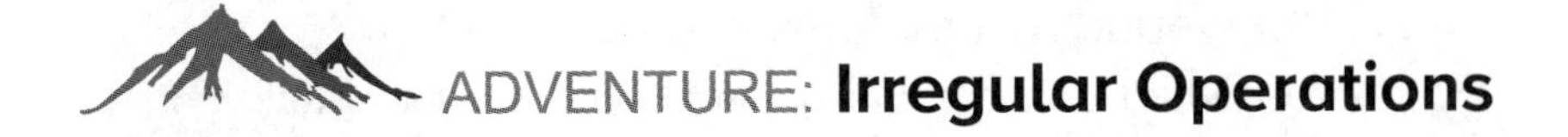

ADVENTURE: **Irregular Operations**

Commercial airlines and airports have different constraints, despite their mutual dependence. Together, they form systems of systems. In the long run, the airline constraint may be aircraft; in the short run, the airline bottleneck may be flight crews when they run out of allowable working hours. In the long run, the airport constraint may be runways; in the short run, the airport bottleneck may be gates when ground holds and flight diversions skew traffic.

Airlines and airports both conduct irregular operations, which is a sign that the regular constraint is not what's limiting throughput at that moment. Thus, weather may prohibit airlines and airports from using all their regular capacity.

Everybody Complains about the Weather . . .

When I worked in corporate strategy, my firm had been researching high-resolution weather forecasting for several years. It was time to get the technology rolled out, and this was a perfect fit for a proof of concept.

Airlines and airports sought better forecasting in order to trigger irregular operations sooner and with higher confidence that a forecast of bad weather wasn't a false positive. The objectives included keeping passengers and flight crews safe and happy, protecting cargo, safeguarding equipment and ground crews, saving jet fuel and deicing fluid, and managing runways, taxiways, and gate utilization.

Forecasts for weather en route were deemed sufficiently accurate, so opportunities for improvement in forecasts were at originating and terminating airports. In addition to forecasting weather severe enough to ground or divert flights, the technology could forecast when a change in wind direction would require every arriving and departing flight to reposition in order to use runways from the other end or use alternate runways.

. . . But Nobody Does Anything about It?

When Charles Dudley Warner quipped that "Everybody talks about the weather, but nobody does anything about it," he couldn't have imagined the technology that today depends on decent weather, nor the technology used to predict the weather. Both would amaze him because he lived before the Wright Brothers' first flight and well before computers were invented.

When commercial flights are scheduled, pre-flight decisions are made based on the weather forecast:

- Will flights depart on time if there is rough weather at the origin, en route, or at the destination?
- Which alternate airports are available if flights must be diverted en route?
- How much extra fuel should be loaded to handle higher consumption due to headwinds, routes around storm cells, and potential diversions?
- Should flights be delayed until rough weather has passed or subsided?
- Should flights be canceled because they are likely to strand passengers at an intermediate airport rather than getting them to their destination?

- Should aircraft be prepositioned out of harm's way, ready to resume regular operations instead of having to be repositioned later?

Weather forecasts don't answer all those questions, but forecasts do provide essential information to support expert judgment. By chance, a severe winter storm hit some airports being forecast during our proof of concept, and the high-resolution local forecasts were demonstrated to be more accurate than normal-resolution regional forecasts. It turns out that today we really *can* do something about the weather.

Lessons Learned

Lessons learned include:

- Irregular service operations are common, but hard to predict.
- Weather forecasting is computationally challenging due to the volume of sensor data and the complexity of modeling the physics behind weather events.

Service Engagements

Technical Service engagements typically begin with a proposal that becomes a contract when signed by the client and the provider. The Statement of Work lays out:

- Objectives
- Background
- Deliverables
- Client responsibilities
- Provider responsibilities
- Price
- Due date

The Statement of Work is based on the Work Breakdown Structure that will be used for project and process management.

Technical Services can be priced several ways:

- **Time and Materials.** Priced by the hour, plus expenses
- **Fixed Price.** Priced based on an estimate for the entire engagement
- **Value Price.** Priced by the value the client expects to receive

A Time and Materials contract can expand as requirements grow, but the client typically signs off beforehand on the change in scope, additional work hours, and extended schedule. T&M contracts frequently use a blended hourly rate when workers with widely varying billing rates will be assigned.

A Fixed Price contract might seem inflexible, but if requirements grow, the client typically signs off on the associated changes. Military contracts are often Fixed Price with milestone payments. That way, the provider earns revenue as a multi-year engagement is completed in phases. The provider thus has a backlog of unbilled work on the service contract.

A Value Price contract focuses on outcome rather than effort and cost. The client might agree to pay the provider a percent of revenue increase, cost savings, or an amount based on market share accomplished by the Technical Service. This is intended to align the interests of the provider with the client.

Given the fees involved and the potential client benefits, clients often send a Request for Proposal to multiple providers. When the field has been narrowed, Technical Service contracts are generally negotiable. Unfortunately, buyer's remorse happens when clients choose the lowest bidder and then complain that they didn't get the A-team for their engagement. The flip side is the winner's curse, which happens when a provider underestimates the amount of work or sets a price too low to achieve its desired profit margin [Anandalingam, 2005].

If the Technical Service deliverable is custom software, the contract may specify a limited-use license rather than perpetual license. That way, the service provider retains the right to reuse intellectual capital on similar engagements with other clients. However, if it's a labor-only contract, the client will negotiate ownership of the software.

If the Technical Service is performed as a process rather than a project, the contract may include a Service-Level Agreement. For instance, it might specify that repairs within a certain distance will be started within two hours. Or the SLA might specify that a certain percent of calls to a help desk will be resolved on the first contact. Realization is the term for revenue received, which can be lower than the amount billed if a project misses a milestone or a process misses an SLA.

If project managers or service managers can open multiple accounting codes for a single engagement, bad news about cost overruns can be buried to make an engagement look more profitable than it really is. If executives cannot determine which clients are profitable and which ones aren't, they are more likely to repeat the winner's curse.

Constraint Management

Can Technical Services be the enterprise constraint? Yes, if the enterprise is a Technical Services provider. Otherwise, Tech Services are not the enterprise constraint because they do not regulate what the enterprise produces. However, if the client's enterprise constraint relies on hardware or software that is broken, Technical Services are the bottleneck at that moment. Technical Services contracts may contain Service-Level Agreements because getting that hardware and software restored is top priority for the enterprise to resume managing its constraint.

As explained in an earlier chapter, Constraint Management was invented for Manufacturing and Distribution industries. In those product-based industries, inventory management is a central concern. But Constraint Management has also been applied to technology-based industries, such as Software as a Service, where the central concerns are availability, performance, and security—not inventory. Constraint Management has also been applied to labor-based service industries, such as Healthcare, which uses Critical Chain to manage patient care. It has also been adapted for expertise-based services, such as Technical Services.

Constraint Management applications that apply to Technical Services were covered in previous chapters, so they are summarized here, but not explained again:

- **Service Production**, such as business process automation or help desk, uses service levels to adjust capacity on demand and pull service requests through the process.
- **Service Distribution** hires on demand based on the pull for resources to replenish skill buffers.
- **Service Projects**, such as consulting and custom software development, following Critical Chain put contingency in the project buffer, enforces

Relay Race work rules, monitors buffer penetration, and thereby pulls the entire project to on-time completion with shorter overall duration. Service projects following an Agile method manage a work buffer instead of a time buffer.

Technical Services potentially benefit all the technology that binds the system of systems within an enterprise. Those services range from development of new technologies to maintenance of old technologies. The technologies may be operational or strategic. Indeed, Technical Services may be instrumental in establishing Constraint Management where it has the most leverage on Enterprise Strategy.

That said, Constraint Management does not have applications for services delivered without a defined project or process. Yet Technical Services has plenty of them. The practitioner must figure out during an engagement how to solve the client's problem. Experienced practitioners draw on similar previous cases to solve current cases, but lots of experimentation and improvisation may be required. A developer must understand the client's business, as well as its technology. A scientist will usually see many failed experiments before some succeed. A technician may be unable to reproduce the client's problem, which is essential to finding a solution. Patentable inventions must be novel, so they can't be a result of a defined process, though they can define a process, ironically. In time, a formal process may evolve; however, it is executed by less experienced, less skilled practitioners. The expert practitioners then move on to new challenges.

Conclusion

Employment in service industries today far exceeds employment in product industries. Fortunately, Constraint Management can be applied to technology-based and labor-based services, too. The service economy is light on hard assets and heavy on intangible assets—especially software.

Technical Services organizations fall in the Professional, Scientific, and Technical Services sector, which is the services sector most different from the product sectors where Constraint Management originated. Although expertise-based services don't have inventory, they do have other artifacts that can be buffered, such as skill groups and project schedules.

Technical Services are, however, challenging because the semi-repeatable nature of services engagements means it can be hard to maintain a stable enterprise constraint. Naturally, the constraint can float from external during slack periods to internal during busy periods. The constraint can slip from one skill group to another, although that's more controllable than the external-internal shift.

In a services business that sells expertise, everybody's an expert on something, and that sometimes gets twisted into being an expert on everything. Good consultants know when to say, "I don't know." Excellent consultants know when it's appropriate to add, ". . . and neither does anybody else."

Here are some anonymous sayings reminiscent of Technical Services:

- There is no compression algorithm for expertise.
- Percussive maintenance means thumping a misbehaving device until it works.
- Anyone who thinks the customer is always right has never worked in Tech Support.

PART 3

SYNERGY

CHAPTER 18

CONSTRAINT MANAGEMENT REDUX

The Information field and Constraint Management have a mutually beneficial relationship. As seen in previous chapters, Constraint Management offers solutions to problems in the Information field. Information is, of course, an essential input to Constraint Management.

That said, the relationship can be adversarial when Constraint Management practitioners see the wrong information being used for management decisions, which happens more often than many managers might imagine. Eli Goldratt wrote two books about problems with software from a Constraint Management perspective [Goldratt, 2005/Goldratt, 2006]. Fortunately, in the ensuing years we have learned how to correct those problems [DDtech, 2019].

Thus, it's time to revisit Constraint Management relative to the Information field. This chapter also compares Constraint Management to other prominent performance improvement approaches, Lean and Six Sigma, to see what they add. Finally, this chapter explores various types of constraints in preparation for the following chapter, Strategy Redux.

Constraint Management for Information

As discussed in the Constraint Management chapter, its principles were first devised for Manufacturing and Distribution, and later extended to other sectors, including Services. As subsequent chapters about various facets of the Information field have shown, Constraint Management principles do apply there, too.

Hardware fits the Manufacturing and Distribution solutions from original Constraint Management. However, the firmware and systems software that ship with hardware may not be as amenable to the Constraint Management solution for projects.

Computer and network operations are similar enough that the original Constraint Management solution for production can work if a stable constraint can be established. However, because computer and network operations happen in real time and sometimes serve demand that cannot be queued, maintaining a stable operations constraint is not always possible. Indeed, architects, developers, and operators in the Information field wrestle with this problem. Flexibility is the antidote to instability, however.

Because software requirements are intangible, unique, and volatile, application software may or may not be amenable to the Constraint Management solution for projects. Agile is more appropriate than Critical Chain in some circumstances. Unfortunately, Agile is sometimes enforced, despite circumstances that favor another approach. On the other hand, Technical Services are often amenable to the Constraint Management solution for projects despite being expertise-based, semi-repeatable, and forever evolving. Indeed, when deliverables include more than software, Critical Chain can excel.

If Data or Knowledge are the primary products, then Constraint Management may be applicable to the projects, processes, or skills that produce those products. Furthermore, whenever Information skills in general are the constraint, such as in Technical Services, Constraint Management can manage skill buffers.

This claim that Constraint Management can be applied to the Information field is controversial in some circles. The strictest definition of an operations constraint is it must be a physical productive element of a system. Except for enterprise systems, which run continuously, most computer hardware is idle more than half its life. Software, on the other hand, violates the physical criterion because software is not the physical medium it's stored on. The same goes for data. The search for a physical constraint behind software is generally not helpful. The assumption that the programmers who write software are the constraint reaches a dead end when said programmers are deceased, retired, or resigned. Plenty of software continues in use long after its programmers have departed.

Information for Constraint Management

While Constraint Management for the Information field is a choice, information for Constraint Management is mandatory. Buffer Management is a universal aspect of every Constraint Management solution, and Buffer Management is driven by information.

For example, the original Constraint Management solution for Manufacturing operations compares the Buffer Level to Buffer Zones. A Buffer Level in the green zone says there's plenty of inventory in process to keep the internal constraint fully occupied, so no action is required. A level in the yellow zone says there may not be enough inventory, so plans should be made to create more. And a level in the red zone says there definitely is not enough, so work-in-process inventory should be increased.

Thus, even in this simple example of an operational decision, three elements of Information are required: One Buffer Level and two thresholds between Buffer Zones. The enhanced Constraint Management solution for Service operations makes a similar comparison, except the Buffer Zones are bidirectional because the Buffer Level should not be too high or too low: the Buffer Zones are red-yellow-green-yellow-red, which adds two zones.

When Constraint Management is not used for Manufacturing operations, one of the most frequently used elements of Information is utilization. Unfortunately, driving every machine or person to have maximum utilization overwhelms the constraint with excess inventory. Thus, this is a counterexample of information leading to suboptimal operational decisions, and it's one of the primary lessons taught to new Constraint Management practitioners.

The Constraint Management solution for Distribution also uses Buffer Levels and Buffer Zones, but each product may have its own. In other words, 1,000 products could have 1,000 Buffer Levels and 3,000 Buffer Zones, so the amount of information required in even a modest-sized company can be substantial. Of course, it is easily within the capability of computers to highlight only the products requiring action.

Unfortunately, Buffer Management is not a universal solution. For example, electricity is a commodity where supply must be matched with demand continuously because there is no practical way to buffer it on a large scale.

Information is also required for tactical decisions, such as which orders deserve priority or whether overtime should be approved to work off a backlog of orders. Buffer Management does not address these decisions, but Constraint Management in the broader sense does. It does so by considering the effect on net profit. More profitable orders might go first (a decision with mostly short-term ramifications) or orders from favored customers might go first (a decision with longer-term ramifications). Overtime might be approved if it generates incremental net profit (a decision with side effects on employee morale and customer satisfaction).

Strategic decisions include (1) acquisition and divestiture, (2) investment in plant and equipment, or (3) investment in research and development. Strategic decisions require quantitative comparison of alternatives. (See the appendix.) Rather than using only point estimates, however, sensitivity analysis can suggest how far actual outcomes can deviate from desired outcomes before an alternative becomes preferable.

The Horizons Model did not originate with Constraint Management, but it is nevertheless a useful addition for labeling the purpose and time frame of strategic decisions. Neglect of current Horizons 2 and 3 jeopardizes future Horizon 1.

Focusing Steps

The Focusing Steps are arguably Constraint Management's single biggest contribution to performance improvement. To review, those steps are:

1. **Identify** the constraint. If it's not in the right place, move it.
2. **Exploit** the constraint. Ensure it is not inhibited.
3. **Subordinate** everything else. Don't let anything overload the constraint.
4. **Elevate** the constraint. Increase its capacity.
5. **Repeat** these steps. Don't stall after each improvement.

The steps work because many natural constraints are not in the best place for managing overall throughput. In some cases, however, the effective constraint is outside the boundaries of the system under consideration. For instance, a Manufacturing constraint may be in quality assurance, not production. Or a Distribution constraint may be in transportation, not warehousing. Thus, it pays to examine presumed boundaries carefully.

Does it matter whether the true internal constraint is identified? Surprisingly, it may not. If Constraint Management is applied to something that's a near constraint, that may be sufficient for production to exceed what the market will buy. Once that happens, it won't matter that the system could eke out more units by managing the true internal production constraint until the firm addresses its external market constraint.

Furthermore, the Focusing Steps address operational constraints, not strategic constraints. For instance, the steps do not question whether the enterprise is manufacturing and distributing the right products, yet a shift in offerings might be many times more profitable than optimizing existing production.

Though the Focusing Steps can be useful in Computer Manufacturing and Distribution, they are harder to apply to Software and Technical Services because (1) the constraint isn't necessarily stable, and (2) non-constraints always have enough work to keep busy and thus overload the constraint. Moreover, the Focusing Steps are harder to apply to Agile Methodology, where the constraint is likely to be skills, than to Critical Chain Methodology, where the constraint is a path through the project plan.

Global Optimization

Constraint Management directs focus onto constraints because the instinct of most managers and executives is to optimize everything. However, as explained previously, driving every element of a system to its maximum production does not optimize what the system overall produces because the constraint gets bogged down.

Thus, global optimization is arguably one of the most important lessons from Constraint Management for the Information field: Don't deliver information in support of local optimization at the expense of global optimization. Instead, work with executives, sponsors, managers, and users to achieve global optimization. Of course, that's easy to say, but hard to do. It ought to be what Business Analysts accomplish, though not every Information organization has such a job role anymore. That's a shame because recognizing that enterprises are systems of systems is a vital first step toward providing information in support of global optimization.

Information requirements submitted by managers and users generally focus on local optimizations when their responsibility is limited to one business function, one geography, or one product/service line. This can lead to excessive business controls, arcane business processes, and missed opportunities for simplification and consistency.

Similarly, many executives think they know intuitively where the enterprise constraint is because their responsibility crosses functions, geographies, or product/service lines. But when their opinions conflict, they can't all be right. Of course, this happens because most are still looking at local optimizations, not the enterprise constraint. Fortunately, guesses about where the constraint lies are testable: (1) subordinate everything else to the suspected constraint, and (2) the real constraint won't run out of work, but non-constraints will.

Conflict Resolution

Disagreement about where the constraint is or what change would improve performance are symptoms of conflict. Some conflicts can only be resolved via a binary decision: For instance, either measure utilization everywhere and achieve local optimization or implement Constraint Management and achieve global optimization. They are mutually exclusive.

If only all decisions were so clear. Here are some examples where the conflicts are shades of grey:

- As CEO, should our strategy be to invest more in H1 productivity or new products/services in the H2/H3 pipeline?
- As CIO, should we invest more in Modern Systems and less in Legacy Systems?
- As CTO, which systems should we migrate to Cloud Computing and away from Traditional Computing?
- As a line-of-business executive, should we invest more in Shadow Information or Central Information?
- As project manager, should we use Planned or Agile Methodology?

The underlying question in each example is how best to allocate finite resources. The decisions are likely to be enacted as strategic initiatives.

Conflict can be expressed in this template:

To achieve our Objective,
we must do Action #1 because Requirement #1 but
we must do Action #2 because Requirement #2.

For instance, "To have a flourishing enterprise, we must improve our current product/service offerings because our costs are too high, but we must also invest in research and development because we're losing our competitive edge."

Because conflicts block progress toward improved performance, Constraint Management includes a method for conflict resolution. The key is to challenge assumptions behind the template. For instance, are costs really too high? Maybe costs are fine, but customers aren't buying because the sales process is broken. Why is the enterprise losing its competitive edge? Maybe more R&D isn't the answer. Maybe it's an image problem that better marketing could repair.

When conflicts span business functions, geographies, or product/service lines, managers and workers within one area cannot resolve the conflict unless they have counterparts in the other areas who will agree to a resolution. Sometimes that happens, but it's rare because the obvious solution is to have a winner and a loser. If there are two conflicts, a barter arrangement may create a win one, lose one outcome; that's a compromise rather than a genuine solution. Conflict resolution more often happens through escalation up to a manager or executive who has responsibility for all the affected areas, but that is still likely to produce a winner and a loser.

A better outcome is win-win by making the conflict disappear altogether. For instance, choose Planned or Agile Methodologies based on best-fit to each project's characteristics. Migrate vital Legacy Systems and Shadow Information to Hybrid Cloud Computing. Disinvest in H1 products/services producing least net profit and customer retention in order to reallocate investment to especially promising H2 products/services.

Much of the uncertainty, instability, and incompleteness in Information requirements can be traced to core conflicts. For instance, Enterprise Resource Planning software didn't solve the operations management problem because it optimizes non-constraints, sometimes to the detriment of the enterprise constraint.

Some problems solve themselves. People will often do the right thing if not blocked by policies and if not measured incorrectly. The key is to remove whatever stands in the way of a simple, sensible solution.

On the other hand, some problems are genuinely unsolvable. In computer science, for example, an unsolvable problem (P vs. NP) has no computable answer. In psychology, an unsolvable problem is a conflict that cannot be resolved, such as a fundamental personality difference—and unsolvable problems outnumber solvable problems 2:1.

Although Constraint Management has a method for conflict resolution, it only works on solvable problems. For example, a highly accurate forecast from noisy data is an unsolvable problem because a forecast will always be inaccurate to some degree. It's only a question of how often and how much. The solution is to reframe the problem from unsolvable to solvable by designing a system that reacts to actual conditions instead of striving to forecast something that is inherently unpredictable.

Decisive Competitive Edge

The Constraint Management approach to strategy is to use its operations solutions to present an "unrefusable offer" to customers—and "unfathomable strategy" to competitors. For example, by implementing the production solution, the distribution solution, or the project management solution, an enterprise can promise much faster, more reliable delivery—and perhaps earn a premium price for it. An unfathomable strategy happens when competitors remain convinced that nobody else can really deliver faster or more reliably than they already do, never mind get a higher price.

This approach assumes that competitors can easily change prices, but it's hard for them to adopt a different paradigm, change policies, and implement Buffer Management. This assumption is less true today than when Constraint Management was new.

One reason for this is that different firms can use the same software and get different results:

- Software can be configured or programmed with good or bad decision rules.
- One firm can have a different physical constraint.
- Even if all firms have the same decision rules and physical constraints, one can have better skills that exploit the constraint (if internal) or adapt to disruption (if external).
- Differences in skills may be outside operations: in marketing, sales, finance, procurement, etc.

- One firm may have a more effective pricing strategy, such as offering faster delivery for premium price.
- One firm may be a member of a well-honed supply chain that benefits all members.
- One firm may have larger market share and customer loyalty.

Strategy and Tactics

As explained in the Constraint Management chapter, a Strategy and Tactics Tree is the diagram of a complete and sequenced plan for change. Each standard Constraint Management solution has such a diagram. It covers about 80% of a complete change, which is enough to create an undifferentiated strategy. Thus, some gaps remain to be filled to create a differentiated strategy.

The Constraint Management approach to strategy has proven effective in capital intensive industries, especially where capacity changes happen in relatively large increments and competitors' relatively slow reaction time opens a sustainable first-mover competitive advantage. Building another factory or opening another store are not small decisions. Even adding one large machine to a factory or one major department to a store can require significant investment and planning.

In contrast, the Constraint Management approach to strategy is harder to apply in organizations that are knowledge-based, technology-enabled, or innovation-driven because competitors react rapidly. Capacity changes can happen in modest increments, and fast-follower responses make first-mover advantage hard to sustain. For example, mobile phone applications are readily matched by competitors. Consequently, an unrefusable offer and unfathomable strategy won't fly when customers already have fast, reliable delivery and competitors are quick to catch on to innovations.

As typically presented, the Constraint Management approach to strategy lays out a logical series of steps that ultimately lead to dramatic process improvement that in turn fundamentally changes how the organization reaches its goal. There are several issues:

- The strategy can take years to implement.
- Big changes require buy-in from everyone.
- Strategic decisions do not necessarily depend on dramatic process

improvement. For example, launching new products, entering new markets, or acquiring companies are big changes, but not necessarily process improvements.
- Uncertainty is not allowed: There are no alternatives or decisions.

Dynamic strategy addresses these issues by taking a more experimental approach. The annual plan still includes strategic initiatives, but proof-of-concept projects serve as a living laboratory. For example, in the Information Technology field, a POC reveals within a few months whether a technology works, whether it solves a business problem, and whether customers will buy it. More about this in the next chapter.

ADVENTURE: **Blue Ocean "Strategery"**

Eli Goldratt, founder of Constraint Management, conducted seminars to engage and educate C-level executives. The centerpiece was an offer to develop an Enterprise Strategy which would generate a decisive competitive edge via an unrefusable offer to the clients' customers based on Constraint Management.

In turn, Eli expected clients to perceive this as an unrefusable offer because it was a Blue Ocean Strategy: No competitor would be able to match the client's decisive competitive edge within a decade. Clients would be able to deliver products and services faster than any of their competitors, plus they would acquire a newfound ability to prosper during highly cyclical demand patterns.

Refusing the Unrefusable

Eli's handpicked Constraint Management experts led executive sessions with the client's entire C-suite, about a dozen executives. My role was program manager and lead author of the final report. After four days of intense client meetings that revealed numerous conflicts faced by the senior executives, we wrote strategy recommendations that would transform their high-tech manufacturing with Constraint Management while it resolved the conflicts.

To Eli's surprise, the client politely declined his unrefusable offer. The change in strategy was too radical—and there were no pathfinders to follow. Later, however, Eli did this kind of meeting successfully with other clients [Kendall, 2005].

Lessons Learned

Lessons learned include:

- Executives not in crisis may see comprehensive change in strategy as unnecessary.
- An unrefusable offer may be seen by executives as a bet-their-jobs commitment.
- The bigger the change, the fewer people who want to be pioneers.
- Management innovations depend on success stories to achieve broad adoption.

Continuous Improvement

Constraint Management is not the only performance improvement method. Indeed, Lean and Six Sigma gained prominence at about the same time as Constraint Management, yet are better known. Some enterprises have had success with a combination.

Here are distinctive features in a nutshell [Nave, 2002]:

- **Constraint Management** leverages the constraint.
- **Lean** reduces waste.
- **Six Sigma** diminishes variation.
- **Trio** selects the best features of the various methods.

Before describing the other performance improvement methods in a little more detail, the following section summarizes the current state of Constraint Management.

Constraint Management

Whenever charismatic founders die, the movements they originated are at a critical juncture. There is no consensus on what has happened to Constraint Management since Eli Goldratt died, but one perspective is that it has diverged into Open and Closed tribes [Marris, 2013]:

- **Open Constraint Management.** Innovations accepted, so upgrades are major.
- **Closed Constraint Management.** Complete as is, so upgrades are minor.

Sometimes, however, a tribe pursues an alternate path. Demand Driven Material Requirements Planning, for instance, doesn't mention Constraint Management, but its heritage is unmistakable [DDtech, 2019].

What does Open Constraint Management bring to the table? It avoids inertia by embracing innovative thinking. More importantly, it identifies where to make improvements (at the constraint) and how to make improvements (with Buffer Management).

Lean

Lean emerged from Just-in-Time (JIT) manufacturing and the Toyota Production System (TPS). But Lean, too, has diverged [Marris, 2013]:

- **Good Lean.** Eliminates waste
- **Bad Lean.** Continuously downsizes

Good Lean eliminates waste by tightly coupling manufacturing steps, minimizing inventory between steps, and balancing capacity throughout. Thus, it requires a stable process and decades of refinement. (In contrast, Constraint Management can improve productivity right away via its Focusing Steps.)

Bad Lean acquired its reputation by being cited as justification for repeated downsizing. It thus improves the efficiency of non-constraints. Its emphasis on cost reduction rather than revenue growth also separates it from Constraint Management.

What does Good Lean bring to the table? It embodies the spirit of continuous improvement wholeheartedly. And it does reduce waste, so it's a useful companion to Constraint Management.

Six Sigma

Six Sigma was born as Total Quality Management (TQM) and popularized by General Electric. It initially achieved acclaim, but its use has declined markedly because its track record is spotty [Staley, 2019]. Thus, it's no surprise that Six Sigma has diverged too [Marris, 2013]:

- **Good Six Sigma.** Uses experiments, data, and statistics to improve quality measurably
- **Bad Six Sigma.** Awards belts and executes many projects, but with few results

The name, Six Sigma, refers to the symbol for standard deviation. Six standard deviations from the mean in a normal distribution represents just one defect per million items, which is an extremely high standard even for manufacturing processes. For software, however, it is an impossible standard because software development is nothing like manufacturing.

What does Good Six Sigma bring to the table? Experiments, data, and statistics can drive quality improvement. Unfortunately, they are too seldom used that way.

Trio

The Continuous Improvement Trio refers to the combination of Open Constraint Management, Good Lean, and Good Six Sigma [Pirasteh, 2006/AGI, 2010]. Curious about which method is best, a 10 billion dollar company decided to conduct its own two-year experiment by using Six Sigma in 11 of its manufacturing plants, Lean in four plants, and the Trio in six plants. Lean contributed 4% of cost savings, Six Sigma did 7%, and the Trio did 89%.

This result is impressive because the company had previous experience with Lean and Six Sigma but not with Constraint Management. If a Constraint Management practitioner had been involved, the target would have been revenue growth rather than cost saving. Despite this, using the Focusing Steps from Constraint Management to guide the changes far outperformed the alternatives.

The following are some synergies among the Trio [Marris, 2013; Jacob, 2010; Sproul, 2012]:

- **Constraint Management** can guide what and how to change. It increases revenue. It handles disruptions with buffers.
- **Lean** reduces waste. It decreases operating expenses. It streamlines stable systems by cutting work-in-process inventory.
- **Six Sigma** diminishes variation. It identifies and quantifies root causes of variation.

Quality

Six Sigma was invented to improve quality because high quality reduces waste and increases product reliability. However, Constraint Management says surprisingly little about quality. There is, for instance, no Constraint Management solution for managing quality. Instead, Constraint Management assumes quality can be achieved as a necessary condition for productivity when processes are repetitive and products are standardized.

In software, however, quality cannot be assumed because development and maintenance processes are not exactly repeatable and deliverables are unique. Although Six Sigma isn't a practical objective for software, quality is nevertheless vital in the Information field. It includes both objective and subjective elements. Objective is "The calculations are correct." Subjective is "I can easily use the system to do my job."

Defect Density is a useful quality measure for software. Defect Removal Rate is a software productivity measure that's useful in predicting when a target quality level will be achieved.

Value

While one view of quality is the absence of defects, value is the presence of business benefits. Though high quality and high value tend to go together, it's possible to have high quality and low benefits, or low quality and high benefits.

Constraint Management recognizes that software drives users to follow processes guided by business rules, and if those rules are wrong, the results will be wrong too [Schragenheim, 2016]. Thus, Constraint Management insists that business rules be given careful consideration before being implemented in software. For example, while users might specify a requirement that every element of a system must have high utilization, Constraint Management declares that to be a misguided policy because it optimizes non-constraints and overloads the constraint.

Things can go wrong in several ways, depending on how software is:

- **Built.** Are Buffer Management rules implemented correctly?
- **Configured.** Are Buffer Zones set correctly?
- **Used.** Do users understand the logic behind the Buffer Management rules?

All those parts must converge for software to deliver value.

Limits

A constraint and a limit are not the same. A constraint restricts what a system can produce, so it can be a control point for managing the entire system. On the other hand, a limit also restricts what a system produces, but a limit cannot be used to manage the system. For example, if a company has a no-overtime policy, its factory will produce only during one shift. How much it produces during that shift is not governed by that limit, so the system operations cannot be managed via the company overtime policy.

Policies are created to address specific issues and scenarios, but they may persist long enough to be applied to situations where they have unintended consequences. For example, 99% of schools in the U.S. still follow a calendar based on an agrarian economy even though less than 1% of the population today is made up of farmers or ranchers.

Limits are also embodied in laws and regulations. When limits originate outside the enterprise, they are "facts of life" that generally cannot be altered by managers. That is different from internal policies, rules, and guidelines that are alterable in theory, but as a practical matter, changing internal limits can be arduous.

Limits are also instantiated in detailed business rules and guidelines. Rules must be followed for routine matters, such as "No refunds on sale merchandise." Guidelines are for non-routine matters and exceptions, such as "Refunds for preferred customers are at the discretion of the store manager." The intent of rules and guidelines is consistent, appropriate decisions and actions even when supervisors are not present.

Limits existed long before computers. Software, however, takes limits into new territory by hiding implementation details and obscuring the logic behind limits. That can entrench limits to the point that nobody really understands where they came from, what their purpose is, how they work, or how they interact.

Sometimes rescinding or modifying a misconceived limit is the most direct way to improve performance and create a competitive edge. Nevertheless, the hardest part of implementing Constraint Management is getting managers to stop doing things that work against the goal, such as forecasting, time reporting, cost allocation, and utilization targeting.

Limits can be used strategically: Abandoning a limit based on misguided conventional wisdom is a strategy that competitors may not match readily. For example, using Critical Chain to complete large, complex projects early may not be something that competitors believe they can match. The same goes for using Constraint Management on skill groups. Eli Goldratt estimated that it takes competitors two years to match a strategic change and up to 10 years to match two strategic changes.

ADVENTURE: **Conflicting Policies**

This system-of-systems adventure is about conflicting policies. One team was a profit center, so its consultants sought as many client billable hours as possible to maximize their revenue. The other team was a cost center, so its technicians wanted as few hours at each client as possible to minimize their cost. This was a clear example of people behaving according to the measures used to evaluate their performance.

Each team had different skills—business and technical consulting versus software configuration and technical support—so their skills were not interchangeable. Both teams were required to complete their respective portions of every client's project. While each client expected one team from the service provider, what they got was two teams with an obvious seam between them—and a beleaguered project executive trying to make it look seamless.

No Constraint in Sight

There was no internal constraint: Both teams had enough skilled staff and all the tools they needed. Clients might have bought more services if the offer were better, so there was no insurmountable external constraint. However, the conflicting policies limited what software and services were delivered to clients. Both teams were operating under accounting fallacies.

The profit-center team strove to maximize billable hours because the accounting system told them that increased their utilization. However, each contract put a cap on revenue unless the client authorized an increase in scope. So, unauthorized hours simply accumulated as unrealized (uncollectable) revenue in the accounting system, and higher utilization did not translate into increased revenue.

On the other hand, the cost-center team strove to minimize hours at client sites because the accounting system told them that lowered their cost. However, unless someone joined or left the team, its staff level and therefore its labor costs were fixed, not variable with billable hours. Missed hours did show up in the accounting system as lower cost on a given contract, but in the aggregate, minimizing hours harmed client satisfaction yet did nothing to reduce total operating expense.

Consequently, clients questioned why so many consultants attended every meeting, while the project executive had to remind technicians to show up on time and stay until their tasks were complete. Quality issues would begin the cycle anew with technicians leaving prematurely and having to be called back.

It's Policies All the Way Down

Remarkably, executives did not think this was a problem because it was the way the business had operated for years. Had its executives been interested in Constraint Management, they would have discovered that layers of problematic policies needed to be untangled.

Ancient cosmology said the Earth is flat and resting on the back of a giant turtle, which rests on the back of a smaller turtle, and so on. Like the familiar saying from cosmology that "It's turtles all the way down," when introducing Constraint Management in large enterprises, "It's policies all the way down."

Lessons Learned

Lessons learned include:

- In large enterprises nobody has a comprehensive view of the system of systems. Not employees. Not managers. Not executives. Not consultants. Not industry gurus.
- Prevailing policies are easy to accept as immutable unless someone in authority points out inherent fallacies.
- Why make a distinction between limits and constraints? Managers frequently must challenge policies before they get to a constraint amenable to Constraint Management.

Duality of Constraints

Constraints can reside in two broad locations:

- **Internal constraint.** Within the system boundary (in production, marketing, sales, procurement, trading, underwriting, consulting, maintenance, etc.)
- **External constraint.** Outside the system boundary (in the market, supply chain, distribution chain, job market, government)

Constraint Management asserts that a system generally has one internal constraint or one external constraint at a time, but the active constraint can shift between those positions. This occurs routinely when the enterprise has seasonal demand: The constraint can be internal during busy season and external during slack season. It also occurs over a longer time horizon as the business cycle ebbs and flows. Furthermore, when a Constraint Management solution is first implemented, it tends to shift the constraint from internal to external rather rapidly.

Because a complex enterprise is a system of systems, however, it can have both internal and external constraints at the same time. How so? One line of business may be thriving while another is dwindling. This often happens as one technology displaces another as market leader.

Internal constraints are not necessarily good. They mean that the market will buy more than the enterprise can produce.

External constraints are not necessarily bad. Some enterprises get their competitive advantage by filling every order from their preferred customers in full and on time, while declining some orders from regular customers.

Constraints get their specific location two ways:

- **Natural constraint.** Located through happenstance
- **Designed constraint.** Decided by management

Regardless of how it arises, the constraint is the element with the least capacity and therefore the most restriction on flow. When a constraint is designed, however, a buffer is established ahead of it, and all the non-constraints are arranged to have spare capacity, which smooths flow.

Constraints can be stable or unstable:

- **Fixed constraint.** Does not change location except by management decision
- **Floating constraint.** Changes location at a random time, possibly to a random place

Fixed constraints make Buffer Management practical because capacity is deliberately unbalanced just enough. Floating constraints are best avoided because the old buffer no longer protects the new constraint and rearranging buffers is not a trivial exercise.

Constraints can be tangible or intangible:

- **Physical constraint.** Corporeal (example: a machine, vehicle, person, computer, or network)
- **Time constraint.** Temporal (example: a project's Critical Chain)
- **Cash constraint.** Financial (example: contingency built into budget)
- **Digital constraint.** Noncorporeal (example: software, data, or knowledge)

Physical constraints and time constraints are the home turf for Constraint Management. However, digital constraints are no less real, and Buffer Management works on them too, even though the work in process is digital. For example, a buffer can contain insurance policies queued in a database for underwriting. Buffer Levels and Buffer Zones can work the same way when a buffer is digital instead of physical.

That said, digital constraints are not limited to the Information field. Any enterprise that depends on digital assets, including Manufacturing and Distribution enterprises, can have a digital constraint. For instance, a firm that's required to maintain meticulous records during R&D, manufacturing, quality control, distribution, maintenance, and disposal can find recordkeeping to be the slowest element of the system. Recordkeeping takes far more healthcare provider time nowadays, too. However, if some data can be gathered automatically, perhaps by bar code or RFID scanning, the digital constraint can be elevated.

Constraints can be actual or potential:

- **Operational constraint.** The current constraint (Horizons 0 & 1)
- **Strategic constraint.** The future constraint (Horizons 2 & 3)

Operational changes increase throughput of the current constraint. Strategic initiatives are a plan to improve or relocate the future constraint.

ADVENTURE: **Floating Constraints**

Constraint Management strongly recommends prevention of floating constraints because the buffer is then in the wrong place relative to the new constraint. In Manufacturing and Distribution, an internal operations constraint is fixed in place by unbalancing capacity and building a buffer ahead of the intended constraint.

In some other industries that's not easily done because conditions triggering a floating constraint are beyond management control. In those cases, the enterprise either must adapt to the change or its customers experience disruption.

Floating Constraints Can Reduce Capacity

Airport #1 had parallel runways built too close together for Instrument Flight Rules. Therefore, whenever visibility dropped below a safe threshold, airport capacity was cut in half as the constraint floated from ground operations to flight operations on a single runway. Once disrupted, flight schedules might not recover until the next day.

Floating Constraints Are Unpredictable

Airport #2 was susceptible to wind shifts that required every plane to get reoriented toward the other end of the runways whenever a storm front passed. Thus, as the orderly flow of inbound flights turned into stacks of flights navigating to new positions, landings were disrupted. Outbound flights could taxi to the other end, which was handled by ground control, so not the overall constraint. However, during bad weather, this airport's inter-terminal transportation could not tolerate icy conditions, so sometimes planes could have flown, but passengers couldn't get to their planes.

Flexibility Can Make the Most of a Floating Constraint

Airport #3 was normally quiet, but when flights were diverted from other airports, it immediately became crowded with aircraft and passengers who were not where they wanted to be. Rather than having gates exclusive to specific airlines, this airport assigned gates on an as-needed basis. Even when stretched to the limit, this policy enabled the airport to use all its gate capacity, thereby exploiting this floating constraint.

Over-elevating a Constraint Creates a Different Problem

Airport #4 vastly overbuilt during the waning days of an economic boom. Subsequently, even on its busiest day, it had over 50% excess capacity and no alternative use for it. Thus, elevating the internal constraint through capital investment eliminated floating internal constraints by creating a persistent external constraint. I imagined a tumbleweed rolling through the airport lounge as I was its only customer.

Lessons Learned

Lessons learned include:

- Some things that trigger floating constraints are beyond management control.
- Flexible capacity improves service levels even when the new constraint is overloaded.
- Capital investment can be an expensive way to elevate a constraint, but in some industries it's the only way.

Ultimate Limit

Goldratt concluded that management attention is the ultimate limit. That is, after elevating all the internal and external constraints, growth of an enterprise is still limited by how many issues its managers can handle [Goldratt, 2010]. Actually, Eli called it the ultimate constraint, but based on the distinction between constraints and limits, management attention is a limit, not a constraint.

Management jobs have lots of bad multitasking from meetings and frequent interruptions. This leads to mistakes, which lead to more interruptions. Thus, managers have finite time, but some of it is wasted. Moreover, managers are caught in a core conflict between current stability and future growth.

Worse still, chaos is contagious. Some managers are crisis-creators when they overload workers on projects, assuming they will accomplish more under pressure. The signature of an executive fire-starter is all managers running around as though their hair is on fire.

This behavior is driven by:

- **Complexity.** Managers dissect complex systems into subsystems with local optima. However, rather than simplify, this creates excessive specialization.
- **Unknowns.** Managers strive to optimize within noisy data. However, insistence on excessive accuracy in no way eliminates normal variation.
- **Conflicts.** Managers forced into compromises can't ignore the underlying conflict, so unresolved conflicts resurface eventually.

Several general trends have converged to create this situation:

- Middle management layers have been thinned, thereby reducing capacity.
- Executives are under extreme pressure to balance existing operations and growth.
- Cost and demand exhibit high variability.
- Enterprises pursue multiple strategic initiatives concurrently.
- Stakeholders introduce new requirements.

Some remedies might seem obvious. Adding managers spreads the workload, but it requires more communication and coordination. Policies are intended to be a proxy for management attention, but exceptions still take management time. Delegating to subordinates also unloads managers, but trust is a fragile commodity. In complex enterprises, escalation is the opposite of delegation: Conflicts that cannot be resolved at lower levels get raised higher and higher in the organization until executives with sufficient authority address the conflict.

Therefore, Constraint Management solutions are designed to require less management attention at the operational level so managers can direct more attention to strategic matters. As the Information field implements those solutions in software, it helps alleviate the ultimate limit.

ADVENTURE: **Digital Constraints**

A Financial Services company licensed a software package to handle 20 new products. It then spent a couple of years and several million dollars implementing just one because the software was not designed or written for easy customization.

Nonetheless, the vendor had enticed the client into licensing its software because that was key to expanding the client's offerings and thereby growing revenue. However, there was a conflict. The software vendor was in the business of developing standard software. Its client wanted customizations that the vendor was unable or unwilling to undertake.

Software Is Hard

The client had deep industry expertise, of course, but it did not have Application Software Understanding or Impact Analysis tools. That was a critical gap, because implementing all the financial products required an understanding of where logic and data were shared so that changes would not have unintended side effects. Sometimes shared logic and data made it easy to change multiple products at once. But shared logic and data also made it hard to customize products with unique attributes.

Customizing the products was a strategic initiative because the client sought a decisive competitive edge. To accomplish it through software, however, the client needed to understand the code, including all dependencies, before modifying it.

Digital Constraints Can Be Sticky

The client therefore approached a consultant about its Application Understanding services. However, the client wanted all the remaining products implemented in about the same time and budget as it had spent internally on just the first product. After negotiation to bring the contract more in line with the scope of work, the consultant started work on the hardest products first. However, as unique customization requirements accumulated, project scope expanded beyond the client's willingness to pay.

Though the consultant did get a few more products implemented, as far as I know they never did implement the entire set. Hence, the client was left with customized software for some but not all its desired products.

Thus, the client did not thoroughly address its digital constraint. Rather than gaining a decisive competitive edge, its products were conventional, and its performance was about average.

Lessons Learned

Lessons learned include:

- Digital constraints aren't necessarily any easier to alter than physical constraints.

- Customizing a software package means the client becomes responsible for maintaining the changes even as the vendor updates the base package unless the vendor agrees to accept that responsibility.
- A standard software package is unlikely to create a decisive competitive edge unless it revokes a policy that competitors will be unable to match.

Paradigms

A paradigm is a way of thinking [Scheinkopf, 1999]. Thus, a paradigm includes beliefs and assumptions that lead to specific policies, business rules, guidelines, laws, or regulations. Paradigms can be helpful or not.

For example, in the Conflicting Policies adventure, the executives were captured by the utilization-maximization paradigm and couldn't think beyond it. In that paradigm, (1) maximum utilization keeps everyone busy, and (2) operating expenses must be minimized. Consequently, clients aren't happy, and the enterprise does not grow.

Constraint Management advocates the revenue-maximization paradigm: (1) if the constraint is internal, maximize only its utilization and subordinate all non-constraints to it, and (2) grow revenue and manage operating expenses accordingly. Consequently, client satisfaction and revenue growth take precedence over operating expense minimization.

Thus, in addition to the limits mentioned previously, paradigms are limits on thinking. Like the other limits, no one can manage a system with a paradigm, but faulty paradigms can limit what an enterprise produces and whether it grows. Indeed, executives cannot set good policies, business rules, and guidelines if they have no way of thinking about something. That's why Constraint Management is a paradigm shift.

The implication of paradigms for the Information field is that virtually all application packages sold commercially and all applications developed internally are based on a paradigm. If that paradigm is deficient and its implementation in software makes it hard to change or replace, software can lock the enterprise into a competitive position that's vulnerable to competition. A change in strategy may need to reveal shortcomings of the prevailing paradigm before changing or replacing software becomes compelling.

Backsliding

Relapse happens when new managers/executives undo strategic initiatives successfully completed by their predecessors. When Constraint Management doesn't match their paradigm, they may conclude that it must be wrong, even if it is working better than what was there before, because conventional wisdom is powerful. If relapse seems possible, redoing the buy-in steps with current managers/executives is advisable.

Conclusion

Constraint Management solutions require the right kind of information, especially Buffer Levels and Buffer Zones. However, Constraint Management also requires that conflicting information be eliminated, such as universal utilization figures. By challenging faulty paradigms, Constraint Management provides sound reasons to abandon misguided rules, guidelines, and policies that have been hardened in software. Moreover, Constraint Management is designed to address not just operational and strategic constraints, but also management attention and paradigms.

The Information field is subject to digital constraints, which have not been a focus of Constraint Management, because Information products and services include software, data, and knowledge. Nevertheless, the Information field can benefit from any of the Constraint Management solutions for production, distribution, and project management because they all can be adapted to the digital world. Indeed, Constraint Management's Focusing Steps, global optimization, conflict resolution, and continuous improvement are all concepts applicable in the Information field.

CHAPTER 19

STRATEGY REDUX

There was a time when strategic planning was an annual event. No more. The average age of companies in Standard & Poor's 500 fell from 60 years in the 1950s to less than 20 years in the 2010s [Sheetz, 2017]. Automation was the leading cause. Events unfolded too quickly for an annual strategy cycle to respond effectively.

With the world awash in excess capacity and firms competing on knowledge, technology, and innovation rather than just on efficiency, competitive advantages are not as durable as they once were. Customers act on good deals. Competitors extract good ideas. Suppliers interact with good partners. Enterprises that innovate leap ahead. Enterprises that cannot react get left behind.

Thus, this chapter examines the relationship between Enterprise Strategy, Constraint Management, and Technical Strategy. Constraint Management provides a way to focus Enterprise Strategy, while the Information field offers ways to align Technical Strategy so that it supports Enterprise Strategy.

Executive Perspectives

Strategy reflects executive concerns and priorities, so when revisiting strategy, it's appropriate to recall these factors from an earlier chapter:

- Chief Executive Officers. Advances in **technology** trigger upheaval.
- Chief Operating Officers. They are more concerned about **digital assets** than physical assets.
- Chief Marketing Officers. Industry **convergence** shifts customer experience to cross-sector products and services from multiple enterprises.
- Chief Financial Officers. **Integrating** information across the enterprise is the biggest challenge.
- Chief Human Resource Officers. **Automation** requires new skills.
- Chief Information Officers. Implementing **vision** for technology is a struggle.

Clearly, Information is seen as both a threat and an opportunity throughout the executive suite. Yet Information Technology and everything needed to make it work well are not easy. For instance, if the product is software, is a minimum viable product better than a fully viable product? The answer may be different for a start-up than for a mature enterprise. The answer will also be different for personal software rather than for enterprise software.

Traditional Strategy

It is said that all Enterprise Strategies boil down to these alternatives [Ovans, 2015]:

- **Growth.** Do something new. For example, devise a Blue Ocean Strategy.
- **Stability.** Build on what you already do. For example, extend into an adjacent market.
- **Renewal.** React to emerging possibilities. For example, act like a start-up even if yours is a going concern.

There are several problems with traditional approaches to strategy [Dettmer, 2003]:

- Planning methods compartmentalize functions and plan them independently. This is a symptom of the system-of-systems problem.
- Planning breeds inflexibility and resistance to change. Expecting change without first achieving buy-in is an uphill battle.
- Enterprises build plans around their current organization instead of adapting. Inertia blocks initiatives at the starting line.
- Implementing strategy is difficult. Planning is hard, but execution is harder.

Nevertheless, approaches to strategy have evolved from new business opportunities funded on a five-year plan to incubators working on a one-year plan to proofs of concept executed in 90 days. Thus, while a strategic plan may span years, nimble enterprises execute rapidly and use the lessons to pivot more often.

The following are a few nuggets from strategy professionals:

- **Awareness.** The problem is not that strategic threats and opportunities go unnoticed. It's that they are not prioritized in time or with adequate resources.
- **Counter-plans.** These cover things you do not want to happen, how you will avoid them if possible, and how you will cope with them if they become unavoidable.
- **Transformation.** The conventional view of transformation is that it's an episode between periods of stability, but continuous transformation is the new normal.
- **Consequences.** The leading cause of problems is solutions because they too often have serious unintended consequences.

As noted earlier, Profit Patterns are strategic changes that shift the business environment and thereby create winners and losers. Two new patterns to consider are:

- **Technology Enables Constraint Management.** On a small scale, the Constraint Management solutions for operations, distribution, and project management can be implemented manually. But for enterprise-scale, software is required.
- **Constraint Management Enables Technology.** Using Constraint Management solutions can improve technology. That includes the Computer Software, Information Technology, and Technical Service industries.

ADVENTURE: **Not Your Typical VC**

Venture capitalists who make headlines invest in start-ups, hoping to cash out at a significant profit when successful start-ups achieve a public offering. But an exit strategy is not the only possible goal. Indeed, strategists may pursue successful transitions from new ventures to going concerns which generate streams of benefits.

Another reason to invest in start-ups is to become their technology partner. Although start-ups can use VC funds to buy technology, the start-up then must manage the technology at the same time it's growing the business. If technology is an enabler rather than the product, having a technology partner offloads that technical responsibility.

Yet another reason for venture capital is to invest internally in a new product line or an entirely new line of business. This super-charged organic-growth approach may be the best alternative when a partnership or an acquisition is impractical or insufficient. Thus, a strategist can work within a large enterprise to foster opportunities that otherwise would be ignored or crushed by conventional wisdom.

These approaches all handle abnormal levels of risk with a mix of due diligence, board membership, executive consulting, portfolio management, and technical support. For me, Technical Strategy was not a single adventure, but an amalgam of adventures in various contexts over a couple of decades.

Planning Is Hard

Strategic Planning teams create business models and business cases. They may include a mix of organic growth, acquisitions or divestitures, and partnerships.

A business model is an explanation of how an enterprise will achieve its goal. If it's a business, how will it make money? If it's a non-profit or government agency, how will it deliver services?

A business case is the financial and operational justification for a business model. If it's a business, how long will the funding last, and can it become a going concern within that horizon? If it's a non-profit or government agency, will its funding enable it to deliver services indefinitely?

My biggest surprise when evaluating start-up proposals was how poorly thought through the business models and business cases were. Although everyone had a concept, few could articulate a coherent business model and reasons to believe it could succeed.

Internal proposals were somewhat better because we provided a standard template with examples to proposal teams and because a POC is much simpler than a start-up. However, assigning a lead VC was critical. Because the lead VC's reputation was on the line, proposals that could not be salvaged were not brought forward for evaluation. On the other hand, the acceptance rate after evaluation was high because the lead VC was committed and the initial round of problems had been solved.

Furthermore, the lead VC stayed engaged through execution, not just funding. That relationship was vital because every POC ran into problems it could not solve on its own.

Execution Is Harder

The Strategy Execution team was formed for several reasons. First, traditional approaches to strategy were too slow: up to five years from initiation to fruition. Second, offerings launched without prior customer involvement oftentimes missed the mark. Finally, some opportunities withered for lack of modest funding because the risk seemed disproportionately high.

Setting aside a pool of funds and assigning VCs to execute proofs of concept solved all those problems. POCs started at 90 days, then extended only if justified. POCs always involved customers, who decided to buy or not buy, so their involvement was a commitment. Finally, the program became self-funding when the income generated matched or exceeded seed funding.

Cities Struggle to Harness Technology

One class of POCs included technology for urban areas. When the U.S. was founded, 95% of its population lived in rural areas. Today, over 80% of its population lives in urban areas. When suburbs are included, that number rises to 84%. Worldwide, a million people a month move to cities. Thus, more of humankind lives in urban areas than rural areas for the first time in history. Urbanization is as much a growth generator in the 21st century as automation was in the 20th century.

This is not meant to dismiss the problems that rural areas and small towns face, such as medical services, retail outlets, and local employment. But solutions to rural and small-town problems differ from solutions to urban and suburban problems. A solution for urban/suburban areas does not necessarily come at the expense of rural areas/small towns.

Cities depend on things outside for survival, such as food, power, water, and waste disposal. Cities also depend on things inside for survival, such as residences, transportation, fire suppression, and policing. All those things benefit from technology—at a cost.

Key performance indicators for city governments are subject to conflicts, such as cost versus response time for city services. And some metrics that make sense in a business context make no sense in a city management context. For instance, utilization of fire trucks and ambulances is the wrong metric. Response time matters more.

A fundamental shift in business and government is that experienced managers cannot know everything, as they did in simpler times. Consequently, business and government have become more instrumented so they can shift to data-driven, fact-based decision-making rather than intuition based on experience.

Decision-makers are awash in data, much of it unstructured. Thus, they need analytics. Real-time dashboards support time-sensitive services such as police, fire, weather, utilities, and traffic. Executive dashboards support routine decisions such as staff deployment, infrastructure maintenance, and departmental budgets.

Lessons Learned

Lessons learned include:

- Great concepts need well-thought-out business models and business cases to justify investment.
- Dedicated Strategy Execution teams can accelerate strategic projects while they increase the win rate.
- City governments need help with technology because they rely on information more than before.

Strategic Initiatives

Strategic initiatives translate vision into projects to fulfill that vision. The flow of strategic initiatives thus includes projects beyond normal operations. Key decisions include which projects to start, freeze, delay, accelerate, redirect, or cancel.

Strategic initiatives can, of course, change where the enterprise constraint is. It may be internal, but it's more likely to be external. When subcontractors or business partners or supply chain members are the constraint, the enterprise may have little control over its own operations. Moving the constraint to be internal brings it under enterprise control.

Strategic initiatives can also change how the enterprise operates. Predict and Prepare relies on forecasts and large inventories or large latent capacities to cope with market fluctuations. In contrast, Sense and Respond can use Buffer Management to cope with market fluctuations by adjusting inventories and capacities on demand. Thus, adoption of Constraint Management can itself be a strategic initiative.

Strategic initiatives increasingly, however, seek to inflict or counter digital disruption. Digital assets dominate some industries, but they exist in every industry nowadays.

Digital Disruption

Digital assets are different because they aren't depleted by use in the same way that physical assets are. An enterprise can sell the same software products to many customers. An enterprise can use the same data, knowledge, or network to deliver services to many customers. However, digital reuse does not prevent obsolescence. Nor does it eliminate the need for updates. When the enterprise cannot keep its digital assets fresh and fully stocked, it may be suffering an internal digital constraint. The pipeline of upcoming software, data, knowledge, or network then is the constraint buffer.

Take digital textbooks, for instance. The publishing industry is well on its way to dematerializing paper textbooks into digital textbooks. That way, the latest edition is available whenever students have connectivity. However, the publishing business model is changing from sales to rentals, and the constraint may be moving from printing presses to digital devices.

Digital disruption happens when a competitor uses digital assets in a manner that threatens another enterprise, if not an entire industry. Sometimes disrupters and the disrupted are in the same industry, such as when a manufacturer adopts Constraint Management. But sometimes disrupters and the disrupted aren't even in the same business. Ride-sharing companies are disrupting taxi companies. Short-term accommodation companies are disrupting hotels. Video streaming companies are disrupting cinemas. In every case, these disrupters use someone else's assets—cars, condominiums, movies—to deliver services.

Although dematerialization encourages digital disruption, technology is more often an enabler than a driver of digital disruption. "Disruption starts with unhappy customers, not technology. New technologies come and go. The ones that stick around are those the consumers choose to adopt. Many fast-growing start-ups have the same technologies as incumbents, but they deliver faster and more accurately what customers want." [Teixeira, 2019]

ADVENTURE: Strategic Guidance

Sales of large, complex service engagements are lengthy and expensive for sellers and for buyers. Therefore, when I was a research manager, we created a digital asset that could be used at any stage in the sales cycle to illustrate (1) how much we wanted to sell our services, and (2) how much the client wanted to buy our services.

Naturally, profitable engagements that would lead to high client satisfaction and follow-on business were most desirable for us. Affordable and timely solutions to each client's specific business and technical problems were most desirable for them.

The stakes were high. An opportunity worth over 100 million dollars could require the seller to invest over one million dollars during the bid-and-proposal stage, which included due diligence and contract negotiation. Thus, our analysis had to cope with sparse data about a small collection of high-value entities.

Where to Invest

The objective was to use this scoring model to decide how to spend our bid-and-proposal budget. Essentially, it was a classification problem: Based on what we knew at the time, was a given sales opportunity most likely to be a Win, a Loss, a client Withdrawal before we bid, or a No Bid decision by us?

We built the model with data from past opportunities where we knew the outcome. Input consisted of answers to a couple dozen straightforward questions, so data collection and updates were easy. Some data was quantitative, but most was qualitative (answered on this scale: very high, high, moderate, low, very low; it was converted to numbers upon input). Half the data described the clients' propensity to buy. The other half described our inclination to sell.

Data was always incomplete, but most sparse during the early stages of the sales cycle, so we designed the model to tolerate missing data while using every bit of available data. This was a significant departure from statistical packages that would delete entire cases or substitute the mean for missing values. This loss of data would have destroyed the classification. Fortunately, as opportunities progressed through the sales cycle, the data became less sparse and classifications improved.

The resulting graph of scores was against Buy and Sell axes, each ranging from –1 to +1. The further right on the Sell axis, the more we wanted to sell. The further up on the Buy axis, the more a client wanted to buy.

By designating quadrant boundaries at zero on both axes, we hoped that scores would separate neatly with Wins in the upper right quadrant. They did not.

The Four-Corners Problem

The business world is replete with quadrant models:

- Strength versus Weakness and Opportunity versus Threat (SWOT)
- Market Growth Rate (Stars and Cash Cows) versus Relative Market Share (Question Marks and Dogs)
- Urgent and Not Urgent versus Important and Not Important

However, items are placed into those quadrants based on judgment, not measurement, so closeness to a quadrant boundary is not defined in those models.

In our uncalibrated quantitative model, the scores clustered around the central four-corners where the quadrant boundaries all met. (See left side of Figure 19-1.) In some instances, corner cases were much closer to cases in adjacent quadrants than to other cases within their own quadrant. For instance, win W_1 is closer to loss L_1 in the lower left quadrant than to win W_2 in the upper right quadrant. Thus, small changes in the input data—well within the margin of error—could shift a score only slightly, yet enough to cross a quadrant boundary. Worse yet, the lower left quadrant (least likely to win) was adjacent to the upper right quadrant (most likely to win) at the four-corners intersection.

Worst of all, Wins and Losses did not fall into separate quadrants as hoped. Thus, the uncalibrated model did not yield useful classification.

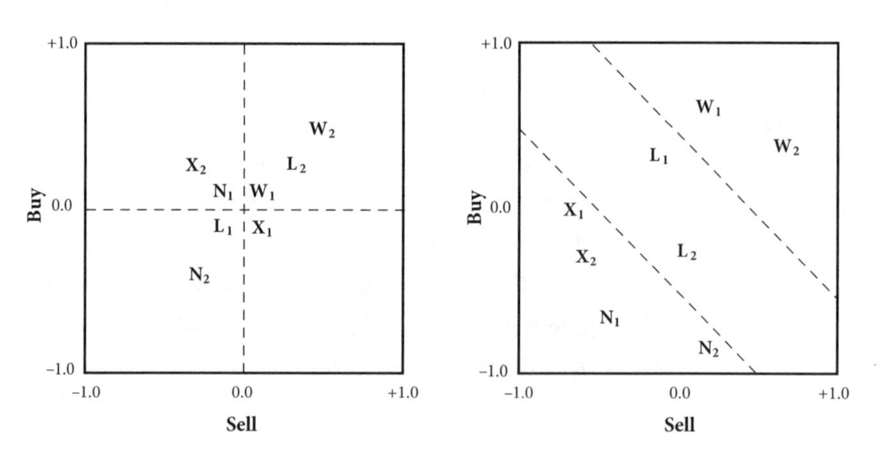

Figure 19-1 Scoring Models.

The Manhattan-Zone Solution

We could not use a standard statistical model like Discriminant Analysis or an Artificial Neural Network because our sample contained too few cases and too many missing values. Therefore, our next step was to invent a weighting algorithm that (1) applied more weight to the variables that better explained past outcomes, and (2) compensated for missing values when predicting future outcomes. This calibrated quantitative model created multiple bands of scores instead of a single cluster of scores.

Our final step was to define zones using Manhattan distances instead of quadrants: d = x + y or Distance = Sell + Buy. (See right side of Figure 19-1.) Each zone was a graph region having the same range of Manhattan distances. For example, all scores where the sum of Buy and Sell was greater than +1.5 fell into the Win zone. All sums between +0.5 and +1.5 fell into the Loss zone. And sums less than +0.5 fill in the No Bid/Withdrawal zone. Visually, zones ran on a 45-degree diagonal from upper left to lower right. A desirable property of zones was when every boundary was between just two zones, unlike the central four-corners of quadrants.

Best of all, the Wins and Losses fell neatly into distinct zones. Thus, the calibrated model did yield reliable classification both for past and future opportunities, even with missing values.

Win-Win Outcome

When we applied the model to future opportunities where we hoped to predict the outcome, this scoring model successfully guided our bids and proposals onto high-odds opportunities. A client wanting to buy as much as the seller wanted to sell was a win-win outcome because interests were aligned.

Just as important, we improved client relations by suspending fruitless discussions. Having early insight into opportunities where enthusiastic sellers were chasing reluctant buyers saved both from lose-lose outcomes.

Lessons Learned

Lessons learned include:

- Useful digital assets sometimes require experimentation and lateral thinking.

- Digital assets can assist business functions that wouldn't ordinarily use digital assets.
- Strategic decisions about high-value entities can be guided by scoring models.
- When the stakes are high, scoring models do not necessarily require a large sample to be useful.

Role of Technology

In the world today, there are no technology-free enterprises. Even small farmers in remote areas of developing countries are using mobile phones to sell their crops. It's more a question of how much reliance there is on technology.

Technology is broader than just the Information field, of course. For instance, it includes machines in manufacturing, robots in warehouses, and autonomous vehicles for deliveries. Nevertheless, Information Technology is a convenient context for understanding the roles that technology may fill and how those roles affect Enterprise and Technical Strategy.

One perspective on technology roles comes from revenue generation:

- **Technology enables the core business.** Does not generate revenue
- **Technology is an adjunct to the core business.** Generates revenue separately
- **Technology is embedded in products and services.** Generates revenue indirectly
- **Technology is the business.** Generates revenue directly

Another perspective on technology roles comes from organizational context:

- **Buyer.** Central versus Shadow Information
- **User.** Front versus Back Office
- **Developer.** Systems of Record versus Insight versus Engagement versus Innovation
- **Operator.** Traditional versus Cloud Computing
- **Seller.** Business to Consumer versus Business to Business

Yet another perspective on roles concerns technology's effect on competitiveness:

- **Advantage.** Leading products and services
- **Parity.** Undifferentiated products and services
- **Disadvantage.** Lagging products and services

Given these diverse roles for technology, it's no surprise that technology questions have been recommended as a prelude to strategy setting. Sets of questions reflect various orientations, as the following lists demonstrate.

Technology Questions from Constraint Management

As noted in an earlier chapter, Constraint Management poses these technology-related questions [Goldratt, 2000]:

1. **What is the power of the technology?** This is the only question about technology itself.
2. **What current constraint or limit does the technology eliminate or vastly reduce?** This question asks obliquely whether the technology is aimed at the constraint.
3. **What policies, norms, measurements, and behaviors are used today to bypass the above constraint or limit?** This question establishes the As Is viewpoint.
4. **What policies, norms, and behaviors should be used once the technology is in place?** This question establishes the To Be viewpoint.
5. **Do the new rules require any change in the way we use the technology?** This question recognizes that the benefits of technology can be blocked by outdated or ill-conceived rules.
6. **How to cause the change?** The answer to this question can be extraordinarily complex, as previous chapters have shown.

These technology questions are a decent starter set because they direct attention to the relationship between technology and constraints. They do not, however, provide much guidance for Technical Strategy.

Technology Questions from Strategy Consultants

Here are technology questions from an Enterprise Strategy perspective [McKinsey, 2013]:

1. How will Information change competition in our industry?
2. What will it take to exceed our customers' expectations in a digital world?
3. Do our business plans reflect the full potential of technology to improve our performance?
4. Is our portfolio of technology investments aligned with opportunities and threats?
5. How will Information improve our operational and strategic agility?
6. Do we have the capabilities required to deliver value from Information?
7. Who is accountable for Information and how do we hold them to account?
8. Are we comfortable with our level of Information risk?
9. Are we making the most of our technology story?

Like the previous set, these technology questions are a decent starter set because they direct attention to the relationship between technology and Enterprise Strategy. Again, however, they do not provide much guidance for Technical Strategy.

Technology Questions from the Information Field

Here are 20 technology questions specific to the Information field:

1. **What technology trends are affecting the enterprise, and the customers or industries it serves?** It probably won't be the only enterprise buying, selling, or developing technology, so look around. Technology may set or comply with industry and company standards.
2. **What risks does technology bring—and is doing nothing an option?** Rapid obsolescence? Technology lock-in? Ransomware? Data breach? Doing nothing is rarely a good option.
3. **How long can the enterprise survive if a technology or a key project fails?** If the answer is "not long," Technical Strategy must be conservative.

If the enterprise can survive such trauma, Technical Strategy can be more aggressive.

4. **Does Technical Strategy align with Enterprise Strategy?** If Enterprise Strategy is growth and Technical Strategy is stability, that's misalignment. If Enterprise Strategy is increasing technical requirements while the budget for technology is headed down, that's misalignment.
5. **Does Technical Strategy cover all strategic horizons?** It's easy to focus on current operations and near-term market opportunities while neglecting future operations and long-term opportunities. Commoditization and value migration are notorious pitfalls for Information Technology.
6. **Does Technical Strategy consider speed?** Positioning buffers carefully can increase speed. But for an enterprise to execute faster than its competitors, its suppliers and customers/clients must keep pace, too.
7. **Does Technical Strategy address technical skills?** Having the right mix of skills requires lead time for recruiting, hiring, training, and deployment. Technology without technical skills is a dead end.
8. **Does Technical Strategy address staffing levels?** Staffing levels are not the same as skill mix and currency. Having too few staff with the right skills can be as bad as having too many staff with the wrong skills.
9. **Does Technical Strategy employ appropriate methodology?** If one method serves projects of all sizes and complexity, some projects are probably riskier than they should be.
10. **Does Technical Strategy employ appropriate metrics?** If no one is measuring Defect Density and Removal Efficiency, quality isn't being managed. The same goes for usability, reliability, and performance metrics.
11. **Does Technical Strategy provide the right tools?** Application Understanding, Data Analytics, and Robotic Process Automation are just examples.
12. **Does Technical Strategy cover repeatable operations?** Matched front- and back-office processes are streamlined. Core processes that create competitive advantage should not be outsourced.
13. **Does Technical Strategy consider Technical Services?** Technical Services typically conform to target service levels that may need to be adjusted occasionally.

14. **Does Technical Strategy recognize Professional Services?** When there are warring factions or business culture is outdated, enterprises sometimes need professional help formulating strategy and managing change.
15. **Does Technical Strategy reflect architecture?** Application Programming Interfaces, Microservices, and Robotic Process Automation are just a few of many technologies with architectural implications.
16. **Does Technical Strategy recognize digital constraints?** Digital assets are more likely to be limits than constraints, but when they are constraints, Buffer Management may work somewhat differently.
17. **Does Technical Strategy address the entire technology portfolio?** Each category in the portfolio may deserve different Technical Strategy:
    - Life cycle stages: Concept, Research, Development, etc.
    - System types: Systems of Record, Insight, Engagement, Innovation
    - Program/project types: large development, small development, maintenance, etc.
    - Enterprise versus Personal Computing
    - Central versus Shadow Information
    - Cloud versus Traditional Computing
    - Modern versus Legacy Systems
    - Fantasy versus Feasible projects
18. **Does Technical Strategy realize where technologies fall in the Hype Cycle: Innovation, Expectations, Disillusionment, Enlightenment, Productivity?** [Gartner, 2019] It's easy to be over-weighted on nascent technologies or on Legacy technologies.
19. **Does Technical Strategy accommodate people-technology role reversal?** Enterprises are shifting from being people-led and technology-assisted to being technology-led and people-assisted.
20. **Where are stakeholders in the Innovation Adoption Framework?** [Rogers, 2003] Pioneers and Early Adopters are key constituencies for technology adoption. On the other hand, saboteurs are anti-change agents.

This set of technology questions is high level, so it is not exhaustive. It's meant to stimulate thinking about diverse elements of Technical Strategy.

ADVENTURE: Spies and Saboteurs

During my first year as an executive consultant, my initial client engagements were notable for high anxiety and low subterfuge. They were examples of misaligned Technical Strategies with no miracle fix.

Mole in a Time Box

The consulting practice was new, so it was not fully staffed or equipped yet. Nevertheless, the sales organization managed to sell a Technical Strategy engagement to a reluctant client. Unfortunately, they decided the scope, schedule, staff, and deliverables without involving the delivery team. The contract was signed before I was assigned.

The reason for client anxiety quickly became apparent. They had deep, long-festering technical problems and had been persuaded to take a leap of faith that we had solutions. So, the client made sure the engagement was in a time box.

The reason for our sales organization's anxiety also quickly became apparent. They had only a superficial understanding of the client's problems and even less understanding of our solutions. Thus, the sales organization offered one of its own as a free resource to assist the delivery team. Of course, every move the delivery team made was immediately reported back to the sales executive, who then fretted to the delivery executive.

As for the client, they were marginally cooperative. But the engagement goals were not achievable even with wrap-around days. No creative Technical Strategy was going to lift them out of the hole they had dug without major investment.

Working in the Dark

When consultants work at client sites, the office space isn't always productivity enhancing. We were once assigned workspace in a file room with no windows. That wasn't a problem, but the lights were controlled by an occupancy switch. After the lights were turned on manually and remained lit for about five minutes, if the switch failed to detect motion because everyone was deep in thought and therefore motionless, it would shut off the lights—and the sensor. Then team members would have to fumble their way to reset the switch in total darkness.

That was an unintentional metaphor for the project itself. The client had no useful data about its software portfolio or its IT projects. As a basis for our Technical Strategy recommendations, we used a combination of automated tools, document examination, and interviews to understand how they got into their current technology predicament; it was a lot like fumbling in the dark.

Turning Things Around 360 Degrees

Another engagement was intended to deliver a Technical Strategy, and it did, despite covert opposition. During my first hour at the client site, two client managers pulled me aside to whisper threats that their personal mission was to make our engagement fail. Consultants rarely get a warm welcome, but threats from strangers is beyond the pale.

I'm still not sure what they expected to achieve, because our work continued without disruption. We recognized that saboteurs emerge in toxic business cultures that have such low morale they foment retaliation. If saboteurs feel their jobs or retirement are threatened, they turn their aggression inward, onto executives, managers, and peers. If saboteurs have company loyalty, they turn their aggression outward, onto customers, suppliers, partners, and consultants.

We mostly worked around the saboteurs until we could propose a viable Technical Strategy. The CIO was pleased enough to hire us to execute key elements of that strategy, which infuriated the saboteurs. They were apparently undeterred, however, because a decade later, long after our Technical Strategy was implemented, the client was spiraling toward bankruptcy.

Lessons Learned

Lessons learned include:

- Technical Strategy can reach a “too far gone” state. That’s one reason average enterprise age is much shorter today.
- Spies emerge when there is a lack of trust. Sometimes the fears are justified.
- Saboteurs are symptoms of a toxic business culture and an unhealthy client–consultant relationship. However, their ability to help or harm an engagement may be negligible.
- Consulting engagements are like a dance: If the client and consultant aren’t in step, someone’s toes will get stepped on.

Strategy Traps

The earlier Strategy chapter mentioned that strategy traps are scenarios that are hard to get out of once you’re in them. Therefore, strategy traps are sometimes called strategy anti-patterns.

Just as Enterprise Strategy traps are best avoided, so are Technical Strategy traps:

- **Dysfunctional Requirements.** Optimizing non-constraints, operating within the noise, forecasting, and local optimization are all Information requirements with unintended consequences.
- **Mega-projects.** Projects well beyond the feasible region based on past accomplishments may have ultra-high benefits, but those projects are ultra-high risk.
- **Horizons Imbalance.** Way too much today, not nearly enough tomorrow. Auto and truck manufacturers are shifting investments from H0/H1 for gasoline/diesel vehicles into H2/H3 for electric/autonomous vehicles.
- **Obsolete Business Models.** When business models are built into Information systems that can’t be changed readily, the business isn’t adaptable or resilient.
- **Obsolete Technical Architecture.** Large, complex systems cannot be rearchitected easily. There is no architect role in Agile and no required

design document, but all Information systems have an architecture, whether designed or accidental.

- **Portfolio Creep.** Whatever is built or bought today will have to be enhanced and maintained tomorrow. Information portfolios get into trouble one day at a time. Legacy Systems consume as much as 80% of some enterprises' Information budgets.
- **Data Mismanagement.** Data can be a digital constraint even if an enterprise isn't in the Information business.
- **Crushing Technical Debt.** When a system is too valuable to retire, too inflexible to meet current requirements, and too expensive to replace, technical debt is a likely culprit.
- **Skill Deficits.** Hiring and retaining skilled staff and managers is difficult when firms offer little security and job seekers come to expect better opportunities elsewhere every couple of years.
- **Industry Benchmarking.** Industry benchmarks are always questionable due to the diversity of enterprises in the sample. Moreover, using benchmarks to justify cost cutting toward the industry average is a race to the bottom. Unfortunately, no industry benchmarks depict a race to the top via Constraint Management. For that, an enterprise must go where no benchmark has gone before.

Because Technical Strategy traps can be terminal, they are reflected in the Information technology questions. Ideally, of course, the questions are asked before an enterprise falls into a strategy trap.

Enterprise Scenarios and Strategies

As illustrated in Figure 19-2, Enterprise Scenarios are an As Is view of the business, while Technical Scenarios do the same for technology. For example, an Enterprise Scenario might include a recent acquisition or merger, and the corresponding Technical Scenario might cover the current state of information system rationalization and data migration. If the enterprise is still running duplicate systems after a reasonable time for consolidation, its Enterprise and Technical Scenarios are misaligned.

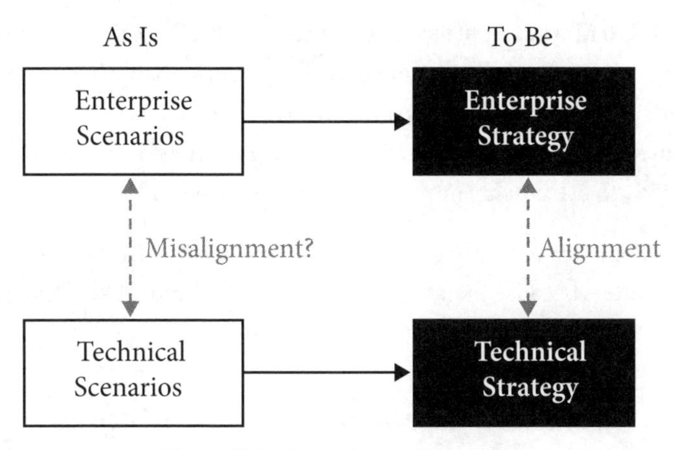

Figure 19-2 Scenarios and Strategies.

Enterprise Strategy and Technical Strategy provide the To Be views of the business and its technology. For example, Enterprise Strategy may plan for future acquisitions or mergers. In that case, Technical Strategy might be to consolidate current systems, plus create tools and procedures to make future work proceed faster. That would bring Technical Strategy into alignment with Enterprise Strategy. Alignment happens in space (products, services, skills) and over time (horizons).

Though far from an exhaustive list, here are some additional examples of Enterprise Strategies:

- **Blue Ocean.** Offering unique products/services can create a barrier to entry if the means of production are hard to replicate.
- **Fail Fast.** Tolerance for risk, plus rapid experimentation, eliminates unsuccessful products/services before huge investments accrue.
- **Premium Product/Service.** Additional functions/features justify higher prices and may capture a larger share of industry profit.
- **Process Innovation.** Changing fundamental operations can reduce investment. For instance, collecting revenue before manufacturing a product creates positive cash flow.
- **Reduce Customer Limitations.** This Constraint Management strategy helps customers manage their own constraints.

- **Speed to Market.** A provider/seller strategy that prioritizes getting products/services to market before competitors may establish an unassailable lead. This includes First-Mover and Fast-Follower variants.
- **Time to Value.** A consumer/buyer strategy that prioritizes putting technology to use quickly minimizes the payback period.

For example, cable/satellite TV providers are not complaining too strenuously about cord cutters because their retained customers are more likely to subscribe to premium programming, which is more profitable.

Technical Scenarios and Strategies

Here is a sample of Technical Strategies and the problems they are meant to solve:

- **Acquisition.** If somebody has already built a product that an enterprise wants to sell, acquiring that product or the enterprise that built it obtains the product and perhaps eliminates a competitor for it and related products.
- **Automation.** When the Enterprise Strategy relies on increased automation, it often requires changes in information systems. However, sometimes automation applies to the Information function itself, such as automating backup and restoration of data.
- **Best of Breed.** Buying separate technical components and integrating them may allow an enterprise to tailor functionality, but at considerable complexity. Middleware and spreadsheets are often used to bridge incompatible components.
- **Build.** When enterprises have unique requirements, they build their own systems. When requirements are common, buying systems is a viable alternative. However, somebody still must build the product.
- **Data Aggregator/Broker.** Some customers prefer to buy data rather than gather and manage it themselves because barriers to doing it in-house can be formidable.
- **De facto Standard.** Achieving enough buzz and market share may lead buyers to select the most popular technical alternatives even when technically superior alternatives exist.

- **Divestiture.** Business units being sold or spun off need their own information systems, and they may have different requirements if the parent's systems are too heavyweight for a smaller organization.
- **Hybrid Cloud.** Born-on-the-Cloud systems are here now. And the easy Legacy Systems have already been migrated to Cloud Computing. It gets progressively harder, so some Legacy Systems will never be moved.
- **Infrastructure as a Service.** Rather than building and operating Information infrastructure, getting it as a service relieves customers of operation and maintenance.
- **Innovation.** In addition to advanced products and services, technical innovations lead to trade secrets and patents, which may create additional competitive advantage.
- **Knowledge.** When an enterprise competes on knowledge, the Information portfolio must extend to Machine Learning, Artificial Intelligence, or Cognitive Computing.
- **Mandatory Changes.** Compliance with laws and regulations is compulsory, and the deadlines are often non-negotiable.
- **Modernization.** If migrating Legacy Systems to a modern platform cannot be justified, improving them in place is an option. It's an opportunity to reduce accumulated technical debt.
- **Optimization.** Achieving high efficiency or high reliability on business functions may require complex optimization models. However, Constraint Management achieves optimization through straightforward methods.
- **Outsource.** When a digital asset or a process is not a competitive differentiator, it may be more economical to pay someone to maintain the asset or perform the process.
- **Planned Obsolescence.** Products are naturally superseded by better technology. This strategy is often disparaged, but the alternative is to have a long tail of aging technologies that require continued maintenance, repair, and technical support. Thus, the cost to keep a technology at some point exceeds the cost of a newer version.
- **Portfolio Migration.** Movement of hardware and software to newer technical platforms is one way to modernize.

- **Pre-integrated Technology.** Vendors may do product integration on behalf of customers, thereby relieving them of that complexity.
- **Security Audit.** The personal and enterprise consequences of data breaches, identity theft, and ransomware are severe, so correction of security vulnerabilities never ends. Sometimes security deserves a dedicated audit.
- **Shadow Information.** When the central Information organization does not meet line-of-business requirements, they are inclined to implement their own technology, which may result in duplication and integration issues.
- **Software as a Service.** Rather than building or buying software, getting it as a service relieves customers of operation and maintenance.
- **Sunset.** Every technology eventually reaches the stage in its life cycle where it's better to remove it than sustain it. This has been more pronounced for computing hardware than software due to Moore's Law, but software has the added advantage of being able to run on older hardware generations, thereby extending its life.

Some of those Technical Strategies, such as Innovation and Optimization, pertain to the enterprise constraint. Many, like Modernization and Portfolio Migration, pertain to Information constraints. But some, such as Mandatory Changes and Security Audit, are necessary just to keep the lights on.

Simultaneously pursuing a mix of Technical Strategies is common within large enterprises for several reasons. First, as shown earlier, the roles that technology can play are diverse. For instance, technology sellers and buyers often employ different strategies. Second, large enterprises have diverse product/service lines with long histories. What works best for one may not be necessary or appropriate for others, even in the same industry. Third, Technical Strategies apply at the enterprise, line-of-business, product, and project levels, and each level has more instances than the one above. Ideally, they would be logically consistent, but the instinct to optimize locally makes that problematic unless governance enforces consistency. Finally, Technical Strategies vary over time. Horizons 2 and 3 may be leading edge, while Horizons 0 and 1 are more prosaic.

ADVENTURE: Friends and Foes

When planning and executing a Technical Strategy engagement, it is standard practice to create an organization chart showing members from the client as well as the consultant and any other participating organizations. For the consultant, however, best practice takes that chart further, into identification of friends and foes.

Bogies to Bandits

A heat map is created by coloring boxes on the chart so that friends appear green, nonpartisans appear white, and foes appear yellow or red, depending on their degree of opposition. It's common to find friends and foes at every level of the client's organization. Though uncommon, it is possible to find foes in the consulting organization too. For instance, someone may disagree with the goal, facts, findings, conclusions, or recommendations.

Heat Maps Reveal Warning Signs

A sea of yellow and red is not a good sign, of course. This might be an adequate reason to decline an engagement or to work harder on buy-in before signing a contract. Other warning signs include:

- **Sponsoring executives are foes.** Client managers and consulting executives may have convinced the client's senior executives to proceed against their instincts.
- **Power mismatches.** Friends may not have sufficient authority to approve necessary changes, such as introducing a new tool.
- **Hidden powers lurk deep in the client organization.** An architecture board or an operations manager with sign-off authority may withhold approvals or resources.
- **Personal priorities out of sync with enterprise priorities.** A manager with strong allegiance (think Agile versus Waterfall) may disagree with the direction of a solution.

- **Someone feels their position is threatened.** Adoption of Constraint Management changes management responsibilities.
- **Trip wires could set off a nonpartisan.** Surprises, such as departmental reorganization or a hiring freeze, may flip a nonpartisan into opposition.

Lessons Learned

Lessons learned include:

- A heat map indicates who is most likely to delay or halt a project, and who is most likely to get it unstuck and on track.
- Use a heat map to guide the buy-in process before, during, and after the engagement.
- Some objections are legitimate, and addressing them makes the project better, so the purpose of a heat map is not to squelch dissent.
- Update the heat map as sentiments change. Foes may become converts. Friends may be fair-weather. Nonpartisans may be unpredictable.

Dynamic Strategy

One of the challenges with Enterprise Strategy is a long time can elapse between when a strategy is formed and when its outcome is known. When a strategic initiative relies on large projects across many departments (legal, finance, marketing, human resources, etc.), a year or more can elapse before useful feedback arrives. For some Enterprise Strategies, such as mergers and acquisitions, there is just no other way. But for other strategies, particularly Technical Strategies, there is an alternative.

Proofs of concept take an experimental approach. As the term is used here, a POC is a joint project between a technology provider and a technology customer. The objective is to demonstrate that the technology works, it solves a business problem, and the customer will implement it. Thus, POCs supplement the traditional “analysis and planning” part of Enterprise Strategy with “execution and learning.”

Unfortunately, POCs can be misguided. One observer of attempts to implement Factory of the Future, or Factory 4.0, cautioned that "Digital Strategy should be the result of a crafted vision and a structured process to achieve it; [Otherwise,] POCs are haphazard uncoordinated bottom-up initiatives that last long without proving much." [Hohmann, 2019] Indeed, the most successful POCs are driven top-down by sponsoring executives on both the customer and provider sides. The key is to remember that the purpose of a proof of concept is to prove something, not just display a technology showcase.

If done well, a POC provides useful feedback within a few days to a few months. That in turn enables Observe, Orient, Decide, and Act—an OODA loop. If a chosen approach isn't working, the POC can rapidly pivot to another; if none of the approaches work, perhaps the strategy itself isn't sound. Overall, dynamic strategy is less vulnerable to unforeseen problems because they can be detected early.

From the technology provider perspective, successful POCs often come after product research and before or during product development. Hence, POCs most often happen during H2 (emerging technology), but could apply during H3 (embryonic technology) if a technology is showing early promise or a customer problem is particularly acute. However, a POC can occur during H1 (current technology) if a different implementation or different customer set would break out of a rut.

ADVENTURE: **Strategy Execution Is an Art**

Over the years, my experience with strategy and execution ran the full extent: project strategy, product strategy, line-of-business strategy, and enterprise strategies for businesses and governments. But the most professional fun I had was as technical strategist in a corporate strategy proof-of-concept program. On average, I led two POCs per year, though some spawned larger projects that lasted more than a year, and I assisted several more POCs per year.

Let me begin by reviewing and affirming some POC characteristics from previous chapters in the context of the entire program:

- The program filled a niche. It fostered worthwhile projects that otherwise would not have been done because they were caught in a conflict. Frequent conflicts included lack of funding, high risk, disparate goals across business units, and customer reluctance to be first adopter.
- The program solved the funding dilemma by establishing an enterprise-level fund, acceptance criteria, and governance procedures. The program looked at long-term revenue potential, not just individual POCs.
- The program offloaded selected risk from business units onto the Enterprise Strategy Execution team, where it could be managed at the program level.
- The program executed rapidly. Funding decisions were made within 48 hours of receiving a qualified proposal. POCs could kick off as soon as approved, sometimes on the same day. Initial results were known within 90 days. Final results were known within a year unless POCs spawned multi-year projects with expanded scope.
- The program had special powers, including dedicated VCs, a methodology for rapid execution, a formal feedback method, an organizational network, executive sponsorship high enough to bridge conflicts across any business units, and a learning culture that meant mistakes weren't repeated.
- The program provided rapid feedback. POCs show what works, what doesn't, and where there are better alternatives.

During the years I was there, the program exceeded its goal by successfully completing more POCs with high client satisfaction than originally envisioned. Although every POC had its challenging moments, the overall program always had reasons to celebrate.

Lessons Learned

Lessons learned include:

- The program succeeded because it was small, nimble, and used unconventional tactics. The senior strategy executive called it his force multiplier.
- Shorter proposals were generally better because fluff was replaced by substance.
- Smaller teams were generally better because fat was replaced by muscle.
- POCs never failed technically, because they weren't science projects. They used known technologies to tackle hard business problems. Some struggled for lack of good data, however.

Conclusion

The military distinguishes a battle plan from battle tactics because a plan only lasts until the battle starts. Or, in the words of boxing champion Mike Tyson, "Everybody has plans until they get hit." While Enterprise Strategy is not that dire, strategic plans are a lot less durable in high-tech industries, such as those in the Information field.

How much do strategic planning and execution matter in business? Among large enterprises, the top 10% captures 80% of the profits, the bottom 10% destroys as much value as the top 10% creates, and the middle 80% is getting squeezed from above and below. However, even leading companies can fall. Over a 20-year period, about half the star companies fell out of the top 10%, some all the way into the bottom 10% [Coy, 2019].

Michael Porter remarked, "The essence of strategy is choosing what not to do." Strategy Traps provide such guidance, both for Enterprise and Technical Strategies. As for what to do instead, Technology Questions from Constraint Management, Enterprise Strategy consultants, and the Information field guide strategic planning and execution.

Planning routes from current Enterprise Scenarios to future Enterprise Strategies is difficult because there are many unknowns, including how customers, suppliers, employees, and competitors will react. Of course, even the best-laid plans are fruitless unless the plans are executed. Inflexible execution may succeed in fulfilling a plan that nevertheless misses the mark because conditions have changed en route.

Given the strong dependence that enterprises have on Information nowadays, Technical Strategies must be aligned with Enterprise Strategies. Otherwise, the enterprise may aspire to a decisive competitive edge that its technology will not permit. Dynamic strategy via a proof-of-concept program is therefore a means to pivot as conditions change.

CHAPTER 20

CONCLUSION

From previous chapters, what conclusions can be drawn from examining Strategy, Information, and Constraints together?

- Constraint Management at scale requires Information.
- The Information field can benefit from Constraint Management.
- Constraint Management focuses Strategy.
- The Information field can better align Technical Strategy with Enterprise Strategy when constraints in both domains are considered.

Here are a few more topics to be wrapped up in this chapter:

- Constraint Management in the Information field
- Capacity Constrained Resources and Bottlenecks
- Technology Rediscovery
- Induced Demand
- Receding Goals
- System of Systems
- Sports Team
- Strategy, Information, and Constraints
- Some Assembly Required
- Exceeding the Goal

Information Constraints

Previous chapters described potential Information constraints. Here is a summary:

- **Hardware.** Like a machine in a factory, Enterprise Computing hardware can be a constraint if it cannot process data in enough volume or with adequate speed during peak demand. Cloud Computing is one way to manage a capacity buffer so that capacity can be expanded and contracted on demand. Then, hardware is more likely to be a constraint for the Cloud vendor than for Cloud customers, but the vendor has the benefit of demand aggregation. Personal Computing hardware is unlikely to be an enterprise constraint because it sits idle much of the time.

- **Software.** Software can be an enterprise constraint if it cannot scale enough to handle peak processing demand. That is, some hardware may be idle while the software is bogged down elsewhere. A different architecture is one way to buffer a software execution constraint. Software can also be a constraint if it is the product and the market has stopped buying the old version in anticipation of a new version.
- **Data.** Structured data can be an enterprise constraint if it is the product or service. However, data may be more perishable than steel and less perishable than fish, and freshness affects buffer levels. The standard buffer management method can be applied to a digital constraint. For instance, streaming television vendors acquire new programs as interest in old ones declines and the buffer level drops into the red zone.
- **Knowledge.** Unstructured knowledge can be a constraint if it is the product or service (e.g., dictionaries, encyclopedias, and Internet searches).
- **Networks.** Communication can be a constraint if an enterprise is in the network business (e.g., telephones, television, Internet).
- **Skills.** Skills can be a constraint in any enterprise. Thus, selected skill groups may be the most logical place for Buffer Management. This is especially true in labor-based services.
- **Projects.** Projects have a constraint in the Critical Chain (time), prioritized backlog (scope), or skills (resources). Furthermore, projects can be the enterprise constraint if the enterprise produces complex products, such as ships. In a multi-project environment, the enterprise may have a skill constraint that limits the number of concurrent projects.
- **Processes.** Manufacturing processes yield physical products. Business processes yield intangible deliverables. Information processes operate hardware and software. The original Constraint Management solution was invented for processes, but its Buffer Management method can be applied to intangible items as well as physical products.

- **Services.** Labor-based services may be constrained by skills. Automated services may be constrained by machines. Although project management works essentially the same way for services as it does for products, process management can be quite different. A services process must adjust capacity on demand when Service-Level Agreements set performance targets, while a product process more often adjusts work in process because capacity is fixed or adjustable in large increments.

The opposite of constraint blindness is constraint fixation. If you think you see constraints everywhere, take a deep breath and focus. They probably aren't all constraints unless the enterprise has leaned itself to the point that just about every resource is capacity constrained. Some are probably limits that cannot be used for Constraint Management, but need to be overcome before a constraint will emerge. A few might be local system constraints, but only one is the enterprise constraint, and it's more likely to be external. So, pick one, establish a buffer, and manage it. If you pick the wrong one, the real constraint will emerge as the chaos subsides.

Constraint Management in the Information Field

The Information field is complex and rapidly evolving. Nevertheless, it is amenable to Constraint Management, which relies on focus rather than all-encompassing solutions. Table 20-1 shows a diverse set of examples.

- **Architecture, Methodology, and Portfolio Management** are strategic capabilities. Although those capabilities can affect the placement of constraints, those capabilities themselves are generally not subject to Constraint Management. An exception might be Portfolio Management, which can use strategic decision-making and Buffer Management on the Information budget. More about that in the Decisions appendix.
- **Hardware Manufacturing, Make to Stock**, can employ the original Constraint Management solution for production. The constraint is a machine or skilled persons, the buffer is work in process, and products are produced on assembly lines or in job shops.
- **Hardware Manufacturing, Make to Order**, can likewise employ the original Constraint Management solution for projects. If the constraint is the

Table 20-1 Constraint Management in the Information Field

Category	Capability/Product/Service	Solution	Type	Constraint	Buffer
Strategy	Architecture			Not applicable	
Strategy	Methodology			Not applicable	
Strategy	Portfolio Management	Delta Analysis	Financial	Budget	Contingency
Manufacturing	Hardware Manufacturing: Make to Stock	Production	Process	Machine/Skills	Work-in-Process
Manufacturing	Hardware Manufacturing: Make to Order	Project	Planned	Critical Chain/ Skills	Time/Skills
Distribution	Hardware as a Product	Distribution	Replenishment	Inventory	Inventory
Development	Software Development: Complete	Project	Planned	Critical Chain/ Skills	Time/Skills
Development	Software Development: Incremental	Project	Agile	Skills	Backlog
Development	Network Development	Project	Planned	Critical Chain/ Skills	Time/Skills
Development	Data/Knowledge Development	Production	Process	Skills	Backlog
Distribution	Software/Data/Knowledge as a Product	Distribution	Replenishment	Inventory	Inventory
Operations	Traditional/Shadow/Network Operations	Production	Process	Hardware/ Software	Capacity
Operations	Infrastructure/Software as a Service	Production	Process	Cloud Computing	Capacity
Services	Professional Services: Management Consulting	Project	Planned	Critical Chain/ Skills	Time/Skills
Services	Scientific Services: Research & Development	Project	Planned	Critical Chain/ Skills	Time/Skills
Services	Technical Services: Help Desk, Repair, etc.	Production	Process	Skills	Backlog

Critical Chain of tasks, the Project Buffer is time. If, on the other hand, the constraint is skills, the buffer is selected Skill Groups.

- **Hardware as a Product** can be distributed using the original Constraint Management solution for replenishment. The constraint and buffer are physical inventory.
- **Software Development of Complete Products/Systems** can be done with the Critical Chain solution for Planned projects. The constraint is the Critical Chain or skills, so the buffer is time or skills, respectively.
- **Software Development with Incremental delivery** can be done with Kanban, an Agile method. The constraint is skills. The buffer is backlog of work.
- **Network Development** can be done with Critical Chain project management. Exterior connections are often done by telecommunication carriers. Internal connections may be done by the enterprise itself or its subcontractors.
- **Data/Knowledge Development** can be done with the production solution. Buffer Management works on digital assets, not just physical inventory.
- **Software/Data/Knowledge products** can be distributed using the Constraint Management solution for replenishment. The constraint and buffer are digital assets.
- **Traditional Information, Shadow Information, and Network Operations** can follow a modified version of the production solution. The original version recognizes a machine or person as the constraint, it buffers work in process, and the result is variable delivery times. Operations in the Information field often must contend with variable demand and completion deadlines. Therefore, the modified production solution varies capacity on demand.
- **Infrastructure/Software as a Service** also uses the modified production solution. However, rather than adjusting hardware/software owned by the enterprise, which may have a limited range of capacity, "as a Service" capacity is adjusted in a Cloud Computing environment specifically designed for a wide range of scalability.

- **Professional, Scientific, and Technical Services** can be managed using solutions for project management, production, and replenishment. Because services cannot be completed in advance, there is no physical inventory in pure service enterprises.

Note that the foregoing are independent scenarios. It's unlikely that an enterprise would have many at once, and it is impossible to have all at once.

For enterprises not in the Information business, Technical Strategy often elevates an Information constraint until a non-Information constraint emerges. On the other hand, for enterprises in the Information business, an Information constraint is likely but not inevitable.

ADVENTURE: **It's Our Policy**

Before going to a local store to make a major purchase, I did research online. The retailer's website offered free shipping on the unit I wanted, but I thought I'd support my local store by purchasing it near my home. I discovered, however, that the retailer's local store would charge a substantial fee to deliver an identical unit weighing 600 pounds (275kg) to my home 10 miles (16km) away.

Blame It on the Computer

When I asked about free shipping, the customer service agent said, "Sorry, I can't change it in the computer." Therefore, because my purchase was not urgent, I instead ordered from the online store, and the company shipped that item 2,000 miles (3,200km).

En route, the unit passed within two miles of my home and within one mile of the local store on its way to a terminal 50 miles (80km) in the opposite direction. The next day it then returned those 50 miles on a different truck but on the same highway, again passing near the store, before it was delivered to my home—without a fee.

Lessons Learned

Lessons learned include:

- This retailer apparently devised independent policies for its online and physical stores. Maybe it had customer segmentation data that supports this, but are online and local customers actually different segments if customers can choose the sales channel? Or maybe the retailer's measurement system drove divergence: Online store revenue was paramount, while physical store cost recovery was paramount.
- When companies set inconsistent policies, customers notice. This retailer's Technical Strategy might have been to limit customer service agent discretion, or it may not have occurred to the software requirements author that a service agent might need to be empowered to make an exception for a big-ticket sale.
- Blaming the computer is not a good tactic when customers have options.

Capacity Constrained Resources and Bottlenecks

When an element of the Information field is not a constraint, it may nevertheless be a capacity constrained resource, which means it has more capacity than the constraint, but not a lot more. Thus, if the constraint breaks down or an unexpected spike in demand arises, the CCR may not have enough spare capacity to cope with the disruption. A CCR that becomes a transient constraint is sometimes called a bottleneck to distinguish it from the persistent constraint.

For example, suppose a Massively Multiplayer Online Game handles thousands of concurrent players per server. Virtual machines and the imaginary worlds they embody are automatically deactivated or activated as players enter and exit. The constraint is external, in the market, because new players are admitted automatically once their payment is verified. However, as the gamer database has expanded, time to retrieve player profiles and initialize virtual worlds has lengthened. On the day that the marketing department offers a coupon for a few hours of free game play to attract new players and elevate the external constraint, the infrastructure that handles network connections, credit checks, and virtual machines begins to falter under the peak load. Time to admit new players becomes intolerable, and some abandon their sessions in frustration. Response time for old players degrades, and they express their dissatisfaction on the customer support portal. Thus, the game infrastructure is a CCR because it was designed to handle typical demand from an external constraint, but on the day the marketing initiative is unleashed, the infrastructure becomes an internal bottleneck because it was not sufficiently scalable.

Information executives must align Technical Strategy with Enterprise Strategy at two levels. First, there's the operational level illustrated in the earlier gaming scenario. Apparently, there was misalignment between the Enterprise Strategy for growing the gamer community and the lack of preparedness of the Technical Strategy for peak demand. Had the executives known that a rogue wave was coming, they might have scheduled extra technicians and reconfigured the infrastructure in advance. Second, there's the strategic level. To anticipate and align Technical Strategy with Enterprise Strategy, the Information executives could have approved projects to reengineer the hardware, software, and network enough to handle higher demand on a sustained basis. Relying on existing scalability to handle an extraordinary demand spike presumed that the infrastructure could adapt to demand well beyond the proven frontier.

There is no shortage of technology advocates. All will gladly tell you about the wonders of their technology. All will claim that it either solves a problem or creates an opportunity. But like a submarine lurking below the surface, Information technologies that typically operate as capacity constrained resources may surface unexpectedly as bottlenecks. That's especially likely when business decisions are made without including Information executives in the discussion.

Technology Rediscovery

Concepts in the Information field often get rediscovered, repackaged, and renamed. Anything touted as revolutionary is more likely a blend of new, borrowed, and rediscovered. That is progress, but it's evolutionary, not revolutionary.

Management and technical innovations touted as the next great thing often fall into disrepute before being eclipsed by something "new." For instance, Object-Oriented Programming was once hailed as the solution to complex code and data.

However, now that time has elapsed, large systems built with that programming model are proving difficult to maintain, despite lofty expectations—a lesson that has been relearned with many Information technologies.

The top reasons Information projects fail is the same list today as it was decades ago. Yet each generation of Information professionals seems destined to rediscover it.

Induced Demand

Make something cheaper or easier and people will naturally consume more of it. Traffic engineers toiled for decades before realizing that adding lanes to existing roads or opening a new road will not alleviate traffic jams because the added capacity attracts even more traffic. While businesses will strive to induce demand for their products and services if each additional unit contributes to revenue and profit, governments more often struggle with induced demand because it generates incremental cost that must be recouped somehow.

Information executives face the same problem, with virtually unlimited demand for a limited supply of Information solutions. The backlog of unfulfilled Information investment requests is often large. Shadow Information enables business units to apply their own discretion, but even that may not clear the entire backlog. It trades one problem (unmet investment requests) for another problem (building and operating critical infrastructure).

One way to cope with excess demand for Information solutions is to stop optimizing non-constraints. Another is to use accounting measures (see the Decisions Appendix) to evaluate alternatives for their effect on the goal, compliance with external mandates, or contribution to strategy. Ultimately, however, it comes down to priority setting.

Receding Goals

In Constraint Management writings, an enterprise goal is typically stated as, "Make money (or produce goal units) now and in the future." This is a directional goal. Business and government goals are usually more specific, however. For instance, a quantitative goal could be to increase profit by 7% every year. That might seem like a practical goal, but due to compounding it amounts to doubling profit in about 10 years.

If the goal is receding, the goal line moves faster than the progress toward it. The notion that a goal can be a moving target is not widely recognized in Constraint Management. Thus, this is a problem that Buffer Management does not solve.

For example, if the goal of a retail business is to grow revenue and profit steadily, but each new store contributes less because the best locations have already been taken, that's a receding goal. Its effects are frequently seen in an enterprise that retrenches after a disappointing expansion strategy. Similarly, if the goal of a city government is to improve services, but the population grows faster than new infrastructure can be built and old infrastructure maintained, that's a receding goal. Its effects are sometimes seen in portable classrooms while school buildings are expanded, or in potholes that remain unrepaired because new ones emerge faster than old ones can be filled.

From a strategy perspective, a receding goal can happen when too much emphasis is placed on H1 (current operations) and not enough on H2/3 (future growth). The worst case is when the goal doesn't just slowly recede; it accelerates away from current capabilities such that the gap gets progressively wider.

In the Information field, the leading edge of technology moves faster than the trailing edge, so the range of technologies in use generally widens. This can create paired receding goals: (1) cutbacks in funding Legacy Systems create technical debt for the current business, and (2) more resources are devoted to Modern Systems, which are nevertheless underfunded relative to the Enterprise Strategy for future business.

ADVENTURE: **Moving the Goal Posts**

Induced demand and receding goals can be two sides of the same coin. Instead of alleviating a constraint, induced demand can further overload it. A receding goal means progress appears slower even though effort hasn't decreased.

Objects Are Further Than They Appear

In research and development, a receding goal is elusive when the closer you get, the harder the next step becomes. For instance, back in my research days, I conducted experiments by providing different types of information to human decision-makers and measuring the effectiveness of their decisions as they managed a simulated factory. Demographic variables included age, education, and experience. Psychological variables included risk-taking and open-mindedness. These attributes of the human subjects were integral to the research design—and a major finding was that some types of information were better for decision-makers with certain psychological profiles.

When the experiments began, every new subject fit into the design somewhere. But as the experiments proceeded, it became increasingly evident that certain combinations of demographic and psychological variables were rare. Thus, more and more subjects had to be recruited, while fewer and fewer subjects actually progressed the studies toward completion.

Growth Curve Flattening

Constraint Management matters more when a goal is receding because the rate of improvement must exceed the rate of growth. However, nothing grows forever. Many entities, natural and human-made, follow an S-shaped growth curve. Scaling up for the steepest part of the growth curve can leave an enterprise with excess capacity as the growth curve levels out at the top. A better Enterprise Strategy may be to hop on a different growth curve (H2 and H3). But governments are often stuck with current circumstances (H0 and H1), and the Information field is tugged in both directions at once.

Lessons Learned

Lessons learned include:

- Directional goals generally do not recede, but quantitative goals sometimes do.

- To achieve a receding goal, the rate of performance improvement must exceed the rate of goal change.

System of Systems

As illustrated in the Hardware, Software, and Data chapters, the Information function in an enterprise is a system of systems. Each local system may pursue a different local goal and have different local constraints, but those local systems overall are meant to support the enterprise goal. It's common for the Information function to contain capacity constrained resources that surface occasionally as transient bottlenecks. Applying Constraint Management principles keeps them subordinate to the enterprise constraint.

Furthermore, a large enterprise overall, not just its Information function, is a system of systems. System boundaries affect the location of local constraints and the direction of local flows. Conflicts between local goals and the enterprise goal should be resolved by subordinating the local system or rethinking the enterprise goal.

Although enterprises may depict themselves as a hierarchy of super-systems and subsystems, most enterprises do not behave in a strictly hierarchical way. For instance, even if Information Systems don't span branches of the organizational hierarchy, their cross-application data flows certainly do.

A matrix organization formalizes cross-system spans, but each dimension of the matrix may be pursuing different goals and managing different buffers. For instance, inventory managers want to minimize inventory, while sales managers want to maximize sales, which requires more than minimal inventory.

Some enterprises organize themselves into multiple lines of business. It might seem that each is effectively its own enterprise, but that's not actually true if there is resource sharing, transfer pricing, risk pooling, or cost recovery.

Regardless of how they are organized, enterprises exhibit emergent behavior which cannot be predicted by examining constituent systems individually because the whole is more than the sum of its parts. Consequently, a change to the local constraint in one system may affect adjoining systems in unexpected ways.

A system of systems can include any combination of manual, semi-automated, and fully automated processes. It's not just computer systems; it's also business units and business functions, along with their projects, policies, and procedures. Moreover, the same element can be a part of more than one system. For instance, one resource manager may supply workers to multiple projects, one project manager may manage more than one project, and one executive can be responsible for more than one business unit.

Constituent systems may need Buffer Management to maintain their status as capacity constrained resources and avoid becoming bottlenecks. When stress takes a system into its red zone and holds it there, a ripple effect on other systems begins. In other words, it's normal for any system to stray into its red zone occasionally. Whether it works its way out in a timely fashion is what matters.

Optimizing a constituent system constraint can suboptimize the enterprise constraint. For example, suppose the enterprise constraint is in make-to-order manufacturing. To fulfill its growth strategy, this enterprise must produce more product, without compromising quality, which the market will readily purchase. However, manufacturing depends on timely and accurate product specifications, which are transmitted by customers in digital files. The enterprise has outgrown its current information system that validates, indexes, converts, stores, and retrieves those specifications. Production is sometimes delayed because the specifications are invalid, index codes must be added manually, storage must be manually allocated, and conversion from the customer's file format and coding scheme introduces anomalies into the files that control manufacturing equipment. Despite this, the Information budget has been cut to the industry average even though the enterprise's market share is well above average due to its reputation for deft handling of custom orders. Unfortunately, that funding is insufficient to fix the current system or purchase a replacement system. Hence, the misguided quest to optimize the Information budget is suboptimizing performance of the enterprise constraint, as well as jeopardizing the enterprise growth strategy.

Intuition of experienced managers, executives, or consultants is often enough to identify the current operations constraint in constituent systems. However, intuition isn't quite as effective at identifying the enterprise constraint because experience is spread across multiple managers, executives, or consultants. Each brings deep insight into at most a few systems, and nobody has a deep understanding of every system.

Furthermore, intuition may not work all that well for selecting future strategic constraints, the places where operations constraints ought to be relocated. Finding the enterprise constraint in a system of systems generally requires investigation instead of intuition. The key question is what overall limits growth of the enterprise? The answer may be different for each horizon.

Although the Focusing Steps—identify, exploit, subordinate, elevate, repeat—were invented for understanding and improving a single system of dependent events, the Focusing Steps can be applied to a system of systems, too. When constituent systems operate semi-autonomously, events can happen in parallel, the sequence isn't fixed, and roadblocks and inspiration can occur anytime. Thus, using the Focusing Steps on a system of systems may require some deep detective work.

An internal constraint restricts the flow of value, so velocity slows and queues form. An external constraint restricts the flow of orders, which can be observed in excess product inventory or lengthening service delivery times. While it's possible to manually trace physical flows of work in process and finished goods, many physical flows are represented in computer systems by corresponding data flows. Tracing the corresponding data flows may be easier, especially if the physical flows span geographic locations. Of course, intangible items, like financial transactions, are themselves data flows, so physical tracing is irrelevant.

Application Understanding tools can uncover data flows, within and between programs, files, databases, and applications. However, those tools typically cannot trace data flows across enterprises in a supply chain because their software portfolios are separate. Hence, simulation models of manufacturing or business processes may assist investigation of constraints within an enterprise or across its supply chain.

Sports Team

When an enterprise's Information function behaves as a chain of dependent events, the original Constraint Management solutions work fine. However, when it consists of interdependent events (for example, the software life cycle) or independent events (for example, technical support), the Information function is less like a factory and more like a sports team because the flow of value is not predefined and many things happen simultaneously.

Players train, practice, and compete together. They obey rules or suffer penalties. They pursue a shared goal. They play within limits. Coaches help them act as a team by following a playbook. Talented players innovate when the playbook isn't working. Many things happen at once. Furthermore, the team has a constellation of support functions, including trainers, groundskeepers, facilities managers, security, physicians, transportation, finance, publicity, statistics, etc.

Where is the constraint in a sports team? There must be one, because teams don't win every game. Where is the buffer? There must be one to enable Buffer Management. What measures matter, if any? Some pundits say intuition beats statistics.

In sports, skills are the constraint. Better players score more and make key defensive plays. Thus, the general manager and scouts recruit the most-skilled players, within the limit of the team budget, and trade or cut the least-skilled players.

Players on the bench or on a farm team are effectively in a skills buffer. While coaches manage the operations constraint (current team roster), the manager and scouts manage the strategic constraint (future team roster).

How players' skills are measured steers decision-making. If every team uses the same measures, teams with the biggest budget will acquire the most-skilled players.

For many years, conventional wisdom in baseball said batting average and runs batted in were the best measures. When statistical analysis showed that on-base and slugging percentages were better predictors of wins, an underperforming team became one of the most successful. They did it by acquiring undervalued players with unconventional skills [Lewis, 2004].

In Constraint Management terms, the general manager executed a strategic initiative to reform the team by elevating the constraint. Despite extensive preparation, execution at game time is when a team wins or loses, and conventional wisdom is the strongest obstacle to buy-in. Therefore, the manager had to overrule the scouts' and coaches' conventional wisdom (subordinate their decisions to the constraint) until the turnaround occurred and buy-in was driven by results.

Though the terminology is different, an enterprise's Information function is like a sports team. The constraint is often skills. The buffer is often skill groups. Measurements steer decisions. Strategic initiatives shift or elevate the constraint. And conventional wisdom is an obstacle to buy-in.

ADVENTURE: **Meta-Learning**

As the author of several lessons-learned documents, I noticed a pattern. Executives ask for lessons-learned documents in good faith. However, the author may spend more hours analyzing events and documenting recommendations than all the readers' hours together. It's as though the documents are intended to be unearthed in a time capsule by some future generation rather than used here and now.

The bigger the failure, the less anyone wants to see lessons in writing because nobody wants to venture there ever again. Those are pariah projects.

The bigger the success, the less anyone wants to read about lessons—even if there's room for improvement or risks that were narrowly averted—because everyone wants to venture there again, regardless. Those are bandwagon projects.

Thus, lessons are mostly read when the endeavor was fairly neutral and will be repeated. Those are mundane projects. Even then, the reading tends to be cursory, and the lessons quickly forgotten.

Easter Eggs

How do I know this? An Easter Egg in a document is a deliberate mistake, an outrageous claim, an implausible conclusion, or just a wry anecdote—all encapsulated with a disclaimer that says, "Call me if you find this." I didn't get many calls.

I was once, however, asked for a deliverable from a Technical Strategy project completed 15 years prior. So, there's that.

The Problem with Pickles

Did you know that pickle consumption correlates highly with diseases, accidents, and crime [Battersby, 2016]? For instance, over 99% of drivers involved in automobile accidents have eaten pickles recently. That's much higher than the percent who drive under the influence of alcohol or drugs. Could pickles be the constraint on human longevity and public safety?

Of course not. Correlation does not mean causality. Pickle consumption correlates with dreaded outcomes because most people like to eat pickles, not because pickles cause those outcomes. Moreover, calculate enough correlation coefficients, and you will find statistical significance just by chance.

Hence, it was surprising that studies of Information investments and company performance showed no correlation during the 1980s. One observer famously quipped, "You can see the computer everywhere but in productivity statistics." However, studies in the 1990s claimed to resolve this productivity paradox by finding significant correlations. This set the stage for the "IT doesn't matter" versus "IT does matter" debate in the 2000s.

Of course, there was no productivity paradox, and if done right, Information does matter. Information investments may not have immediate benefit because in a system of systems the chain of events can be long. Furthermore, some Information investments are effective, and some are not. Some manage constraints, and some optimize non-constraints. Did the studies take these distinctions into account? Nope.

Legerdemain

In fiction, a McGuffin is a goal that a protagonist pursues with little or no explanation. Postmortem documents are sometimes written with an agenda other than learning, such as persuasion: "Based on the success/failure (pick one) of our past strategic initiative, we are launching our new strategic initiative to (fill in the goal), and if we don't do it now, then (fill in a consequence)."

Burying Bad News

I've also read my share of lessons-learned documents, and I have noticed that some authors count on impatient readers not to read past the executive summary. It's possible to bury bad news in the body while extolling minor triumphs in the executive summary. Try reading the conclusions and recommendations first, then see if the executive summary bears them out.

Confidential Sources

As a consultant, I have seen clients slide a document across the table that is clearly labeled confidential. The only safe reaction is to slide it back until its provenance and purpose are established. Or, if it's an attachment or a link, don't open it.

If the client wrote it, they can explain how it's pertinent and what the limits are on its use. In high-tech industries and consulting, intellectual capital is a major asset, and access to confidential documents can muddy those waters.

If the document was written by a third party, does the client have the right to disclose it? The client's assertion that the third party was paid to write it does not mean that the client owns it.

These are situations where it is in everyone's interest to get intellectual property lawyers involved. That is best done before confidential documents are opened because there is no putting that genie back in the bottle.

Lessons Learned

Lessons learned include:

- Lessons-learned documents are too often locked in file drawers or digital archives, rarely, if ever, to be seen again.
- Research into the effect of Information investment on enterprise performance is muddied by (1) confounding effective and ineffective implementations, and (2) failing to recognize that major Information investments take time to bear fruit.
- McGuffins and executive summaries are ways to twist lessons learned.
- Confidential materials containing intellectual capital are a potential minefield.
- Intellectual humility means openness to new information.

Strategy, Information, and Constraints

Figure 20-1 is a conceptual diagram that brings all the elements of this book together:

- **Enterprise Strategy** plans and executes strategic initiatives meant to achieve the enterprise goal. Enterprise initiatives may or may not depend on Information, but often do.
- **Technical Strategy** plans and executes strategic initiatives aligned with the Enterprise Strategy whenever it does depend on technology.
- **Constraint Management** focuses management attention and brings a new paradigm.
- **Operational Constraints** are amenable to Buffer Management.
- **Strategic Constraints** are better places for future operational constraints.
- **Information Elements** (hardware, software, data, etc.) implement strategic initiatives and manage operational constraints—plus capacity constrained resources. Technical Architecture and Portfolio Management affect constraint location.

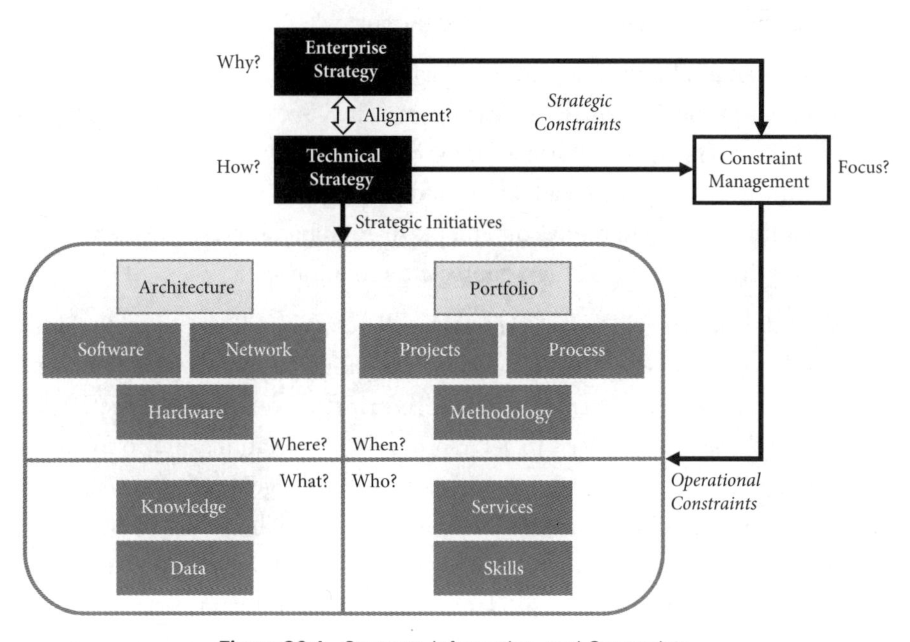

Figure 20-1 Strategy, Information, and Constraints.

Some Assembly Required

Strategy, Information, and Constraints are not topics you often find together in one book. The title of this book, *Exceeding the Goal*, refers to the potential benefits from considering them together instead of as separate subjects. It also refers to making a large enterprise accomplish more by focusing on constraints. When managing large enterprises, or consulting with them on Strategy and Information, leverage points are the places where limited resources have their greatest effect, and Constraint Management is a good way to focus.

Constraint Management works best when the goal is clear, but not as well when the goal is ambiguous or conflicted. Thus, a prerequisite for Constraint Management is getting agreement on the goal. What are we trying to achieve? Are we reaching it? Are we exceeding it?

Constraint Management also works best when the enterprise constraint is clear, and it may or may not be in the Information function. Even if Information is not the enterprise constraint, Buffer Management may nevertheless be prudent so that capacity constrained resources in the Information function do not become bottlenecks.

Constraint Management helps exceed the goal with:

- **Focusing.** Where to improve performance
- **Constraints.** Enterprise versus local systems
- **Buffers.** How to maximize what the constraint produces
- **Limits.** Rules, policies, attention, and paradigms
- **Conflicts.** How to resolve differences that inhibit progress
- **Buy-in.** How to help others appreciate a solution

The Information field has progressed from two major releases a year to minor releases every two weeks. The unit cost of computing has plummeted. That sets aggressive expectations. If an enterprise's Information function has metrics or benchmarks, they can be used to set appropriate expectations and to focus on constraints. However, induced demand and a receding goal can overwhelm an Information function. The Information function can itself be a constraint, though in many enterprises it's used to manage physical or digital constraints elsewhere.

No manual activity can sustain large scale reliably, so enterprises substitute software for labor whenever possible. Firms performing repeatable services are trending toward heavy automation, like manufacturing did decades ago. As the role reversal unfolds from people-led and technology-assisted to technology-led and people-assisted, the solution to system of systems problems is increasingly a matter of software programming rather than training people to perform manual tasks.

Although Central Information can create economy of scale, Shadow Information puts responsibility for the Information function into business units. Unless business decisions are coordinated with the central Information function, the enterprise may act uncoordinated. For example, as this chapter was written, a new building was approved without coordination with the Information function, so that building will have no network access unless the plans are amended. That's an omission best discovered well before construction begins because it is expensive to correct later.

The Information field helps exceed the goal with:

- **Architecture.** Design and implementation of technology that meets requirements
- **Skills.** Technical, business, and managerial
- **Methodology.** Iterative versus Planned methods
- **Portfolio.** Investment mix between Legacy and Modern Systems
- **Services.** Expertise-based versus labor-based versus automated
- **Metrics.** Productivity, quality, and technical debt
- **Frontier.** Track record of what has actually been attained

Information systems have traditionally been built upon the Predict and Prepare paradigm. It says the better the forecast, the better an enterprise can anticipate uncertainties in supply and demand. However, unconventional Information systems enable Constraint Management to implement the Sense and Respond paradigm. It says forecasts are always wrong to some degree, so buffers and rapid replenishment lead to less disruption as demand and supply vary.

Operations are prone to inertia—sticking with what works instead of switching to something that works better. Strategy is a fight against inertia to find that better way. However, strategy has its own form of inertia documented in strategic plans. If plans don't change during execution, an enterprise can easily fall behind on technology. Dynamic Strategy uses proofs of concept to pivot while executing strategy.

Strategy helps exceed the goal with:

- **Enterprise Strategy.** Planning and executing over multiple horizons with varying degrees of risk and reward
- **Technical Strategy.** Balancing cutting-edge technologies that will enable the future business against mainstream and trailing-edge technologies that sustain the current business
- **Alignment.** Removing inconsistencies and gaps between Enterprise Strategies and Technical Strategies
- **Profit Patterns.** Strategic changes that shift the business environment enough to create winners
- **Decisive Competitive Edge.** Strategies durable enough to resist normal competition

Exceeding the Goal

As physical assets have dematerialized into digital assets, Information systems have virtualized. Constraint Management is now confronted with intangible products and services that alter where constraints lie and how buffers should be managed. Although Information systems may not be the enterprise constraint, they can easily be a capacity constrained resource that pops up as an intermittent bottleneck.

Optimizing every local system within a system of systems will not necessarily optimize what an enterprise produces or how robust it is against competitive threats. Indeed, there's ample reason to believe that universal local system optimization cannot optimize an entire enterprise because optimizing every element within a single system generally fails to optimize that system. However, if an enterprise is understood as a system of systems, cross-system interactions can be orchestrated just as within-system interactions are. This orchestration relies on mutual understanding among executives, managers, and workers within local systems, as well as general managers whose chief role is to coordinate across local systems.

Large enterprises are complicated. Finding the enterprise constraint, maximizing its productivity, and subordinating everything else to it is not typical. Using local systems to elevate the enterprise constraint is also not typical. Nevertheless, considering the sports team analogy may help cut through the clutter.

If the enterprise constraint is external—truly in the market, not in sales—subordinating to it means not producing more than the market will buy. What Enterprise Strategy drives local systems to do that? Being demand driven instead of utilization driven would do it. What Technical Strategy is aligned with that Enterprise Strategy? Implementing Information systems that embody Constraint Management make it real.

If the enterprise constraint is external, elevating it means finding ways to get more market share, increase the market size, or sell to a new market. What Enterprise Strategy drives local systems to do that? There are many ways to pursue a decisive competitive edge, including product R&D, market research, and service offerings. What Technical Strategy is aligned with that Enterprise Strategy? Proofs of concept for Information systems can enable innovation and growth instead of inertia. Thriving enterprises do this simultaneously across all strategic horizons.

If the enterprise constraint is internal, subordinating to it means creating only as much work in process as the enterprise constraint can handle effectively. What Enterprise Strategy drives local systems to do that? Adopting Buffer Management of production maximizes delivery of products and services. What Technical Strategy is aligned with that Enterprise Strategy? Prioritizing Information projects that make the enterprise constraint more productive and Information projects that reduce bottlenecks at capacity constrained resources are the place to start. However, subordinating to the enterprise constraint also means rejecting Information projects that optimize local systems to the detriment of the enterprise constraint. It means throttling Information projects that would outstrip the skills buffer and budget buffer.

If the enterprise constraint is internal, elevating it means finding ways to produce more with current assets, acquiring more assets, or improving skills. What Enterprise Strategy drives local systems to do that? A classic example is reactivating latent assets, such as deactivated machines. Another classic example is training and overtime. What Technical Strategy is aligned with that Enterprise Strategy? Moving to Cloud Computing for scalability is one example. Thriving enterprises pursue these strategies aggressively for H1, but simultaneously consider whether different strategies would be better for H2 and H3.

Therefore, how to exceed the goal? Combine Strategy, Information, and Constraints.

Conclusion

This book has not covered every aspect of Strategy, Information, or Constraints. It has, however, covered enough to show how each can benefit the others. In other words, they have synergy: Together they are more than the sum of their parts.

Strategy, Information, and Constraints all strive for simplicity. Designing simplicity is hard. Simplification is harder because you must undo and redo things. Business rules and Information systems never get simpler on their own.

In keeping with the human tradition of storytelling around the campfire, I hope that the adventures recounted here have brought to life points made in the tutorial sections. Furthermore, I hope the lessons learned guide your own adventures and enable you to exceed your goal.

EPILOGUE

ADMINISTRATION, ACADEMIA, AND CONSTRAINT MANAGEMENT LITE

In the Prologue I described how, as a new operations manager, I stumbled into an enterprise constraint without recognizing its significance. Once in academia, I realized that a personal constraint may be such that the only winning move is not to play. When I switched my career into the Information field, I encountered digital constraints without fully appreciating their effect on Technical Strategy.

By the time I began writing this book, however, I had traversed Strategy, Information, and Constraints. This Epilogue describes some adventures off the main path, starting with administration.

ADVENTURE: Administrative Woes

Decisions seem rational to decision-makers in their specific contexts at a moment in time even when they look irrational to persons with a different vantage point. That's not the problem. The problem is when corrections are not made after policies are revealed to decision-makers as irrational and harmful.

Checking the Boxes

When I was promoted into an executive role, I was prohibited from attending new executive training. I never got a credible explanation, but I suspect it was because I was a technical executive (program management, plus engineering responsibility) rather than a business executive (people management, plus profit and loss responsibility). At the time there were many more business executives than technical executives.

Yes, I appealed the decision, because I really was a new executive, but it was denied. Degrees in business and many years of experience apparently didn't matter once I was classified as technical.

Fast forward 10 years and I was told that new executive training was mandatory, and I was noncompliant—tsk, tsk. By then, the number of technical executives had grown considerably, so we were no longer a novelty. Hence, I appealed the later decision, because I really was not a new executive, and was again denied. Therefore, despite being an experienced executive, I sat in an auditorium for three days with brand new executives to hear about policies, procedures, and initiatives that I had been following for a decade.

Administrators got to check the box twice, 10 years apart, for opposing decisions even though neither decision made sense at the time. At least I got to meet the newest executives and share a few war stories.

A Form a Day Keeps the Visitors Away

Meanwhile, terrorism concerns justifiably led to more scrutiny of visitors to division headquarters. Thus, hosts had to complete an online form for every foreign site visitor. However, every day of the visit required a separate online form. This meant that a five-day training course with 20 foreign attendees required 100 forms. No form? No admission. No exception.

Why? There was no budget to enhance the software so that a range of dates could be entered on one form per person per visit—or better yet, one form per event with a list of persons attached. Furthermore, a simple paper form was deemed unacceptable. This online system was a cascade of failure, starting with defective requirements and culminating with indifference to their adverse effect.

That facility closed some years later due to low utilization. I wonder whether the facility administrators ever made the connection.

Insider Outside Looking In

Later, because I lived over 1,000 miles away from my department in corporate headquarters, I worked mostly from my office at home or at client sites, and occasionally from satellite offices. Security policy dictated that my ID badge would unlock the employee entrance to any company building in the country—except headquarters, where my actual department was located.

Once in the HQ building, I could roam freely and borrow an office for the day. However, until cleared by security every day, I was stuck in the lobby, so there was no point arriving early. And I could never use the employee entrance closest to the parking lot.

Lessons Learned

Lessons learned include:

- When there are boxes to be checked, the facts don't matter.
- When there is no software budget, absurd inefficiency is accepted.
- When not assigned an office, your own department can be the least accessible.
- The satirical comic, *Dilbert*, exaggerates uncomfortable truths. For instance, "The ultimate goal of every engineer is to retire without having been blamed for a major catastrophe."

Career progression sometimes takes unexpected turns. My own career took a 15-year detour down the rabbit hole.

ADVENTURE: **Accidental Academic**

My entry into academia was coincidental. If I had arrived on campus one semester earlier or later, circumstances would not have lined up as they did.

Because I had manufacturing experience plus computer experience, I was hired as a teaching assistant and then as a research assistant while I finished my master's degree. As I interviewed for jobs in industry a few months before graduation, my school started a doctoral program in the Information field, and the faculty invited me to stay on and be its first doctoral student.

I had not sought an academic career, but it was an opportunity to go in an entirely different direction. I wrestled with the decision, and eventually made the leap.

Getting In

Because I was first, I had some special consideration. For instance, I was able to study and teach in a traveling scholar program that took me through a rotation at another university with a highly regarded PhD program and research center. When I returned to my home university, I established and managed a behavioral/computing research laboratory. I was awarded graduate fellowships that paid my way during the dissertation stage.

About half of doctoral students drop out without completing their degrees, and the dissertation stage is especially hard because it's entirely unstructured. During the course-work stage, progress toward the goal is steadily measured in credit hours. During the dissertation stage, however, the key question is, "How many pages did you write today?" If the answer is zero, there was effectively no progress toward the goal.

Though relentlessly challenging, I thrived in graduate school. It was the first time in my education when I was able to conduct my own studies rather than follow a prescribed curriculum. My dissertation was the first in my field to win an award.

The job market for new professors in the Information field was hot at the time, so I had multiple offers. On the surface, things were off to a splendid start: I married, relocated, earned a research grant, received high teaching evaluations, and did my duty on committees. However, hours in the day to do research, teaching, and service are the constraint for untenured professors. The constraint for tenured professors is often grant funding.

Getting Out

Work–life balance can be hard to maintain in academia. Privately, I was miserable, so I eventually left and took a job as an R&D manager.

As lucky as I had been in my entry into academia, I was even luckier on my exit. Some former professors find that it's a grueling transition because the academic and practitioner worlds are so different [Bartram, 2019/Wood, 2019], but for me it was more like a homecoming.

A Golden Age

My stint in academia came during its golden age [Shirky, 2019]. Conditions today, however, are sometimes described as a higher-education bubble:

- **Faculty jobs.** When I entered, academia was 60% full-time tenure-track positions. Today it is 80% part-time adjuncts. About 40% of PhDs graduate without any job.
- **Fake degrees.** Half the people claiming a new PhD today have a fake degree, and fake colleges that sell degrees are proliferating [Ezell, 2005].
- **College admission.** Criminal conspiracy influenced admission decisions at some universities.
- **Grade inflation.** Instructors double their odds of favorable teaching evaluations by awarding grades higher than earned [Johnson, 2003].
- **Student debt.** Student loan debt is greater than credit card debt.
- **Data dredging/p-hacking.** Statistics can be misused to create publishable results that are probably false positives [Head, 2015].
- **Fake journals.** Hundreds of thousands of scientific papers have been published without peer review, the standard for credible research [Hern, 2018].
- **Academic relevance.** Practitioners don't read academic journals, but academia doesn't value articles in practitioner journals as much.

Is it fair to call these crises? Maybe not, because they aren't universal. Maybe not, because they have been creeping up for decades. Maybe not, because college enrollment is still high, albeit declining. However, there is no doubt that these are significant issues, and when considered altogether, they're troubling.

Recovering Academic

I've taught thousands of students, practitioners, and faculty. I still use research skills honed in academia to solve business and technical problems. Despite having left a long time ago, however, I'm still a recovering academic.

Lessons Learned

Lessons learned include:

- Personal constraints affect individuals and their families independent from enterprise constraints.
- I was lucky to get into academia—and luckier to get out.
- Being a recovering academic is much better than being a miserable academic.

Constraint Management promotes simplicity, but sometimes does it in complicated ways. It introduces its own concepts, terminology, measurements, and diagrams. Sometimes, however, a different approach is practical.

ADVENTURE: Constraint Management Lite

While shepherding dozens of consultants through complete Constraint Management training, I was pleased that their newfound paradigm enabled them to think differently. But bringing their clients along for the ride was not easy even when clients sat through a multi-day executive briefing by experts. For instance, when a dozen C-level executives from the same company drew diagrams of their core problem, none were the same.

When we explained Theory of Constraints in detail, it took considerable time and effort to get buy-in. On the other hand, when we briefly explained only the relevant principles behind Constraint Management, we quickly and easily got enough buy-in to proceed. Once that happened, knowledge transfer shifted from push to pull. When a solution works, people naturally want to know (1) why it works, and (2) how to get more.

Buy-in Without Baggage

Therefore, when introducing Constraint Management, first in Technical Services and later in Technical Strategy, we took a "lite" approach. That is, we used plain language and straightforward principles to explain what we were doing and why.

Doctors don't try to tell you everything they know. They tell you just enough so that you can follow their directions for medication, bandages, diet, exercise, etc. Constraint Management can work the same way.

Lessons Learned

Lessons learned include:

- Constraint Management promotes simplification, but its body of knowledge is complicated, which discourages newcomers.
- Pilots can fly planes without being aeronautical engineers, and managers can manage constraints without being constraint engineers.
- Less is not more, but lite is enough.

Conclusion

The Goal is a best-selling business novel about Constraint Management [Goldratt, 2014]. Although it's a common reading assignment, academia doesn't cover much beyond it. The complicated body of knowledge doesn't help. Neither does the gulf between practice and academia.

Administration in large enterprises does not accommodate exceptions willingly. Policies may be defended vigorously even when their absurdities are revealed.

Constraint Management Lite is simple explanations of basic principles followed by a successful small-scale implementation. Ideally, that triggers natural curiosity so that the implementation can be scaled up or out.

PART 4

APPENDIX

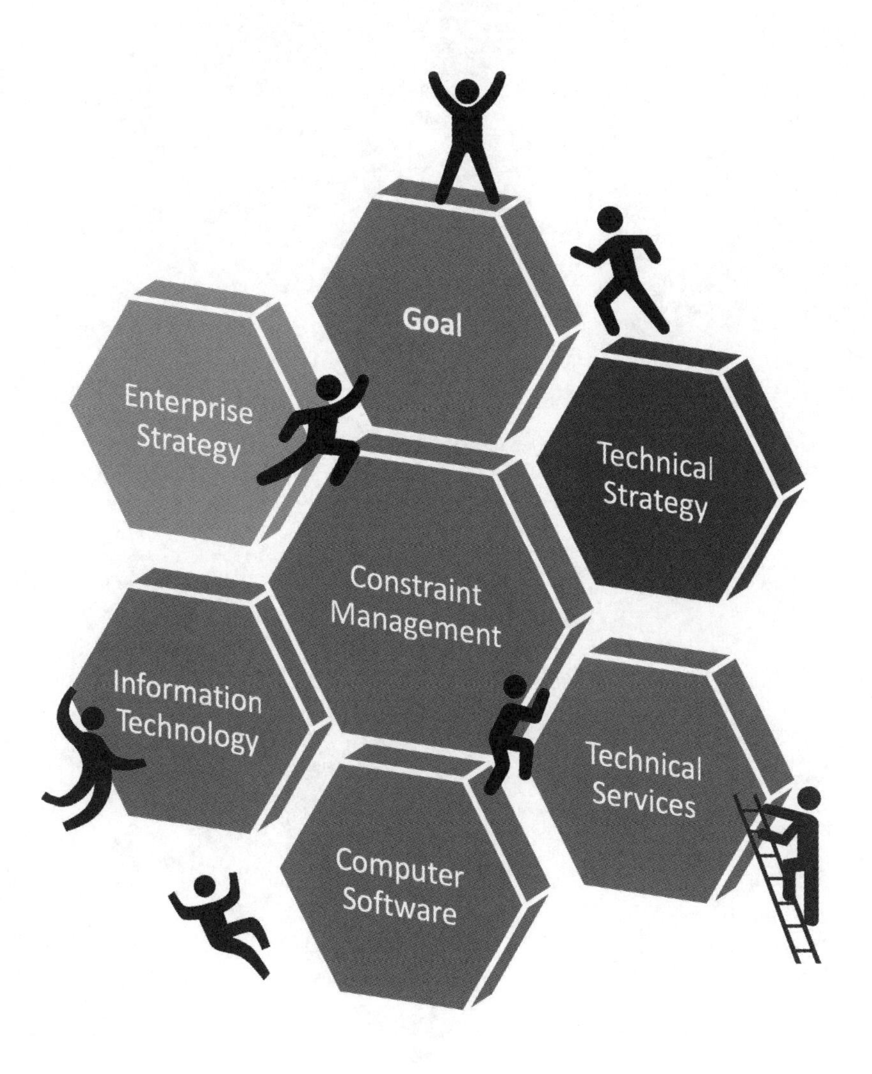

Principles are fundamental propositions that support reasoning and behavior. Principles should be timeless, while strategy and technology are dynamic. Nevertheless, some issues and behaviors are consistent enough for principles to be derived that transcend fads, fashions, and folderol.

In addition, principles derived originally for one domain may apply elsewhere. And so it is with Constraint Management principles that originated in operations. Some can also be relevant to Strategy and the Information field.

These appendices do not describe every principle. They cover the ones pertaining most directly to Strategy, Information, and Constraints. For more information, see the *TOC Dictionary* [Cox, 2012], *TOC Handbook* [Cox, 2010], and *The Online Guide to Theory of Constraints* [Youngman, 2018].

The principles are organized in alphabetical order for ease of reference. You can read them top to bottom, of course. However, if you want to start with the core Constraint Management principles, read Goal, Chain, Constraint, and Win-Win. Or if you want to see the Focusing Steps, read Constraint, Exploitation, Subordination, Elevation, and Improvement.

APPENDIX A

STRATEGY PRINCIPLES

Alignment Principle

Enterprise and Technical Strategy must work together to achieve the enterprise goal, so they cannot be created and executed separately. Alignment means commitments in the Enterprise Strategy should be feasible in the Technical Strategy and whatever gets done in Technical Strategy should enable the Enterprise Strategy.

Blue Ocean Strategy Principle

Blue Ocean Strategy establishes a decisive competitive edge by creating and serving a market in a manner so differentiated that there are no competitors, yet the product or service is so highly valued that customers are willing to pay premium prices.

Red Ocean Strategy (conventional wisdom) says companies are forced to compete on price when they do not have a decisive competitive edge. As they do so, they lose whatever else differentiates them from competitors.

(Contrast with Profit Patterns.)

Dynamic Strategy Principle

Dynamic strategy means the enterprise can pivot rapidly. It's impossible to plan for all contingencies due to combinatorics, so "sense and respond" is a vital supplement to "predict and prepare." If a fixed strategy cycle is too inflexible, make the strategy dynamic with techniques such as the OODA Loop and Proofs of Concept.

Execution Principle

Strong execution can salvage a less-than-perfect strategy, but even a strong strategy may not survive weak execution. Execution can take various forms, including:

- **First mover.** Start-ups often seek competitive advantage by getting to market first.
- **Fast follower.** If there are no barriers to entry, competitors may recognize an emerging opportunity before it's mature.
- **Critical share.** As the number of loyal customers grows, so may their reluctance to switch to another vendor.
- **Networking effect.** Establishing connections between suppliers, distributors, and customers may be hard for competitors to replicate, especially when members would have to adopt different technology.
- **Technical advantage.** Superior technology may be persuasive, but technical superiority is only sustained through concerted effort.
- **Proof of concept.** Customers who take a "show me" stance can be persuaded by a live but small-scale implementation in their own environment.

Horizons Principle

The Horizons Model divides every strategy into time periods:

- **Horizon 3 (H3)** is the seeds of future business.
- **Horizon 2 (H2)** is the emerging stars.
- **Horizon 1 (H1)** is the current business.
- **Horizon 0 (H0)** is the past business.

Strategy must orchestrate all these horizons in a coordinated manner because decisions affecting each horizon will affect subsequent horizons eventually.

Innovation Principle

Innovations include new products and services, new business models, and new procedures. Technical Strategy often depends on technical innovation. Likewise, Enterprise Strategy often depends on recognizing the current Profit Pattern and steering toward the desired future Profit Pattern.

Disrupters are innovators, but incumbents can be innovators, too. The former seek to leap ahead, while the latter seek to stay ahead.

OODA Loop Principle

When Observe, Orient, Decide, Act are performed in a loop for Enterprise or Technical Strategy, the first three steps are strategy creation, and the last is strategy execution.

In a nutshell, OODA exploits a competitor's mistakes. Thus, if your competition can execute their own OODA loop faster than you can, beware.

Profit Patterns Principle

Profit Patterns are strategic changes that shift the business environment and thereby create winners and losers. For instance, technology can shift the strategic landscape so that new entrants have an advantage—or newcomers may be thwarted by the incumbent's advantage.

(Contrast with Blue Ocean Strategy.)

Strategy Traps Principle

Strategy Traps are anti-patterns for strategy because they lock enterprises into unsustainable strategies. For instance, minor investment in business or technical innovations while competitors are making major investments creates risk that they will pull ahead. Or, shipping work offshore because labor is cheaper—without consideration of actual skill levels, service level expectations, and culture clash with customers—is a commodity strategy that reduces differentiation.

APPENDIX B

INFORMATION PRINCIPLES

Agile Principle

When requirements are unknown or unstable, rapidly designing, building, and implementing information assets in small deliverables generates timely feedback.

(Compare to Waterfall principle.)

Commoditization Principle

As information products and services become commodities (i.e., lose their competitive differentiation), value migrates and cost diminishes. Thus, commoditization should be a trigger for strategic change. Enterprises can embrace the commodity path or switch to the differentiation path.

Dematerialization Principle

Physical products are becoming more reliant on software. Thus, even enterprises that don't think of themselves as software companies are facing some of the same challenges as pure software companies.

DevOps Principle

DevOps is a contraction of development and operations. Traditionally, information systems development and operations have been separate responsibilities by separate groups. DevOps puts them together so that operations issues can be addressed quickly and future instances reduced.

Legacy Principle

Effective information systems outlive the technology they were built with/on. Hardware, operating systems, middleware, and software development tools evolve rapidly enough that software applications readily accumulate technical debt.

Longevity Principle

Effective information systems outlive or outlast their creators, sometimes by multiple generations of developers. Succeeding generations of developers may not have the skills or interest to maintain Legacy Systems. In the worst case, those systems fall into an inadvertent stasis because replacement is the only option, yet it's prohibitively expensive.

Metrics Principle

Metrics are the foundation of estimating and benchmarking. They are essential for establishing a baseline that can predict whether a proposed project is feasible. Likewise, metrics are essential for tracking projects and making mid-course corrections.

Polarization Principle

Personal Computing and Enterprise Computing have disparate requirements, so those markets are diverging. For example, a mobile phone strictly for personal use does not require the same level of security as a mobile phone with access to confidential enterprise data. Likewise, a single network firewall may be sufficient for a residence, but layers of firewalls are needed for enterprise data centers.

Role Reversal Principle

Organizations are shifting from being people-led and technology-assisted to being technology-led and people-assisted. This affects investment in both technology and skills.

Technical Debt Principle

Technical debt is decisions you make now but pay for later. For example, insufficient testing will let defects escape into production where they can cause not just an outage but also business disruption. This principle applies to entire portfolios of information assets, not just individual computer programs or complete systems.

Some technical debt is inevitable because technical work and information systems management require tradeoffs and because demand for systems always exceeds what can be supplied affordably. Too much technical debt locks enterprises into products, services, and markets that may not be strategic.

Variety Principle

As the Information field evolves, the leading edge moves faster than the trailing edge. Therefore, the variety of technologies in active use expands unless there is an effort to contain it.

Developers think a lot more about how to build new components than how they will be operated, maintained, enhanced, used, and supported in the future. Under DevOps, however, developers are responsible for whatever they build, so the inability to "throw it over the wall" changes their behavior.

Virtualization Principle

Physical servers, storage, and networks in data centers are being reconfigured into Virtual Machines plus Software Defined Storage and Software Defined Networks in Cloud Computing. This decouples virtual entities from physical hardware, but the hardware must be more capable to run additional software layers.

Waterfall Principle

When requirements are large and complex, carefully designing, building, and implementing information assets in large deliverables ensures that requirements are thoroughly met.

(Compare to Agile principle.)

APPENDIX C

CONSTRAINT PRINCIPLES

Accounting Principle

The effect of a proposed change can be assessed by comparing its anticipated net profit or return on investment to their baselines. This is true for both operational and strategic decisions. However, product cost is a misleading concept banished from Constraint Management.

Aggregation Principle

Fluctuation in demand from multiple independent sources tends to smooth out when aggregated. For example, sales at retail stores will naturally vary from day to day and location to location. If, however, all stores get their inventory from a warehouse, aggregated demand there will be less variable.

Conventional wisdom says inventory should be pushed out to stores because that's where it's sold. This principle (and the Replenishment principle) say holding it centrally reduces inventory overall while not sticking stores with obsolete inventory.

Attention Principle

Management attention is the ultimate limit. Constant firefighting wastes it because managers are paying attention to matters that should be handled without management intervention.

Solutions that minimize distractions (see the Pull principle) allow managers to focus their attention. Instead of managing operations, managers can then think about setting the goal and executing strategy to achieve that goal.

Buffer Principle

A buffer is units that decouple dependent items so that the item being buffered is less likely to suffer disruption. In a manufacturing setting, raw stock is a buffer against supply uncertainty, finished goods are a buffer against demand uncertainty, and the constraint buffer is work-in-process inventory ahead of the constraint. That buffer enables the constraint to keep working steadily for a time if upstream tasks are disrupted.

As just described, some buffers consist of physical units. But buffers can also consist of time units. For instance, a project buffer is time planned at the end to accommodate late completions of tasks earlier in the project.

Buffers are sized to accommodate normal variation. Simple rules of thumb are generally sufficient. For instance, if it takes five days to resupply, and normal demand is 10 units per day, then the buffer could be sized at 50 units.

Constraint Management divides buffers into zones. In the simplest scheme, when the buffer level is in the green zone, no action is needed. In the yellow zone, preparations are made in case the buffer drops further. The red zone means action is required to keep the buffer from being fully consumed. Additional zones may be used to manage extraordinary conditions.

Information Technology began using buffers long before Constraint Management was invented. In hardware, software, and networks, buffers allow separate elements to work at different speeds. For example, because main computer memory is fast and expensive, while data storage is slower and cheaper, a transfer buffer (special memory) between main memory and storage accommodates the speed difference. Likewise, networks are typically slower than main memory, so a buffer accommodates that speed difference, too.

Capacity Principle

Some managers (at least in manufacturing) think they have a capacity shortage, but they often have excess latent capacity. They cannot see the latent capacity because it is hidden in the chaos caused by pursuit of high utilization. Once Constraint Management is implemented, excess work-in-process inventory drains away, waste on the constraint disappears, and the factory can produce more than managers would have predicted.

It is counterintuitive, but unbalanced capacity is more productive because managers know exactly where the constraint is. With balanced capacity, a problem anywhere causes disruption everywhere because there are no buffers between dependent steps. Thus, Constraint Management strives to balance flow rather than capacity.

Chain Principle

Systems are like a chain if they operate as a series of dependent events. That certainly describes manufacturing, but it describes many business processes as well. If a system operates step by step, where each step depends on its predecessors, that's a series of dependent events.

The old saying that "a chain is only as strong as its weakest link" implies that such a system is only as strong as its constraint. Indeed, the constraint not only limits what the system can produce, if the constraint is disrupted, the entire system is affected. Thus, the constraint is the weakest link in the chain, and efforts to improve what the system can produce should start with the constraint.

Change Principle

Every improvement is a change, but not every change is an improvement. It is often said that people resist change. Let's test that. Suppose someone offered you cash to take a different route to work tomorrow. You probably wouldn't resist that change because you can easily see that the benefit exceeds the cost. In other words, buy-in would be effortless.

On the other hand, when managers announce changes, such as different work rules, resistance is inevitable if workers cannot see the benefit exceeding the cost. Yet that is how many managers approach change. The result is nobody wants change, but nobody is happy with the status quo.

Getting buy-in to changes, such as the introduction of Constraint Management, can be fostered by following these steps:

1. Agree on the problem.
2. Agree on the direction of the solution.
3. Agree that the solution solves the problem.

4. Agree that the solution will not lead to significant negative effects.
5. Agree on ways to overcome obstacles.
6. Overcome unverbalized fears.

Competitive Edge Principle

Constraint Management says a competitive edge comes from market offers so good that customers can't refuse them. For manufacturing at least, this edge comes from implementing Constraint Management applications for production, distribution, and project management.

With one or more of those applications in place, the firm can make offers that initially seem too good to be true. For instance, if the standard lead time in the industry is 12 weeks and the firm can reduce that to two weeks and no competitor can match that speed, the firm has a competitive edge in speed of delivery.

Conflict Principle

In business enterprises and government agencies, conflicts are common. In factories, for instance, the desire for high utilization everywhere conflicts with the desire for low inventory because high utilization drives high inventory—and inventory is expensive. If unresolved, conflicts like this lead to oscillation across sides of the conflict. One month the focus is on high utilization. The next, it's on low inventory. Even when conflicts are resolved through compromise, the outcome is rarely optimal: High utilization may be achieved by producing excess inventory, which then remains unsold.

The Constraint Management solution is conflict resolution. By uncovering mistaken assumptions, the conflict is eliminated. For example, the belief that high utilization everywhere maximizes what a factory can produce has been disproven: High utilization only matters on the constraint.

Conflicts arise in Information Technology, too. An enterprise may have both Central IT (owned by the CIO) and Shadow IT (owned by line-of-business executives), but Central IT may be less responsive to the needs of the business, while Shadow IT creates islands of information within the enterprise.

Constraint Principle

In common usage, a constraint is any limit that cannot or should not be exceeded. For instance, the budget is a constraint. The deadline is a constraint. Headcount is a constraint. Office space is a constraint. And the hours in a day is a constraint. Are they all active limits at once? No, but we still think of them as potential constraints.

In mathematics, each constraint is a condition of an optimization problem that the solution must satisfy, and as in common usage, there can be multiple constraints. For instance, an input must be less than a specified value and the solution must be an integer.

In Constraint Management, however, "constraint" has a narrower meaning. The constraint is the one thing that can be managed to govern what an entire system produces. That system could be a factory, a project, a consulting practice, or a business function such as sales. This constraint is also called the weakest link (see the Chain principle) or the bottleneck.

Thus defined, a constraint has these properties:

- An internal constraint is within the system, while an external constraint is outside the system. (See the Internal/External principle.)
- Ideally, the constraint is stable, but if several elements of a system have nearly the same capacity, the constraint can float among them. This is undesirable because floating constraints are harder to plan and less effective at the controlling system.
- In manufacturing and distribution, it's reasonable to assume that all constraints are physical things like machines and people. In other industries, especially services and Information Technology, it's reasonable to assume that some constraints are not physical. For example, the constraint on a project is a set of tasks, which are not physical. Likewise, a computer program is not physical. The medium it's stored on is physical, but the program itself is not.

Decision Principle

Constraint Management decisions are simple. They tell workers and managers when to act and how to act. For example, to manage a buffer, take no action when the level is green, make plans to act when it's yellow, and act to restore the buffer when its level is red. To improve system performance, follow the Focusing Steps. (See Focus principle.)

Though decisions should be simple, that does not mean Constraint Management is simplistic. If it were, a lot more people might be doing it.

Finding and managing the constraint can be done on paper for trivial systems, but practical applications of Constraint Management require computers. This puts the recordkeeping and calculations where they belong—in the computer—while enabling managers to spend less time firefighting.

Demand Principle

Forecasts are usually wrong. They are often very wrong. Lengthening the horizon to allow more lead time makes forecasts less accurate. Thus, Constraint Management in general doesn't use forecasts. It uses "sense and respond."

One exception is irregular service operations. Because labor-based services cannot be buffered, forecasting disruptions to regular service operations enables "predict and plan."

A frequently asked question goes like this: "Aren't buffer zones a forecast?" Nope. Zones are sized based on historical data, as explained in the Buffer principle, but they are not a forecast because they are not used to plan production.

Elevation Principle

The fourth Focusing Step is "Elevate the constraint." This means finding ways to increase the constraint's capacity. For example, in a factory, managers can add another machine to do some of the same work as the constraint, or they can schedule a second shift, or hire more workers.

Exploitation Principle

The second Focusing Step is "Exploit the constraint." If time on the constraint is being wasted, this means improving utilization of the constraint. For example, if the constraint sometimes runs out of work due to chaos upstream, implementing a constraint buffer will keep the constraint productive despite occasional upstream disruptions.

Flow Principle

Constraint Management seeks balanced flow, not balanced capacity. When capacity is balanced throughout a system, a disruption in one place becomes a disruption everywhere. But when capacity is unbalanced—the constraint clearly has less capacity than other elements of the system—only a disruption at the constraint is likely to disrupt the entire system. The constraint buffer minimizes those disruptions.

Focus Principle

Focus is the central idea behind Constraint Management. The Focusing Steps are:

1. Identify the constraint.
2. Exploit the constraint.
3. Subordinate everything else.
4. Elevate the constraint.
5. Repeat these steps.

With these steps in mind:

- Workers should focus on top priority tasks.
- Managers should focus on top decisions.
- Executives should focus on strategy and tactics.

Goal Principle

Constraint Management starts with the system goal. It requires the system being managed to have just one goal, otherwise goal ambiguity and goal conflict arise.

For a business, a typical goal is "Make money now and in the future." For a government, a typical goal is "Benefit citizens now and in the future."

Some practitioners consider goal setting to be step zero in the Focusing Steps because you can't manage a constraint if you can't agree on the goal. In other words, changing the goal may change the constraint.

Holistic Principle

Constraint Management applications are intended to be holistic in the sense that they provide complete solutions for the system being managed. Furthermore, there should be no conflict between the solutions—either internally or externally.

For example, a production solution that works internally should not be to the detriment of other members of the supply chain because they all need to benefit from Constraint Management. A project management solution should improve all projects.

Improvement Principle

The fifth (and final) Focusing Step is "Improve performance by repeating the preceding steps." Don't get stuck at one improvement. Keep the cycle going. Continuously improve. Don't let inertia set in.

Leverage Principle

Unlike conventional wisdom, which attempts to optimize everything in a system and often winds up optimizing nothing, Constraint Management is selective. It says find and manage the constraint because it governs what the system overall produces.

Thus, the constraint is a leverage point in the sense that a modest change there can create substantial changes elsewhere. For instance, using the constraint to set the pace for production typically drains significant work in process out of the system, which not only has financial benefit, it also reduces clutter and confusion on the shop floor.

Necessary conditions can also be leverage points. A necessary condition is something that must happen to achieve the goal. For instance, an improvement in the business rules embodied in an IT system can enable better Constraint Management. In this case, the IT system isn't the constraint, it's a necessary condition and leverage point.

Location Principle

The enterprise constraint can be located inside the system or outside it, but not both. If it's inside, it governs how much the system can produce, which is less than market demand. If it's outside, it governs how much the enterprise can sell, which is more than it can produce.

Note, however, that a sales constraint (for example, too few salespeople) is internal if the enterprise does its own selling, but external if the enterprise sells through dealers or business partners. (See Constraint principle.)

Measurement Principle

Measures drive behavior. Mismeasures drive misbehavior.

For example, measuring holes in the constraint buffer sets valid production priorities, but measuring utilization drives overproduction of unnecessary inventory—and a balanced scorecard often promotes high utilization of non-constraints. Similarly, measuring on-time performance at the task level of projects leads to overestimates, wasted contingency, and late delivery; however, measuring penetration into the project time buffer promotes on-time completion of the entire project.

Multitasking Principle

Conventional wisdom says multitasking increases productivity. Some people are proud of their ability to multitask. Texting while driving is a dangerous example borne out by gruesome statistics.

In tasks requiring concentration, such as computer programming, multitasking hinders productivity. The Alpha-Number-Shape exercise in Chapter 13 shows how single tasking is twice as productive as multitasking because every switch to another task wastes time.

Though it is not helpful to try to do two or more tasks simultaneously, there is one kind of multitasking that is beneficial. When someone on a project is unable to make further progress, they can switch to another task on the same or a different project. They should, however, strive to complete each task before switching to another, and only switch mid-task if blocked in some way.

Noise Principle

Constraint Management cautions against trying to "manage within the noise." That is, normal variation is just noise, so it doesn't pay to optimize within it. Buffers are the Constraint Management way of encapsulating noise so that workers and managers only have to make decisions when conditions fall outside the noise.

Forecasts are based in part on noise, and it's hard to separate noise from the signal. That's why Constraint Management doesn't use forecasts. It uses actuals instead.

Minimalism means a process does not have to be modeled in all its detail—just enough to manage performance. Excess details add complexity and obscures the objective.

Likewise, overly detailed project plans (more than 300 tasks) don't improve on-time performance. It doesn't make sense to spend three hours trying to figure out how to do a two-hour project in one hour.

Optimization Principle

The only way to improve performance of an entire system is to improve the performance of its constraint. This achieves global optimization of the entire system.

Local optimizations to individual non-constrained elements create only the illusion of improvement because the system performance is still limited by the constraint. Attempts to drive high utilization on non-constraints just clogs the system with work in process.

Pull Principle

Conventional wisdom says managers should push work through a system. Frequent or constant expediting is a symptom of this. So are large batches and large volumes of work in process.

Constraint Management instead establishes rules, practices, and measurements that pull work through a system on demand. Expediting is done only on those infrequent occasions when something unusual happens to deplete a buffer that must then be restored. Small batches are common. Lots of work in process is not.

Relay Race Principle

The Relay Race Principle is the basis for the Critical Chain project management method. It says when someone has a task to perform, they should do it as fast as possible, then hand off the deliverables to whoever has the next task to perform. If everyone works as though running a relay race, the project overall is more likely to be completed on time or early.

This principle is designed to counteract the Student Syndrome, which is so named due to students' tendency to wait until the last moment before completing assignments or studying for exams. Common practice on projects is to start a task according to the project plan, but then work on other things, such as non-project work, until there's just enough time left to complete the task on time. This has the unfortunate consequences of (1) squandering whatever contingency was built into the task estimates, and (2) increasing the likelihood that the task will be completed late and possibly make the entire project late.

Replenishment Principle

Because variation in demand is highest at retail shops, the Distribution solution says most inventory should be held at a central location, such as a warehouse, where aggregate demand varies least. Retail shops only need to hold enough inventory to cover sales during the time to resupply, which can be daily in some cases.

This principle can also be applied to services. In that case, the buffer is skilled workers or their productive hours. Each buffer is applied to a skill group consisting of people with similar skills.

Sales Principle

When an organization is not selling everything it can produce, the classic Constraint Management conclusion is the constraint has moved external, in the market, because customers aren't buying. For instance, recession may have dampened demand for the customers' own products and services, thereby reducing its purchasing.

There is, however, a sales side to every purchase transaction. If the organization is not selling enough because it has too few sellers or an ineffective sales process, the constraint is actually internal, in the sales organization. Determining whether the constraint is in the market or in sales is vital because the solutions are different.

Segmentation Principle

Segmentation is a time-tested marketing technique. In many cases, different products are produced for different market segments because they are seen as having incompatible requirements.

Constraint Management argues that it's better to segment the market than the products. In other words, rather than producing separate products for civilian and military customers, or for domestic and foreign customers, it may be sufficient to change the packaging and pricing—and activate or deactivate features—rather than producing entirely different products. This simplifies production and reduces inventory.

Simplicity Principle

Eli Goldratt, founder of Constraint Management, brought a physicist's perspective to his study of organizations. He observed that the universe is inherently simple, and we humans just make things complicated because we don't truly understand cause and effect.

He therefore encouraged Constraint Management practitioners to keep things as simple as possible. The more complex the problem, the simpler the solution should be. Here are some examples:

- Simple rules of thumb for buffer sizing and buffer management are good enough to get started, and may not need adjustment thereafter.
- Some authors have proposed that simulation and moving buffers around yield better project on-time performance, but that's a lot more complicated, and the gains are small.

Strategy and Tactics Principle

In Constraint Management, strategy answers "What?" and tactics answer "How?" A Strategy and Tactics tree shows the sequence of changes necessary to achieve the goal.

For example, if the strategy is to create a decisive competitive advantage via shorter delivery time, the corresponding tactics could include implementing a Constraint Management application for production or project management.

Subordination Principle

The third Focusing Step is "Subordinate everything else to the constraint."

If non-constraints are producing more than the constraint, reduce their production until it matches flow through the constraint. Anything more than that just generates excess work in process.

To go faster, oftentimes it's best to stop some things. Like people crowding exits during an emergency evacuation, processes and projects get jammed up. Thus, choking release of work into a factory or freezing a quarter of projects already started can alleviate the logjam and allow the remaining work to proceed to completion faster.

Supply Chain Principle

Supply chain members get more leverage if they act together than if they act only in their own interest. Creating a win-win-win with customers, suppliers, and business partners makes the entire supply chain more competitive.

This principle came about because there is a strong tendency for members of a supply chain to optimize only their own performance. But in a supply chain, nobody really benefits until the final consumer buys. Rather than tallying sales to the nearest neighbor in a supply chain, Goldratt advocated tallying sales only when the consumer buys. That way, all members of the supply chain have incentive to improve overall performance by reducing inventory and speeding delivery.

Technology Principle

If technology isn't the enterprise constraint, it is nevertheless probably a frequent bottleneck. Given the importance of Information Technology in all organizations, not just technology companies, Constraint Management asks these questions:

- What is the power of the technology?
- What current limitation does the technology eliminate or vastly reduce?

- What policies, norms, measurements, and behaviors are used today to bypass the above limitation?
- What policies, norms, and behaviors should be used once the technology is in place?
- Do the new rules require any change in the way we use the technology?
- How to cause the change?

Thinking Principle

Conventional wisdom is often ineffective because it is based on an incomplete or erroneous understanding of causes and effects. For example, Reorder Point Systems with Economic Order Quantities lead to large inventories and false economy. Likewise, a traditional Material Requirements Planning system with a full Bill of Materials explosion is slower and requires more inventory than Demand-Driven MRP.

Constraint Management includes a Thinking Process that attempts to re-create the procedures that Eli Goldratt went through to create solutions. However, the Thinking Process is controversial because it is not known for finding solutions to major problems. It has been thoroughly covered in other publications, and it's not central to this book, so we won't spend much time on it.

Time Principle

Time lost on the constraint is time lost to the entire system. Every other element of the system has some extra capacity that enables it to stay ahead of or catch up with the constraint. However, if the constraint is internal, it does not have this capability, by definition.

Touch time happens when a machine or a person performs a task and thereby contributes to production of a product or deliverable. Wait time, on the other hand, happens when work in process is suspended until a machine or person becomes available to work on it. Wait time on the constraint is bad, but on non-constraints it's inevitable—unless management is pursuing high utilization everywhere—but utilization is not the right goal.

Rework is anti-work. Rework on the constraint is particularly damaging because it induces even more lost time.

Utilization Principle

Utilization is the percent of capacity used, typically measured for a single machine type or skill category, if not a specific machine or person. Conventional wisdom says utilization is a means to drive cost-saving. But cost-saving is not just lower bounded, once an organization starts down that path as a strategy, it often becomes a race to the bottom and other strategies get locked out.

In Constraint Management, utilization only matters at the constraint because non-constraints cannot be 100% utilized without overloading the constraint—unless non-constraints are using their sprint capacity to refill a buffer after an outage. When the constraint is internal, 100% utilization of that constraint is ideal, but when the constraint is external, even the constraint has less than 100% utilization because demand does not pull more.

Win-Win Principle

Every problem has a win-win solution if cause and effect are understood and assumptions are validated and challenged when appropriate. Rather than assigning blame, Constraint Management uses conflict resolution and the buy-in process to create win-win solutions. (See Conflict Principle.)

Constraint Management is not a zero-sum game. In fact, this principle means that Constraint Management is a positive sum game because non-zero sums include negative sums.

APPENDIX D

STRATEGIC DECISIONS

When making strategic decisions and pursuing strategic initiatives, such as investing in technology, Throughput Economics is a method for comparing alternatives [Schragenheim, 2019]. The baseline for comparison is current circumstances. Any number of alternatives can be explored for their ability to improve performance.

Fundamentals

Here are fundamentals of Throughput Accounting from the Constraint Management chapter:

- **Revenue:** R = money generated by sales, interest, royalties, etc.
- **Totally Variable Costs:** TVC = money spent on materials and parts
- **Throughput:** T = R – TVC
- **Investment:** I = money spent on land, buildings, machinery, inventory, and technology
- **Operating Expense:** OE = money spent turning I into T
- **Net Profit:** NP = T – OE
- **Return on Investment:** ROI = NP/I

Although the fundamentals are consistent across enterprises, the specifics are not. For example, work for external customers generates revenue, but work for internal stakeholders does not. Likewise, products need inventory management, while services need capacity management. Other differences are covered next.

Decision-making

In addition to Throughput Accounting fundamentals, Throughput Economics uses Delta Analysis:

- **Delta Net Profit:** $\Delta NP = NP_P - NP_B$
- **Delta Return on Investment:** $\Delta ROI = ROI_P - ROI_B$

Delta is the Greek symbol used in mathematics to mean "difference" or "change." The P subscript designates the effect of proposed alternatives. The B subscript designates the baseline. If a delta is positive, the proposal is an improvement over the baseline.

Change in NP is attributable to changes in T or OE. Change in ROI is attributable to change in NP or I. If simultaneous alternatives are proposed, their combined effect is analyzed because they may not all move the needle collectively.

Cause and Effect

When an enterprise is in the Information business, standard Delta Analysis applies. However, when the Information function supports the enterprise's core business, standard Delta Analysis can be misleading because the technology itself generates no T and changes in OE attributable to technology may be minor compared to changes in OE of the enterprise. Therefore, we will use prime notation to indicate technology-related figures.

- **Technology Throughput:** T'
- **Technology Investment:** I'
- **Technology Operating Expense:** OE'
- **Technology Net Profit:** $NP' = T' - OE'$
- **Technology Return on Investment:** $ROI' = NP'/I'$

With these quantities, decision-makers can explore the effect of technology decisions on the enterprise.

Information Technology

At a minimum, Information Technology includes hardware, but it may embed software too. Table D-1 shows accounting for IT.

- **Internal IT** creates hardware for internal use, so it generates no revenue.
- **IT product business** generates revenue from hardware and associated warranties.

- **IT service business** delivers operations, maintenance, and repair.

IT service businesses include renting Personal Computers, printers, cable TV set-top boxes, digital video recorders, Internet cable modems, security cameras, and other networked devices known as the Internet of Things.

Table D-1 Accounting for Information Technology

	IT	IT Business	
	Internal	**Product**	**Service**
Revenue (R)	Zero	Cash from hardware + warranty	Cash from hardware + operations + maintenance + repair
Totally Variable Cost (TVC)	Zero	Materials + parts + packaging	Zero
Throughput (T)	Zero	R – TVC	R
Investment (I)	Hardware + software + skills	R&D + manufacturing + inventory + skills	Hardware + parts + software + skills
Operating Expense (OE)	Compensation + training	Compensation + training + SG&A + support	Compensation + training + SG&A + support

Manufactured products typically have Totally Variable Cost, which varies directly with production volume. TVC includes materials, parts, and packaging for each unit. An IT service business also consumes parts during Maintenance and Repair, but they do not vary directly with the number of units serviced, so they are instead counted in Investment.

Investment by internal IT often buys hardware and software—and maybe skills. An IT product business invests in Research and Development, Manufacturing, and physical inventory. An IT service business may buy or build hardware and software.

Operating Expense by internal IT includes compensation and training for employees. Compensation includes wages, salaries, incentive pay, insurance, and retirement funding. Thus, it's more than just direct labor. IT product or services businesses also spend on Sales, General, and Administrative expenses plus Technical Support.

Throughput Accounting keeps TVC separate rather than allocating it because product cost is a misleading concept. As illustrated in the Prologue, it leads to misguided strategic decisions about which products to manufacture and in what volumes.

Computer Software

Accounting for Computer Software varies depending on whether the activities are limited to just software engineering or expanded to encompass an entire software business. See Table D-2. Although all the variants produce software deliverables, none of them generate Totally Variable Cost. Therefore, Throughput equals Revenue.

Table D-2 Accounting for Computer Software

	Software Engineering		Software Business	
	Internal	**External**	**Product**	**Service**
Throughput (T)	Zero	Zero	Cash from deliverables	Cash from deliverables + infrastructure
Investment (I)	Ideas + tools	Ideas + tools	Ideas + tools + inventory	Ideas + tools
Operating Expense (OE)	Compensation	Compensation	Compensation + training + SG&A + support	Compensation + training + SG&A + support + infrastructure

- **Internal software engineering** happens when enterprises hire their own software engineers. Because the software deliverables (source code, documentation, test cases) are used only internally, they generate no Throughput.
- **External software engineering**, such as independent contractors, writes, documents, and tests code, but the deliverables are not for sale because it is custom code.
- **Software product business** generates Throughput from software deliverables because they are for sale. The customer runs packaged software on its own or a third party's infrastructure, such as Cloud Computing. Packages may be configured and customized by the customer or the vendor or a third party.

- **Software service business** generates Throughput from software deliverables plus infrastructure to run the software on behalf of clients. This is Software as a Service. It may be strictly standardized across clients or configured uniquely for each client.

Throughput from a software business may be derived from licenses (renewed periodically) or subscriptions (charged for usage). Technical Services contracts, described below, often accompany enterprise software agreements.

All software investment includes ideas that are turned into code and software tools for writing, documenting, and testing the code. If a product business delivers shrink-wrapped software, that business manages inventory consisting of storage media, printed documentation, and packaging. A software inventory therefore consists of multiple units of identical software packaged for sale. This is distinct from a software portfolio which consists of unique software components or products.

Operating Expense includes compensation. Software businesses also spend on training, SG&A, support, and possibly infrastructure.

Technical Services

Technical Services are a subset of the Professional, Scientific, and Technical Services sector, which relies on expertise, not just labor. Professional Services include management consulting. Scientific Services include research and development. Technical Services include installation, configuration, maintenance, repair, training, and help desk. Although these services are distinct, accounting for them is consistent. See Table D-3.

- **Internal Technical Services** do not generate Throughput.
- **External Technical Services** do generate Throughput from engagements.

Clients may sign a Technical Services contract when buying IT devices. Or they may acquire services from another provider. Or they may do services themselves.

Technical Services require investment in skills, tools, and parts. An external Technical Services provider may invest in its own facilities, as well as its clients' facilities.

Table D-3 Accounting for Technical Services

	Technical Services	
	Internal	**External**
Throughput (T)	Zero	Cash from engagements
Investment (I)	Skills + tools + parts	Skills + tools + parts + facilities
Operating Expense (OE)	Compensation + training	Compensation + training + SG&A

Operating Expense includes compensation and training. Technical Services businesses also spend on SG&A.

Strategic Decision Scenarios

When applied to Manufacturing and Distribution, strategic decision-making considers everything loaded on the constraint. For instance, if the constraint is currently 90% loaded with work on current products, it wouldn't make sense to overload it to 120% by introducing new products. However, adjusting the product mix to stay within current capacity or increasing the constraint's capacity might be sensible initiatives.

With fundamentals out of the way, the following scenarios illustrate how decisions shape Technical Strategy. Note that decisions which improve NP or ROI have their basis in adjustments to T, I, and OE. Also, note that strategic decisions may affect different horizons.

Moving to a Strategic Constraint

A common strategic decision involves a transition from the current constraint to a new constraint that is better positioned. For instance, in a manufacturing plant, if Machining is the current constraint but a convergence point such as Heat Treat would be a better place to regulate overall production, buying more machines (increasing I) and/or hiring more machinists (increasing OE) could increase capacity until Heat Treat becomes the new constraint and an appropriate buffer can be managed there.

A similar scenario from Technical Strategy could involve shifting the constraint from data entry to data processing. That is, if month-end reports are chronically late because processing awaits data, automating data acquisition (increasing I) and/or hiring more data entry staff (increasing OE) could increase capacity until processing becomes the new constraint and an appropriate buffer can be managed there.

To justify these decisions with Constraint Management calculations:

- ΔNP would need to rise because T exceeds OE
- ΔROI would need to rise because NP more than offsets additional I

Although this justification might not be immediate in H1, repositioning the strategic constraint could prepare the enterprise for future growth in H2.

Strategic initiatives—investment, pricing, staffing—may change the location of the constraint even if that's not their intent. Therefore, when making strategic decisions, strategists should not just assess capacity of constraint, but should assess all capacity constrained resources so that they don't become bottlenecks unexpectedly. If the constraint is loaded to near 100% of its capacity, market fluctuations will float the constraint between internal and external, and that may disappoint some customers.

Relaxing a Capacity Constrained Resource

Another common strategic decision is to establish and maintain enough spare capacity on capacity constrained resources so that they are less likely to become temporary bottlenecks. Recall the Massively Multiplayer Online Game example from the Conclusions chapter.

For the current business, H1, if marketing executives had discussed their coupon promotion with technical executives, they could have prepared for a temporary demand spike, which would have increased OE. What happened instead was T declined due to insufficient capacity on the constraint, which in turn decreased NP.

For the future business, if the coupon promotion had attracted and retained new game players, technical executives could have prepared for a sustained increase in demand, which would have increased I during H1 in anticipation of somewhat higher OE and much higher T during H2. The net result could have been improvement in both NP and ROI.

Reducing Technical Debt

Technical debt is the price paid later for something done (or not done) today. It's a chronic problem that can afflict any Information system, but it is especially common in Legacy Systems because (1) their own requirements have evolved over time, (2) they may be hampered by outdated underlying technology, and (3) incompatibility with newer systems limits overall functionality.

To their credit, software engineers often address some technical debt on their own initiative during normal maintenance and enhancement activities. For instance, they often rewrite code that is particularly error-prone or difficult to modify.

This honor system may not resolve the biggest technical debt, however. Defect metrics based on discovered defects, Maintainability metrics based on coding flaws, and Testability metrics based on code complexity can all highlight problem areas. Those metrics are especially useful when the technical debt spans multiple systems.

If the scope of technical debt is too large to resolve with an incremental approach, a new project may be required to implement a different technical architecture. Or that objective may be added to the scope of a project whose mission is to implement new functionality. Or it may be time to replace the Legacy System.

Table D-4 is a hypothetical scenario that compares those alternatives. Investing in technology by reducing technical debt can reduce annual OE' for the Legacy System, but it's a small fraction of the enterprise OE. However, if the system is replaced, enterprise T will grow because the new system supports new products, but the vendor's annual maintenance is 10% of the initial license price. Some investment alternatives result in negative ΔROI now, but all yield positive ΔROI in the future.

Hence, the key questions are: (1) how much can the enterprise invest in this technology now, and (2) would other investments yield even higher ΔROI in the future? Given the figures in the table, a dedicated project appears best, but business stakeholders may remain skeptical because that project would only fix technical debt, not add new functionality. Stakeholders may view it as an IT problem, rather than a business benefit.

Table D-4 Technical Debt

	Technology			Enterprise		
	T'	I'	OE'	T	I	OE
Do nothing	$0	$0	$100,000	$10,000,000	$5,000,000	$8,900,000
Project add-on	$0	$100,000	$80,000	$10,000,000	$5,000,000	$8,900,000
Dedicated project	$0	$250,000	$60,000	$10,000,000	$5,000,000	$8,900,000
Replacement system	$0	$750,000	$95,000	$11,000,000	$5,500,000	$9,790,000
			Year 1 (Now)			
	NP	NP	ROI	ΔNP	ΔROI	
Do nothing	$1,000,000	10.0%	20.0%	$0	0.0%	
Project add-on	$1,020,000	10.2%	20.0%	$20,000	0.0%	
Dedicated project	$1,040,000	10.4%	19.8%	$40,000	-0.2%	
Replacement system	$1,115,000	10.1%	17.8%	$115,000	-2.2%	
			Year 2+ (Future)			
	NP	NP	ROI	ΔNP	ΔROI	
Do nothing	$1,000,000	10.0%	20.0%	$0	0.0%	
Project add-on	$1,020,000	10.2%	20.4%	$20,000	0.4%	
Dedicated project	$1,040,000	10.4%	20.8%	$40,000	0.8%	
Replacement system	$1,115,000	10.1%	20.3%	$115,000	0.3%	

For simplicity, this example uses point estimates. If they were replaced by best-case, worst-case, and most-likely values, decision-makers would have more insight into risk. On technology projects, it's common for scope creep to drive up investment while schedule slip delays payback, so thorough analysis should consider both quantities and timing. In this example, the investment generates a multi-year stream of benefits that are not captured in a net present value.

This example also ignores non-quantifiable factors, such as the software vendor's capability. Is its software package really better, or does it create new problems while solving old ones? Will error correction impede data migration? Will there be training issues as users learn the new system? And so forth.

Portfolio Management

The Information function in enterprises includes myriad hardware, software, data, and other assets. To understand some of the complexities behind Portfolio

Management, consider Table D-5. This portfolio represents an enterprise not in the Information business, but one where the Information function is a critical enabler of its core business.

Table D-5 Portfolio Management

ID	Life Cycle Stage	System of...	Method	Platform	Horizon	Risk	Priority	Status
Ops1	Maintenance	Record	Planned	Traditional	H1	H	VH	Active
Ops2	Maintenance	Record	Planned	Traditional	H1	M	M	Active
Dis1	Development	Record	Planned	Traditional	H1	H	L	Freeze
Fin1	Maintenance	Engagement	Agile	Cloud	H1	M	NA	Approve
Mkt1	Reengineering	Engagement	DevOps	Hybrid	H1	L	H	Expedite
CIO1	Maintenance	Record	Planned	IoT	H1	VL	L	Active
CIO2	Maintenance	Record	Planned	Hybrid	H1	L	NA	Active
CIO3	Development	Engagement	Agile	Cloud	H2	L	H	Throttle
CIO4	End of Life	Record	Planned	Traditional	H0	VL	VL	Terminate
CTO1	Concept	Insight	Agile	AI/ML	H2	L	VL	Active
CTO2	Concept	Insight	Agile	Cloud	H3	M	M	Approve
CTO3	Research	Innovation	Agile	AI/ML	H3	M	H	Throttle

	Technology			Enterprise					
ID	**T'**	**I'**	**OE'**	**T**	**I**	**OE**	**NP**	**NP**	**ROI**
Ops1		$100,000	$50,000	$6,000,000	$3,000,000	$5,280,000	$670,000	11%	22%
Ops2		$30,000	$100,000	$1,000,000	$500,000	$880,000	$20,000	2%	4%
Dis1		$1,000,000		$1,250,000	$625,000	$1,100,000	$150,000	12%	9%
Fin1		$100,000	$50,000		$0	$0	–$50,000		–50%
Mkt1		$200,000	$150,000		$0	$0	–$150,000		–75%
CIO1		$25,000	$10,000		$0	$0	–$10,000		–40%
CIO2		$75,000	$25,000		$0	$0	–$25,000		–33%
CIO3		$800,000			$0	$0	$0		0%
CIO4			$30,000		$0	$0	–$30,000		
CTO1	$300,000	$150,000			$0	$0	$300,000	100%	200%
CTO2	$100,000	$100,000			$0	$0	$100,000	100%	100%
CTO3		$80,000			$0	$0	$0		0%
	--------	--------	--------	--------	--------	--------	--------	-----	-----
	$400,000	$2,660,000	$415,000	$8,250,000	$4,125,000	$7,260,000	$975,000	11%	14%

The table shows a dozen hypothetical application projects, each representing a different strategy problem. However, large enterprises have hundreds if not thousands of such projects. Therefore, in the aggregate, there is no easy calculation that leads to a globally optimal solution. Judgment is required.

Columns describe attributes of each project:

- **Identifier (ID).** Unique designation
- **Life Cycle Stage.** Relative maturity
- **System of....** Application type
- **Method.** Techniques for building, maintaining, and operating
- **Platform.** Underlying technology
- **Horizon.** Timeframe
- **Risk.** Likelihood that business will be disrupted
- **Priority.** Importance assigned by application owners
- **Status.** Current condition
- **T', I', OE'** and **T, I, OE, NP, ROI.** As defined above
  - Building hardware is I'. Running hardware is OE'.
  - Writing software is I'. Executing software is OE'.
  - Collecting data is I'. Using data is OE'.
  - Constructing networks is I'. Operating networks is OE'.

Ops1 (manufacturing) is a mature system enabling the business operations that generate the most T. It handles routing of work in process, plus Buffer Management. When Ops1 is down, the business is down, so it's high risk and very high priority. I' and OE' are modest compared to I and OE, so ROI is high, and continuing to fund system maintenance and operation is not controversial.

Ops2 (warehousing) is another mature system supporting business operations, though its T generator is smaller. Its risk is moderate because the business can tolerate some downtime. But because its ROI is small, this application is vulnerable to pressure to reduce I' and OE'. If OE' were halved, ROI would more than double. But reducing OE' would require more I' to increase efficiency.

Dis1 (distribution) is a proposed system that would enable a new T generator. However, I' for system development would be substantial, and future OE' is currently unknown. This proposal is currently frozen while the investigation and debates continue. A bank loan or other funding would be required, but it has been in the backlog since last year, so unless approved this year, it could automatically age out of the portfolio.

Fin1 (finance) is another mature system, but one that enables the business without a direct connection to a T generator. For instance, it might handle time and attendance, plus payroll. Its maintenance is mandatory due to union contracts and government regulations. Thus, continued funding is approved even though its ROI is negative.

Mkt1 (marketing) is a Legacy System slated for reengineering to transition from its original Traditional Computing platform to a Hybrid Cloud platform wherein part of the application runs on Cloud and part remains Traditional. Reengineering will address the backlog of Functional Requirements while reducing OE', which is highest of any current application. Thus, this project is being expedited.

CIO1 (Internet of Things) is another maintenance project. Risk, I', and OE' are all very low, so this system has been routinely funded for years. It's operated and managed by the CIO organization, and generally seen by other executives as just a cost of doing business because its sensor data is used for safety and security, as well as production.

CIO2 is yet another maintenance project, this one on a Hybrid Cloud to enable third-party access. It is mandatory under terms settling a lawsuit, so it is routinely funded with few questions asked.

CIO3 is a proposed project that would build Private Cloud infrastructure to support future application development and operations. It is therefore of keen interest to the CIO team because it would make applications more scalable. However, it is currently throttled, meaning it cannot start until its skills constraint is alleviated. Meanwhile, business process owners have expressed skepticism about its business benefit, fearing that it may be a technology boondoggle.

CIO4 is a Legacy System that has reached its end of life. The Functional Requirements it met have either become obsolete or they have been satisfied by user-developed software—mainly spreadsheets. Therefore, this application will be terminated as soon as data migration is complete.

CTO1 is a Proof of Concept for using Artificial Intelligence to construct a System of Insight. Its I' is higher than typical of POCs, but if successfully implemented as a full-blown system, it could become a new T' generator soon. Work recently began, so the team is still wrestling with unfamiliar technology.

CTO2 is another POC, this one for a Cloud-based System of Insight. If successful, it could streamline manufacturing, warehousing, and distribution. What's more, it could become a new T' generator. The POC is approved, but not yet active. It is intended to demonstrate that the concept is viable before a full-blown development project is approved.

CTO3 is a research project using Machine Learning to construct a System of Innovation. That innovation could radically reshape competition in this enterprise's industry. But there are no guarantees that the technology will work or that customers would buy it. Furthermore, the project is throttled until some developers from CTO1 and CIO3 become available. Their combined experience is vital, and the necessary skills cannot be hired.

Overall, the enterprise has 2.7 million dollars to invest and 400,000 dollars to operate projects that are Active, Expedited, or Approved. The rest are Frozen, Throttled, or Terminated. I' of Throttled projects will be spread over more than one strategic planning cycle unless more budget for the Information function is approved.

Here are some additional takeaways from this illustration:

- The projects span all strategic horizons. However, in an actual enterprise, Maintenance projects and application operation for H1 could consume the majority of the Information budget.
- Research, Concept, and Development projects have inherent uncertainty even if their business-interruption risk is low.
- Frozen and Throttled projects may be beyond the frontier of successful past projects, so the final decision should be made on capability, not just financial measures.
- Technical Strategy decisions are hard to justify with ΔNP and ΔROI when I' and OE' are not directly linked to T' or a T generator, but some technologies are necessary conditions for a viable enterprise.
- Proofs of Concept with T' potential suggest that this enterprise is pursuing a digital services strategy to complement its core Manufacturing and Distribution business.

- Calculating ROI using $\Delta OE'$ instead of ΔNP is not in the spirit of Constraint Management, which strongly advises T growth more than OE reduction.
- The Portfolio makes no mention of the enterprise constraint, which is typical. If that constraint is in manufacturing, Ops1 could be a capacity constrained resource.
- The Portfolio is presented in a single table, but if the enterprise has both Central and Shadow Information, some decision-making would be decentralized. Business functions/units might be tempted to optimize local constraints to the detriment of the enterprise constraint.
- As for the Information function itself, demand exceeds the budget, and skills appear to be a pervasive constraint.

Conclusion

Constraint Management promotes a race to the top instead of the bottom by focusing on increasing Throughput more than reducing Operating Expense. Cost cutting is often a symptom that executives aren't managing the constraint.

Agile projects are designed for rapid, incremental delivery of Functional Requirements for H1 projects. However, the lag between investment and business benefits for H2 and H3 projects can be measured in years—and some of those projects may not bear fruit.

Every goal comes with necessary conditions. Delta Analysis doesn't address them, so supplemental measures and insights are advisable for topics such as customer loyalty, employee satisfaction, and vendor reliability. If the necessary conditions are not being met, Delta Analysis can lead to ineffective decisions.

REFERENCES

[AGI, 2010] AGI-Goldratt Institute, "Combining Lean, Six Sigma, and the Theory of Constraints to Achieve Breakthrough Performance," *Theory of Constraints Handbook*, 2010, pp.1067–1080.

[Alexander, 2013] Ruth Alexander, "The Student Who Caught Out the Profs," *BBC News Magazine*, April 20, 2013.

[Ambler, 2014] Scott W. Ambler, "The Non-existent Software Crisis: Debunking the Chaos Report," *Dr. Dobb's Journal*, February 4, 2014.

[Anandalingam, 2005] G. Anandalingam and Henry C. Lucas Jr., "The Winner's Curse in High Tech," *IEEE Computer*, March 2005, pp. 96–97.

[Anderson, 2004] David J. Anderson, *Agile Management for Software Engineering: Applying the Theory of Constraints for Business Results*, Prentice Hall, 2004.

[Anderson, 2010] David J. Anderson, *Kanban: Successful Evolutionary Change for Your Technology Business*, Blue Hole Press, 2010.

[Anderson, 2015] David J. Anderson and Andy Carmichael, *Essential Kanban: Condensed Guide*, Lean Kanban Inc, 2015.

[Asay, 2019] Matt Asay, "Is Java the next COBOL?" *InfoWorld*, September 11, 2019.

[Baghai, 2000] Mehrdad Baghai, David White, and Stephen Coley, *The Alchemy of Growth: Practical Insights for Building the Enduring Enterprise*, Basic Books, 2000.

[Bartram, 2019] Erin Bartram, "How PhDs Romanticize the Regular Job Market," *The Chronicle of Higher Education*, January 8, 2019.

[Battersby, 2016] Mark Battersby, *Is That a Fact?: A Field Guide to Statistical and Scientific Information*, Second Edition, Broadview Press, 2016.

[BCG, 2018] Boston Consulting Group, *An Atlas of Strategy Traps*, www.bcg.com/bcg-henderson-institute/strategy-traps/, 2018.

[Beck, 2001] Kent Beck et al., *Manifesto for Agile Software Development*, agilemanifesto.org, February 11–13, 2001.

[Boehm, 2000] Barry Boehm, "Project Termination Doesn't Equal Project Failure," *IEEE Computer*, September 2000.

[Brewer, 2012] Eric Brewer, "CAP Twelve Years Later: How the 'Rules' Have Changed," *IEEE Computer*, February 2012, pp. 23–9.

[Brooks, 1995] Frederick P. Brooks Jr., *The Mythical Man-Month: Essays on Software Engineering*, Addison-Wesley, Anniversary Edition (2nd Edition), 1995.

[Carr, 2003] Nicholas G. Carr, "Why IT Doesn't Matter," *Harvard Business Review*, May 2003.

[Chessell, 2013] Mandy Chessell and Harald Smith, *Patterns of Information Management*, Pearson, 2013.

[Ching, 2013] Clarke Ching, "TOC and Software Development," *TOCPA Conference*, May 2013.

[Ching, 2014] Clarke Ching, *Rolling Rocks Downhill: Accelerate Agile Using Goldratt's TOC*, CreateSpace Independent Publishing Platform, 2014.

[Cooper, 2010] Marjorie Cooper, "Traditional Strategy Models and Theory of Constraints," *Theory of Constraints Handbook*, Chapter 17, 2010, pp. 501–17.

[Cox, 2010] James Cox III and John Schleier (editors), *Theory of Constraints Handbook*, McGraw-Hill, 2010.

[Cox, 2012] James Cox III, Lynn Boyd, Timothy Sullivan, Richard Reid, and Brad Cartier, *Theory of Constraints Dictionary*, Second Edition, TOCICO, 2012.

[Coy, 2019] Peter Coy, "We Are the 1%," *Business Week*, May 13, 2019, pp. 10–12.

[DDtech, 2019] Demand Driven Technologies, "A Beginner's Guide to Demand Driven Material Requirements Planning (DDMRP)," www.demanddriventech .com, 2019.

[Dettmer, 2003] H. William Dettmer, *Strategic Navigation: A Systems Approach to Business Strategy*, ASQ Quality Press, 2003.

[Emam, 2008] Khaled El Emam and A. Gunes Koru, "A Replicated Study of IT Software Project Failures," *IEEE Software*, September–October 2008, pp. 84–90.

[Eveleens, 2010] J. Laurenz Eveleens and Chris Verhoef, "The Rise and Fall of the Chaos Report Figures," *IEEE Software*, January–February 2010, pp. 30–36.

[Ezell, 2005] Allen Ezell and John Bear, *Degree Mills: The Billion-dollar Industry That Has Sold Over a Million Fake Diplomas*, Prometheus Books, 2005.

[Fowler, 2018] Martin Fowler, "The State of Agile Software in 2018," *Agile Australia Conference*, August 25, 2018.

[Fry, 2019] Erika Fry and Fred Schulte, "Death by a Thousand Clicks: Where Electronic Health Records Went Wrong," *Fortune*, March 18, 2019.

[Gartner, 2019] Gartner, "Top 10 Strategic Technology Trends," *Gartner.com*, 2019.

[Glass, 2005] Robert L. Glass, "IT Failure Rates—0% or 10–15%?," *IEEE Software*, May–June 2005, pp. 110–112.

[Glass, 2008] Robert L. Glass, Johann Rost, and Matthias S. Matook, "Lying on Software Projects," *IEEE Software*, November–December 2008, pp. 90–95.

[Goldratt, 2006] Eliyahu Goldratt, *The Haystack Syndrome: Sifting Information Out of the Data Ocean*, North River Press, Revised edition, 2006.

[Goldratt, 2010] Eliyahu Goldratt, *The Science of Management*, www.toc.tv, 2010.

[Goldratt, 2014] Eliyahu Goldratt and Jeff Cox, *The Goal: A Process of Ongoing Improvement*, Third revised edition, 2014.

[Goldrat, t2005] Eliyahu Goldratt, Eli Schragenheim, and Carol Ptak, *Necessary But Not Sufficient*, North River Press, Revised edition, 2005.

[Head, 2015] Megan Head and others, "The Extent and Consequences of P-Hacking in Science," *PLoS Biol,* 13:3, 2015.

[Hern, 2018] Alex Hern and Pamela Duncan, "Predatory Publishers: The Journals that Churn Out Fake Science," *The Guardian*, August 10, 2018.

[Hohmann, 2019] Chris Hohmann, "A Collection of Proofs of Concept Doesn't Make a Digital Strategy," *hohmannchris.wordpress.com*, June 29, 2019.

[Holt, 2010] James R. Holt, "The Alpha-Number-Shape Game: The Real Impact of Multi-Tasking," Washington State University Engineering Management Program, 2010.

[Holub, 2014] Allen Holub, "The Death of Agile," *Software Architect*, 2014.

[Humble, 2011] Jez Humble and David Farley, *Continuous Delivery*, Pearson, 2011.

[Hutanu, 2015] Andrei Hutanua, Gabriela Prosteana, and Andra Badea, "Integrating Critical Chain Method with AGILE Life Cycles in the Automotive Industry," 7th World Conference on Educational Sciences, February 2015.

[IBV, 2018] Institute for Business Value, *Insights from the Global C-Suite Study*, 2016–18.

[IDG, 2015] IDG, Introverts vs Extroverts: Is there an IT personality?," *IDG Connect*, January 7, 2015.

[Ioannidis, 2005] John Ioannidis, "Why Most Published Research Findings Are False," *PLOS Medicine*, 2:8, August 2005.

[Jacob, 2010] Dee Jacob, Suzan Bergland, and Jeff Coxx, *Velocity: Combining Lean, Six Sigma, and the Theory of Constraints*, Free Press, 2010.

[Johnson, 2003] Valen Johnson, *Grade Inflation: A Crisis in College Education*, Springer, 2003.

[Jones, 2000] Capers Jones, *Assessments, Benchmarks and Best Practices*, Addison-Wesley, 2000.

[Kahnweiler, 2009] Jennifer Kahnweiler, "Why Introverts Can Make the Best Leaders," *Forbes*, November 30, 2009.

[Kendall, 2005] Gerald Kendall, *Viable Vision: Transforming Total Sales into Net Profits*, J Ross Publishing, 2005.

[Kim, 2014] Gene Kim, Kevin Behr, and George Spafford, *The Phoenix Project: A Novel About IT, DevOps, and Helping Your Business Win,* IT Revolution Press, 2014.

[Kim, 2016] Gene Kim, Patrick Debois, John Willis, Jez Humble, John Allspaw, *The DevOps Handbook*, IT Revolution Press, 2016.

[Kim, 2005] W. Chan Kim and Renee Mauborgne, *Blue Ocean Strategy: How to Create Uncontested Market Space and Make Competition Irrelevant*, Harvard Business Review Press, 2005.

[Kurzweil, 2006] Ray Kurzweil, *The Singularity Is Near: When Humans Transcend Biology*, Penguin Books, 2006.

[Leach, 2014] Lawrence P. Leach, *Critical Chain Project Management*, 3rd edition, Artech House, 2014.

[Lewis, 2004] Michael Lewis, *Moneyball: The Art of Winning an Unfair Game*, W.W. Norton & Co, 2004.

[Lovric, 2019] Darko Lovric and Greig Schneider, "What Kind of Chief Innovation Officer Does Your Company Need?," *Harvard Business Review*, November 11, 2019.

[Marris, 2013] Philip Marris, "TOC + Lean + Six Sigma," *TOCICO Conference*, 2013. https://youtu.be/XTxiOkh35cU.

[McFarlan, 2003] F. Warren McFarlan and Richard L. Nolan, "Why IT Does Matter," *Harvard Business Review*, August 2003.

[McKenna, 2016] Laura McKenna, "The Ever-Tightening Job Market for PhDs," *The Atlantic*, April 21, 2016.

[McKinsey, 2013] McKinsey & Company, "The Do or Die Questions Boards Should Ask About Technology," *McKinsey.com*, 2013.

[Moser, 1999] Ted Moser, Kevin Mundt, James Quella, Adrian Slywotzkiy, *Profit Patterns: 30 Ways to Anticipate and Profit from Strategic Forces Reshaping Your Business*, Crown Business, 1999.

[Muro, 2016] Mark Muro, "Manufacturing Jobs Aren't Coming Back," *MIT Technology Review*, November 8, 2016.

[Mueller, 2010] Ernest Mueller, "What Is DevOps?," *TheAgileAdmin.com*, 2010.

[Nave, 2002] Dave Nave, "How to Compare Six Sigma, Lean and the Theory of Constraints," *Quality Progress*, January 2002, pp. 73–8.

[Neumann, 2012] Peter G. Neumann, "Rethinking the Computer at 80," *New York Times*, October 29, 2012.

[Newbold, 1998] Robert C. Newbold, *Project Management in the Fast Lane: Applying the Theory of Constraints*, CRC Press, 1998.

[Ovans, 2015] Andrea Ovans, "What Is Strategy, Again?" *Harvard Business Review*, May 12, 2015.

[Pirasteh, 2006] Reza Pirasteh and Kimberly Farah, "Continuous Improvement Trio," *APICS Magazine*, May 2006, pp. 31–3.

[PMI, 2013] Project Management Institute, *A Guide to the Project Management Body of Knowledge*, 5th edition, 2013.

[PMI, 2017] Project Management Institute, "Success Rates Rise," *Pulse of the Profession*, 2017.

[Power, 2014] Brad Power, "How the Software Industry Redefines Product Management," *HBR Blog*, 2014.

[Reges, 2018] Stuart Reges, "Why Women Don't Code," *Quillette*, June 19, 2018.

[Resnick, 2016] Brian Resnick, "A Bot Crawled Thousands of Studies looking for Simple Math Errors [and] The Results Are Concerning," *Vox*, September 30, 2016.

[Ricketts, 1990] John Arthur Ricketts, "Powers-of-Ten Information Biases," *MIS Quarterly*, 14:1, March 1990, pp. 63–77.

[Ricketts, 2008] John Arthur Ricketts, *Reaching the Goal: How Managers Improve a Services Business Using Goldratt's Theory of Constraints*, Pearson, 2008.

[Ricketts, 2010] John Arthur Ricketts, "Theory of Constraints in Professional, Scientific, and Technical Services," *Theory of Constraints Handbook*, Chapter 29, McGraw-Hill, 2010.

[Rogers, 2003] Everett Rogers, *Diffusion of Innovations*, 5th Edition, Simon and Schuster, 2003.

[Saylor, 2012] Michael Saylor, *The Mobile Wave: How Mobile Intelligence Will Change Everything*, Vanguard Press, 2012.

[Scheinkopf, 1999] Lisa Scheinkopf, *Thinking for a Change: Putting the TOC Thinking Processes to Use*, St. Lucie Press, 1999.

[Schindler, 2013] Esther Schindler, "Why Your Users Hate Agile Development (and What You Can Do About It)," *IT World*, June 4, 2013.

[Schragenheim, 2018] Eli Schragenheim, "Raw Thoughts on the Management Attention Constraint," *elischragenheim.com*, January 21, 2018.

[Schragenheim, 2016] Eli Schragenheim, "TOC and Software—The Search for Value," *elischragenheim.com*, April 3, 2016.

[Schragenheim, 2019] Eli Schragenheim, Henry Camp, and Rocco Surace, *Throughput Economics: Making Good Management Decisions*, Routledge Press, 2019.

[Schwaber, 2017] Ken Schwaber and Jeff Sutherland, *The Scrum Guide*, Scrum.org, November, 2017.

[Science, 2018] *Science Code Manifesto*, sciencecodemanifesto.org, 2018.

[Sheetz, 2017] Michael Sheetz, "Technology Killing Off Corporate America: Average Life Span of Companies Under 20 Years," *CNBC Markets*, August 24, 2017.

[Shirky, 2019] Clay Shirky, "The End of Higher Education's Golden Age," *Independent Educational Services*, September 14, 2019.

[Sposi, 2014] Michael Sposi and Valerie Grossman, "Deindustrialization Redeploys Workers to Growing Service Sector," *Dallas Fed Economic Letter*, September 2014.

[Sproul, 2012] Bob Sproull and Bruce Nelson, *Epiphanized: Integrating Theory of Constraints, Lean and Six Sigma (TLS)*, North River Press, 2012.

[Staley, 2019] Oliver Staley, "Whatever Happened to Six Sigma?," *Quartz at Work*, September 3, 2019.

[Thomas, 2008] Joseph C. Thomas and Steven W. Baker, "Establishing an Agile Portfolio to Align IT Investments with Business Needs," *IEEE Agile Conference*, 2008.

[Teixeira, 2019] Thales Theixeira, "Disruption Starts with Unhappy Customers, Not Technology," *Harvard Business Review*, June 6, 2019.

[Thompson, 2019] Clive Thompson, "The Secret History of Women in Coding," *The New York Times Magazine*, February 13, 2019.

[Ujigawa, 2016] Koichi Ujigawa and David Updegrove, "Agile CCPM: Critical Chain for Software Development," *Theory of Constraints International Certification Organization White Paper Series*, 2016.

[USDA, 2019] United States Department of Agriculture, "Farming and Farm Income," *Economic Research Service*, 2019.

[Woeppel, 2014] Mark Woeppel, "The ONE Thing to Deliver More Projects, On Time," *LinkedIn*, October 4, 2014.

[Wood, 2019] L. Maren Wood, "Odds Are, Your Doctorate Will Not Prepare You for a Profession Outside Academe," *ChronicleVitae*, July 12, 2019.

[Youngman, 2018] Kelvyn Youngman, *The Online Guide to Theory of Constraints*, www.dbrmfg.co.nz, 2018.

[Yourdon, 2003] Edward Yourdon, *Death March*, Prentice-Hall, Second Edition, 2003.

INDEX

Page numbers followed by *f* and *t* refer to figures and tables, respectively.

Acceptance tests, 136
Accessibility, software, 125
Accounting principle, 467
Acquisition, 414
Adventures (case examples):
  about, 19–20
  administrative, 447–449
  of architecture, 223–224, 228–231
  career, 449–453
  Cognitive Computing, 199–202
  of Constraint Management, 72–74, 376–377, 382–384, 386–391
  data, 160–164, 168–174, 179–181
  data cleansing, 197–198
  of disasters, 211–214
  executive priorities, 21–22
  fairness, 194–196
  goal, 432–434
  of hardware, 103–105, 109–111
  Information field, 51–52
  meta-learning, 437–440
  methodologies, 262–263, 274–275, 281–282
  with network policies, 215–216
  portfolio, 331–333, 336–337, 344–345
  of processes, 315–319, 321–325
  projects, 295–298, 300–308
  of services, 353–354, 357–359
  skills, 235–236, 238–242, 244–245, 248–252, 258–259
  software, 132–134, 141–148
  of strategy, 34–39, 395–398, 400–404, 409–411, 417–421
  Surprise Expert, 11–14
  Technical Services, 54–56
Affordability, software, 124
Age, 256
Aggregation principle, 95, 467
The Agile Manifesto (Beck), 275–276
Agile Methodologies:
  about, 275–280
  Iron Triangle in, 266
  issues with, 283–284
  replanning in, 302
  scope creep in, 299
  SDLC in, 328
  tracking, 294–295
  work rules of, 285
Agile Portfolio, 330
Agile principle, 463
AI (Artificial Intelligence), 61, 193
Alerts, 172
Algorithms, 225
Alignment principle, 46, 444, 459
Alpha tests, 136
Alpha-Number-Shape Exercise, 270–271
Ambiverts, 245–247
Analytics, data, 172–173
Announcement, 135, 327
Anti-patterns, 121

Anti-productivity, 254
APIs (Application Program Interfaces), 122
Application architecture, 220, 225–226
Application Program Interfaces (APIs), 122
Application software, 136
Application Understanding, 140–141
Applications, 101–102
Architecture, 219–233
  about, 219
  Application, 225–228
  case example of, 223–224, 228–231
  Data/Information/Knowledge, 224–225
  Enterprise, 231
  frameworks for, 231–232
  Hardware, 221
  information technology, 220
  IT, 221–222
  as operations constraint, 232–233
  Technical, 219–221
Archiving, data, 176–177
Artificial Intelligence (AI), 61, 193
  pseudo, 203
Aspirationals, defined, 28
Associative array, 159
Atkin's Law of Demonstrations, 111
Attention principle, 467
Auditability, software, 125
Authentication, 210
Automation, 414
Availability:
  architecture prioritizing, 227
  general, 135, 327
Awareness, 395

B2B (business-to-business) markets, 59
B2C (business-to-consumer) markets, 59
Back-end developers, 243
Backlog, product, 277
Back-office, 317
Backsliding, 392
Backups, 139, 176
Backward compatibility, 106
Bad Lean, 378
Bad Six Sigma, 379
Balance loading, 227
Benchmarking, 131–132, 335, 412
Benchmarking fallacy, 132
Benchmarking trap, 132
Beta tests, 136
Biases:
  information, 170
  and knowledge fairness, 194
  power-of-ten information, 170–171
  selection, 173
Binary, 159
Blue Ocean Strategy (BOS), 42, 46, 413, 459
Boolean, 159
BOS (*see* Blue Ocean Strategy)
Bottlenecks, 278, 429–430
Boundary conditions, 130
Boyd, John, 43
BPO (business process owner), 184–185
Braess' Paradox, 216
Brooke's Law, 58, 266
Buffer Management, 82
Buffer penetration, 272
Buffer principle, 79–81, 95, 468
Buffers, types of, 79–80, 84–85, 257–258, 272
Bugs, 130
Burndown chart, 277, 294–295, 340
Business opportunity, 34–35
Business platforms, 30–31
Business process owner (BPO), 184–185
Business processes, 317–319
Business-to-business (B2B) markets, 59

Business-to-consumer (B2C) markets, 59
Buyers, hardware, 114–115
Buy-in, 91–92

Cable TV, 209
Caching, 227
Cancelled projects, 290
Capacity buffer, 80
Capacity Constrained Resources (CCR), 429–430, 489
Capacity management, 314
Capacity principle, 95, 468–469
Cash buffer, 80
Cash constraints, 385
Cause and effect, 484
CC (Cognitive Computing), 61–62, 193
CCR (Capacity Constrained Resources), 429–430, 489
CCTV (Closed Circuit Television), 209
CDO (chief data officer), 26, 27, 183–184
Central information, 27, 53
Centrists, 247–248
CEO (chief executive officer), 23, 27
CFO (*see* Chief financial officer)
Chain principle, 95, 469
Challenging projects, 290
Change principle, 469–470
Chief data officer (CDO), 26, 27, 183–184
Chief executive officer (CEO), 23, 27
Chief financial officer (CFO), 24, 27, 346
Chief human resources officer (CHRO), 24–25
Chief information officer (CIO):
  and local constraints, 346
  as nontechnical executive, 305
  role of, 25, 27
Chief innovation officer (CINO), 26
Chief marketing officer (CMO), 24
Chief operating officer (COO), 23
Chief privacy officer (CPO), 184
Chief science officer (CSO), 26–27, 184
Chief technology officer (CTO), 26–27, 346
CHRO (chief human resources officer), 24–25
CINO (chief innovation officer), 26
CIO (*see* Chief information officer)
Clarke, Arthur C., 205
Classical Computing, 221
Clear interfaces, 226
Client-server, 226
Closed Circuit Television (CCTV), 209
Closed constraint management, 378
Cloud Computing:
  hardware for, 107
  IT Architecture vs., 222
  Traditional vs., 60–61
  value migration of, 59*f*
CMO (chief marketing officer), 24
Code:
  learning, 238
  modifying vs. rewriting, 243
  for software, 120–121
Code modules, 121
Cognitive Computing (CC), 61–62, 193
Cold start, 203
Collaboration, concentration vs., 250
Commoditization principle, 57, 66, 463
Commodity skills, 257
Communication satellites, 209
Compatibility, hardware, 106
Competitive edge principle, 470
Complexity, software, 128–129
Compliance, 232
Compression, data, 165
Compromise, conflict resolution and, 91

Computer software:
about, 15–16
accounting for, 486–487, 486*t*
Concentration, collaboration vs., 250
Configurability, software, 125, 139
Conflict principle, 95, 470
Conflict resolution, 90–91, 372–374
Congruence, 232
Conservatives, 247–248
Constraint Management, 69–98, 367–392
about, 69
backsliding in, 392
benefits of, 442
buffers in, 79–80
and buy-in process, 91–92
case example of, 72–74, 376–377, 382–384, 386–391
and conflict resolution, 372–374
conflict resolution with, 90–91
continuous improvement in, 377–378, 380
data as, 181–182
and decision-making, 94–95
defining constraints in, 78–79
flow of, 75
flowing water analogy of, 70–71
Focusing Steps of, 89–90, 370–371
and global optimization, 371–372
goal of, 74–75
for Information, 97–98, 367–368
in Information field, 425–428, 426*t*
information for, 368–370
and knowledge, 204–205
limits in, 77–78, 381–382
paradigms in, 391–392
principles of, 95–96
Sense and Respond in, 89
for software, 148
solutions from, 80–89
Strategy in, 92–93, 374–376
systems boundaries in, 75–76
and technology, 93–94
technology questions from, 405
weakest link analogy of, 70
Constraint Management production of goods, 82
Constraint principles, 96, 467–481
Constraints:
case example of, 3–9
cash, 385
digital, 385–386, 389–391
duality of, 384–385
fixed, 385
floating, 385–388
internal, 384
limits vs., 77, 381–382
natural (*see* Natural constraints)
operational, 386, 441
physical, 385
process, 183
project, 183
strategic (*see* Strategic constraints)
Content Delivery Networks, 214
Contingency, 269
Continuous Improvement Trio, 379–380
Contracts, 360
Conventional production of goods, 82
COO (chief operating officer), 23
Coordinated Universal Time (UTC), 109
Core skills, 257
Cost cutting, 95
Counter-plans, 395
CPO (chief privacy officer), 184
Crashing, 299
Critical Chain Method:
about, 85, 87, 88*f*, 268, 271–274
Constraint Management in, 285, 308
and hardware, 113
replanning in, 301–302
tracking in, 293–294, 293*f*
work rules of, 291

Critical Path Method:
  about, 85, 87, 88*f*, 267–271
  Constraint Management in, 285
  replanning in, 301
  tracking in, 292, 294
  work rules of, 291
Critical skills, 257
Crunching, 299
CSO (chief science officer), 26–27, 184
C-suite (CXOs):
  about, 18
  agility of, 31
  ambiguous boundaries of, 29
  business platforms and, 30–31
  central vs. shadow information, 27
  and central vs. shadow information, 53
  and data, 182–188
  with data, 182–188
  digital reinvention of, 28
  and disruptive innovation, 28–29
  nontechnical executives, 305
  roles in, 23–27
  on strategy, 393–394
  technology affecting, 30
  as ultimate limit, 388–389
CTO (chief technology officer), 26–27, 346
Current-value flow, 76
Custom software, 122
CXOs (*see* C-suite)

DA (data analyst), 185
Dark patterns, 121
Darwin, 337–338
Data, 157–190
  analytics of, 172–173
  case example of, 160–164, 168–174, 179–181
  compression of, 165
  and data administration, 164
  and DataOps, 178
  de-duplication of, 165
  encryption of, 166
  as enterprise constraint, 181–182, 424
  errors in, 167–168
  extracting, transforming, and loading, 164–165
  formulas and expressions of, 160
  information, knowledge, and, 157–158
  and information biases, 170
  logging (*see* Logging data)
  master, 224
  mismanagement of, 412
  misuse of, 158
  open, 178
  privacy with, 175
  reference, 224
  retention, backup, recovery, archiving, and disposal of, 176–177
  security of, 175
  staff functions with, 182–188
  streaming, 166
  structures of, 159
  and system of systems, 188–189
  and technical debt, 177–178
  transaction, 224
  types of, 158–159
Data administration, defined, 164
Data aggregator/broker, 414
Data analyst (DA), 185
Data at rest, 158
Data cleansing, 196–202
Data errors, 167–168, 167*t*
Data in motion, 158
Data in process, 158
Data modeling, 164
Data networks, 208–209
Data producers, 187–188
Database administrator (DBAs), 186
Data/Information/Knowledge architecture, 220, 224–225

DataOps, 178
Date, 159
DBAs (database administrator), 186
Deadlock, 130
Death march, 266, 337–338
Decision principle, 471–472
Decision-making principle, 94–95, 483–484
Decisive competitive edge, 374–375, 444
De-duplication, 165
Defect Density, 380
Defects, 111–112, 129–131
Degredation, graceful, 227
Delivery monitor, 314
Demand principle, 95, 472
Dematerialization principle, 56–57, 66, 400, 463
Design manager, 185–186
Design patterns, 225
Design thinking, 276
Designed constraints, 385
Development:
  in product life cycle management, 327
  in software development life cycle, 135
  in software system of systems, 152, 153*f*
Development manager, 149–150, 186–187
Development team, 277
Developmental requirements, 124–125, 225
DevOps, 276, 279–280, 285, 463
Digital constraints, 385–386, 389–391
Digital disruption, 399–400
Digital reinvention, 28
Digital robots, 198–199
Digitization, 217
Direct sales process, 114
Disaster recovery, 210, 321
Disinformation, 158
Disposal, data, 176–177
Disruptive innovation, 28–29
Distribution:
  constraint management solution for, 84–85
  of hardware, 112
  in system of systems, 116, 117*f*
Distribution manager, 115
Divestiture, 415
Documentation, 199
Dominance, conflict resolution and, 91
Drift, 232
Drucker, Peter F., 33
Dumps, 151
Dynamic strategy principle, 46, 418–419, 459
Dysfunctional requirements, 411

Early finish punishment, 270
Edge cases, 130
Education, skills, 236–238
Efficiency, software, 125
Elevation principle, 96, 472
Embedded software, 122
Encapsulation, object-oriented programming, 138
Encryption, 166, 210
End of life, 135, 327
End of service, 135
End of support, 327
End-to-end tests, 135
End-user software, 122
Engineering:
  architecture vs., 219
  in system of systems, 116, 117*f*
Engineering manager, 113
Enterprise agility, 31
Enterprise architecture, 221, 231
Enterprise Computing:
  and Cloud Computing, 60
  Personal vs., 59, 105
  value migration of, 59*f*

Enterprise Strategy, defined, 15, 441
(*See also* Strategy)
Error-defer, 167
Error-end, 167
Error-hold, 167
Errors, 130, 167–168, 167*t*
Error-substitution, 167
Error-through, 167
Estimating, 131, 269–270, 272–273
ETL (Extract, Transform, Load), 164
European Union's General Data Protection Regulation (GDPR), 175
Evaporating Cloud, 90
Executable code, defined, 120
Execution principle, 46, 460
Executive priorities (*see* C-suite (CXOs))
Exit criteria, 278
Expediting, 269, 273, 299–300
Exploitation principle, 96, 472
Expressions, 160
External constraint, 384
Extract, Transform, Load (ETL), 164
Extreme Programming (XP), 276
Extroverts, 245–247

Fail fast, 131, 413
Failure, success vs., 289–290
Fairness, 194
Fantasy information, 63–64
Fast tracking, 299
Fault tolerance, 176
Feasibility, 334–335
Feature, functional requirement vs., 124
Feeder buffers, 272
Fever Chart, 293–294, 340
Final Assembly, 78
Financial obsolescence, 64
Financial sector, industries in, 182
Firewalls, 175, 210
Firmware, 102
Fixed constraints, 385
Fixed price contract, 360
Flexible capacity, 340, 343
Floating constraints, 385–388
Floating point, 159
Flow principle, 95, 473
Flowing water analogy, 70–71
Flows, 76
Focus principle, 96, 473
Focusing Steps, 89–90, 96, 370–371
Forecasting, 172
Formulas, 160
For-profit organizations, 74
Frameworks, 231–232
Freezing, 337–338
Front-end developers, 243
Front-office, 317
Full kit, 291
Full-stack developers, 243–244
Functional requirements, 123–124, 225
Fuzzy search, 160

Game Theory, 91
Gamma tests, 136
GDPR (European Union's General Data Protection Regulation), 175
General availability, 135, 327
General Intelligence, 193
Global optimization, 371–372
Global Positioning System (GPS), 209
Globalization, 124, 217
The Goal (Goldratt), 453–454
Goal principle, 74–75, 96, 431–432, 473
Goldratt, Eli, 9, 69, 75, 78, 378, 388, 478
Good Lean, 378
Good Six Sigma, 379
Governance, 339–340
Government organizations, 74
GPS (Global Positioning System), 209

Graceful degredation, 227
Graham, Paul, 250
Graph, 159

Hard disk drives (HDD), 107
Hardware, 101–118
  about, 101
  availability of, 105–106
  case example of, 103–105, 109–111
  compatibility, 106
  defined, 49, 102
  development of, 111–112
  distribution of, 112
  as enterprise constraint, 112, 423–424
  layers, 101–102
  management viewpoints of, 113–116
  Moore's Law affecting, 58
  performance of, 105
  security, 106–107
  and system of systems, 116–118, 117*f*
  types of, 107–109
Hardware architecture, 220–221
Hash, 159
HDD (hard disk drives), 107
Heat Treat, 4, 78
Higher education, 237
High-tech Manufacturing, 111–112
HIPAA (U.S. Health Insurance Portability and Accountability Act), 175
Holistic principle, 474
Homework, 241
Horizons Model:
  about, 40–41, 41*f*, 47, 98, 460
  and hardware, 115
  imbalance in, 411
Hot standby, 176
Human-made disasters, 210
Hybrid Cloud, 415
Hybrid methodologies, 280
Hype Cycle, 202–203
Hypervisor, 102

IaaS (*see* Infrastructure as a Service)
Impact Analysis, 141
Imposter syndrome, 242
Improvement principle, 96, 474
Indexed array, 159
Induced demand, 431
Industrial Revolution, 349, 350*f*
Industry convergence, 23–24
Inequality bugs, 130
Inertia, 227
Inference, 172
Information, 224
  about, 15–16
  central vs. shadow, 53
  CFOs role in integration of, 23
  constraints of, 97–98
  fantasy, 63–64
  importance of, 50
  possible constraint influencers with, 63–66
  principles of, 66–67, 463–465
  role of, 53
  system types, 53–54
  trends in, 56–63
Information biases, 170
Information field:
  Constraint Management in, 425–428, 426*t*
  defined, 16
  exceeding goals in, 443
  industries in, 181–182
  S&T Tree for, 93
  technology questions from, 406–409
Information principles, 66–67, 463–465
Information processes, 319–321
Information projects, methodologies for, 264–266
Information spending, 27

Information system, defined, 49, 75
Information Technology, accounting for, 484–486, 485*t*
Information technology architecture, 220
Infrastructure as a Service (IaaS), 320, 415, 427
Inheritance:
  legacy, 65
  object-oriented programming, 138
Innovation principle, 47, 413, 460
Installation, software, 139
Institutional knowledge, 199
Integration, 227
Integration bugs, 130
Integration tests, 135
Internal constraint, 384
Internationalization, software, 138
Internet, 208
Internet of Things (IoT), 62, 108–109
Interviews skills, 240–241
Intranets, 208
Introverts, 245–247
Investment, defined, 94
IoT (Internet of Things), 62, 108–109
Iron Triangle, 266, 286
IT Architecture, 221–222

Jevons Paradox, 58
Justification, 337

Kanban:
  about, 276, 278–279
  Constraint Management in, 285, 308
  replanning in, 302
  tracking, 295
  work rules of, 291
Kanban boards, 340, 343
Kernighan, Brian, 130
Knowledge, 191–206
  about, 191
  and architecture, 224
  Artificial Intelligence, 193
  case example of, 194–198
  Cognitive Computing, 193
  cold start problem with, 203
  and data cleansing, 196–202
  digital robots, 198–199
  as enterprise constraint, 204–205, 424
  fairness with, 194
  and Hype Cycle, 202–203
  institutional, 199
  and knowledge work, 192
  Machine Learning, 192
  and pseudo-AI, 203
  and Technological Singularity, 203–204
Knowledge work, 192

Languages, software, 137
LANs (Local Area Networks), 208
Late finish accumulation, 270
Layers:
  of architecture, 226
  computing systems, 101–102
Lean, 276, 378, 380
Lean Manufacturing, 76
Leapfrogging, 66
Leased lines, 208
Legacy inheritance, 65
Legacy migration, 65
Legacy replacement, 65
Legacy skills, 255
Legacy Systems principle, 65–66, 139–140, 155, 464
Legacy transformation, 65
Leverage principle, 474
Liberals, 247–248
Lieberg's Law of the Minimum, 78
Life cycle management, 327–328
Limits, constraint vs., 77, 381–382
Local Area Networks (LANs), 208
Location principle, 475

Logging data, 151, 166, 210
Longevity principle, 66, 464
Loss-less compression, 165
Lossy compression, 165

Machine Learning (ML), 61, 192
Main memory, 107
Maintainability:
  defined, 105
  of software, 124
Maintenance, 152, 153*f*
Malinformation, defined, 158
Malware, 158
Manually quantifying requirements, 128
Manufacturing processes, 312–313
Master data, 224
Master record, 159
Measurement principle, 132, 475
Mega-projects, 411
Memory, main, 107
Mentoring, 248–250
Metadata, 159
Methodologies, 261–287
  agile, 275–280, 283–284
  case example of, 262–263, 274–275, 281–282
  choosing, 267
  and Constraint Management, 285–286
  domains of, 264
  hybrid, 280
  importance of, 261–262
  for Information projects, 264–266
  Iron Triangle, 266
  planned, 267–274
  service, 355–356
  and technical services, 284–285
Metrics principle, 127–133, 464
Microservices, 122–123
Middleware, 102
Migration, 135, 327
Milestones, 269, 272, 337
Military software, 136
Minimum viable product, 266
Misestimation, 295
Misinformation, 158, 295
Misreporting, 295
ML (Machine Learning), 61, 192
MLOC (millions of lines of code), 127–128
Mobile phones, 208
Monitoring networks, 210
Moore's Law, 57–58
Multi-project management, 340–343, 341*f*
Multi-project portfolios, 264
Multitasking principle, 96, 270, 291, 475
Multi-tenancy, software, 124
Myers-Briggs Type Indicator, 245

NAICS (North American Industry Classification System), 181–182
Natural constraints, 78, 90, 385
Natural disasters, 210
Networks, 207–217
  about, 207
  and Braess' Paradox, 216
  case example of, 211–216
  defined, 49
  and disasters, 210
  as enterprise constraint, 216–217, 424
  policies with, 214–215
  security of, 210
  software defined, 108, 214
  types of, 208–210, 214
Never used software, 125–126
No deal, conflict resolution and, 91
Noise principle, 476
Non-profit organizations, 74
Non-relational (NoSQL) databases, 225

Nontechnical executives, 305
Non-technical work, 253
North American Industry Classification System (NAICS), 181–182
NoSQL (non-relational) databases, 225
Number numbness, 170
Numbness, number, 170

Obfuscation, 144
Object-oriented programming, 137–138
Observe, Orient, Decide, Act (*see* OODA)
Obsolescence, 64–65, 415
OE (operating expense), 94
Off-by-one errors, 130
OODA (Observe, Orient, Decide, Act), 43, 47, 461
Open constraint management, 378
Open data, 178
Open source software, 122
Operating expense (OE), 94
Operating system, 102
Operational constraints, 386, 441
Operational requirements, 125, 225
Operations:
  of software, 139
  in software system of systems, 152
Operations manager, 150, 187
Optimization principle, 96, 172, 415, 476
Orchestration, 228
Organizational system, defined, 75
Outsourcing, 415
Owner, product, 277

Packaged software, 122, 152
Paradigms, 391–392
Parkinson's law, 270
Partitioning, 227
PBXs (Private Branch Exchanges), 208
Percent complete, 269
Performance, hardware, 105
Perimeter security, 175
Personal Computing:
  and Cloud Computing, 60
  Enterprise vs., 59
  value migration of, 59*f*
Physical constraints, 385
Pipe-filter, 226
Pipeline projects, 340, 342
Plain Old Telephone Service (POTS), 208
Planned Methodologies:
  about, 266–268
  scope creep in, 299
  tracking, 292–294, 293*f*
  work rules of, 284
Planned Portfolio, 330, 348
PMO (Project Management Office), 345–346
POC (*see* Proof of concept)
Pointer, 159
Polarization principle, 66, 464
Polymorphism, 138
Porous defenses, 130
Portability, software, 124
Porter, Michael, 421
Portfolio, 327–348
  case example of, 331–333, 336–337, 344–345
  as enterprise constraint, 347
  and feasibility, 334–335
  and governance, 339–340
  and justification, 337
  and life cycle management, 327–328
  management of, 328–330, 491–496, 492*t*
  methodologies for, 264
  multi-project management of, 340–343, 341*f*
  prioritization in, 334
  and project initiation, 337–338

Portfolio (*continued*)
  and Project Management Office, 345–346
  and project termination, 338–339
  and system of systems, 346–347
Portfolio creep, 412
Portfolio migration, 415
POTS (Plain Old Telephone Service), 208
Power-of-ten information bias, 170–171
Practical obsolescence, 64
Practitioners, 28
Prediction, 172
Principle of Least Astonishment, 124
Principles, 457–481
  about, 457
  constraint, 467–481
  information, 463–465
  strategy, 459–461
Prioritization, 334
Privacy, 125, 175
Private APIs, 122
Private Branch Exchanges (PBXs), 208
Process constraints, 183
Processes, 311–326
  business, 317–319
  case example of, 315–319, 321–325
  as enterprise constraint, 325–326, 424
  Information, 319–321
  manufacturing, 312–313
  methodologies for, 264
  projects vs., 311–312
  service, 314–315
Product backlog, 277
Product cost, 95
Product owner, 277
Production:
  Constraint Management in, 81–84, 83*f*, 312–313
  in system of systems, 116, 117*f*
Production manager, 113
Productivity, 131, 254
Professional/Scientific/Technical services (PSTS), 351–353, 428
Profit Patterns, 42–43, 47, 395, 444, 461
Profit shift, 43
Programs, methodologies for, 264
Project:
  as constraint, 424
  feasibility of, 334–335
  governance of, 339–340
  initiation of, 337–338
  justification for, 337
  processes vs., 311–312
  termination of, 338–339
Project buffer, 272
Project constraints, 183
Project Management Office (PMO), 345–346
Projects, 289–310
  case example of, 295–298, 300–308
  Constraint Management solution for, 85–89, 86*f*, 88*f*, 308–309
  expediting, 299–300
  methodologies for, 264
  with nontechnical executives, 305
  replanning, 301–302
  and scope creep vs. scope surge, 298–299
  and spin, 295
  status of agile, 294–295
  status of planned, 292–294, 293*f*
  success vs. failure of, 289–290
  and work rules, 290–292
Promotion, software, 138–139
Proof of concept (POC), 37–38, 418–419, 460
Protected APIs, 122
Pseudo-AI, 203
PSTN (Public Switched Telephone Network), 208

PSTS (Professional/Scientific/
Technical services), 351–353, 428
Public APIs, 122
Public Switched Telephone Network
(PSTN), 208
Pull principle, 96, 476

Quality, 380
Quantum Computers, 221
Query, 172
Queue, 159

Race condition, 130
RAID (Redundant Array of
Inexpensive Disks) drives, 176
Ransomware, 166
Reassignment, 299
Recovery:
data, 176–177
disaster, 210, 321
of software, 139
Red Ocean Strategy, 42, 459
Redundant Array of Inexpensive Disks
(RAID) drives, 176
Reference data, 224
Regression tests, 135
Reinventors, defined, 28
Relay race principle, 96, 272, 477
Release, 278
Release independence, 227
Reliability, defined, 105–106
Remote sensing satellites, 209
Remote work, 250–251
Replanning, 269, 273, 301–302
Replenishment principle, 84, 96, 477
Reporting, 172
Requirements:
developmental, 124–125, 225
dysfunctional, 411
functional, 123–124, 225
manually quantifying, 128
operational, 125, 225

Research, 327
for hardware, 111
in software development life cycle,
135
in software system of systems, 152,
153*f*
Research institutions, 237
Research manager, 149
Resolution, conflict resolution and, 91
Resource leveling, 272
Resource management, 130
Restart, software, 139
Retention, data, 176–177
Return on investment (ROI), 94
Revenue, defined, 94
Review, 278
Risk, 38
Robots, 108
ROI (return on investment), 94
Role reversal principle, 62–63, 66, 464

SaaS (*see* Software as a Service)
SAFe (Scaled Agile Framework), 340,
343
Sales, system of systems and, 116, 117*f*
Sales manager, 114
Sales principle, 477–478
Salting, 166
Satellite Broadcast, 209
Scalability, software, 125
Scaled Agile Framework (SAFe), 340,
343
Scope creep, 298–299
Scope surge, 298–299
Scrum, 276–277, 285, 291
The Scrum Guide (Schwaber), 277
Scrum master, 277
SDLC (System Development Life
Cycle), 268, 328
Security:
audit of, 416
data, 175

Security (*continued*)
of hardware, 106
networks, 210
perimeter, 175
of software, 125
Segmentation principle, 478
Selection bias, 173
Sense and Respond, 89, 356
Sensitivity analysis, 335
Separation of concerns, 226
Server-less apps, 108
Service Level Agreements (SLAs), 150, 360–361
Service manager, 116
Service processes, 314–315
Services, 349–363
about, 349–351, 350*f*
Buffer Management in, 82
case example of, 353–354, 357–359
defined, 49
disruptions to, 356–357
engagements of, 359–361
as enterprise constraint, 361–362, 425
methodology of, 355–356
Professional/Scientific/Technical, 351–353
in system of systems, 116, 117*f*
Shadow information, 27, 53, 416
Shadow work, 253
Side projects, 241
Signed integer, 159
Simplicity principle, 478
Simulation, 172
Six Sigma, 379, 380
Size, code, 127–128
Skill buffer, 80, 257, 258
Skill constraints, 183
Skills, 235–260
and age, 256
case example of, 235–236, 238–242, 244–245, 248–252, 258–259
with collaboration vs. concentration, 250
for conservatives/liberals/centrists, 247–248
deficits in, 412
education for, 236–238
and imposter syndrome, 242
for interviews, 240–241
for introverts/extroverts/ambiverts, 245–247
Legacy, 255
modifying vs. rewriting code, 243
for non-technical work, 253
as operations constraint, 256–258, 424
and productivity vs. anti-productivity, 254
and remote work, 250–251
with solo vs. team work, 253
specialists vs. full stack developers, 243–244
and stack ranking, 254–255
and wrap-around days, 251
SLAs (Service Level Agreements), 150, 360–361
Smartphones, 107
SME (Subject Matter Experts), 196–197
Software, 119–155
about, 119–120
Application Understanding/Impact Analysis with, 141
benchmarking for, 131–132
case example of, 132–135, 141–148
and code, 120–121
defined, 49
as enterprise constraint, 148, 424
estimating for, 131
languages for, 137–138
and Legacy Systems, 139–140
life cycle of, 135
management viewpoints of, 149–152
metrics of, 127–131

for microservices, 122–123
and Moore's Law, 58
never used vs. unexpected use with, 125–126
open source, 122
operations of, 139
origins of, 122
packaged, 122, 152
promotion of, 138–139
requirements of, 123–125
and software engineering, 126–127
and software industry, 121
and system of systems, 152–154, 153*f*
technical debt with, 143–144
testing of, 135–136
unexpected use, 125–126
user-developed, 136
work breakdown structure with, 136
Software as a Service (SaaS), 122, 152–153, 416, 427
Software defined networks, 108, 214
Software defined storage, 108
Software engineering, 126–127
Software industry, 121
Solid-state drives (SSD), 107
Solo work, 253
Source code, defined, 120
Space buffer, 79
Special Intelligence, 193
Specialists, 243–244
Specialty networks, 209–210
Speed to market, 414
Spin, 295
Spreadsheets, 136
Sprinting, 79, 277
SSD (solid-state drives), 107
S&T Trees (*see* Strategy and Tactics Trees)
Stack, 159
Stack ranking, 254–255
Staff buffers, 85
Stand-up meeting, 277
Starts, 269, 272
Statement of Work, 359–360
Stock buffer, 79
Storage:
hardware for, 107
software defined, 108
Story points, 277
Strategic constraints, 386, 441, 488–489
Strategic decisions, 483–496
Strategic-initiatives flow, 76
Strategy, 33–47, 393–422
about, 14–15, 39–40
alignment of, 44
Blue Ocean, 42
business opportunity, 34–35
case example of, 34–39, 395–398, 400–404, 402*f*, 409–411, 417–421
constraint management approach to, 92–93
decisive competitive edge, 374–375
digital disruption, 399–400
dynamic, 418–419
Enterprise Scenarios and, 412–414, 413*f*
exceeding goals with, 444
executive perspectives of, 393–394
Horizons Model, 40–41, 41*f*
initiatives for, 399
OODA, 43
principles of, 46–47
Profit Patterns, 42–43
proof of concept, 37–38
Technical Scenarios and, 414–416
technology partnership, 36
technology research affecting, 33–34
technology role in, 404–408
traditional, 394–395
traps in, 44–46, 45*t*, 411–412, 461

Strategy and tactics principle, 96, 478–479
Strategy and Tactics (S&T) Trees, 92–93, 375, 478–479
Strategy principles, 459–461
Strategy Traps Principle, 44–46, 45*t*, 47
Streaming, data, 166
Streaming TV, 209
Stress interviews, 240
String, 159
Student syndrome, 270
Subject Matter Experts (SME), 196–197
Subordination principle, 96, 479
Success, failure vs., 289–290
Super project, 340, 342
Supply chain principle, 479
Support, software system of systems and, 152, 153*f*
Support manager, 151
Swarming, 299
Symbolic overtime, 251
System boundaries, 75–76
System Development Life Cycle (SDLC), 268, 328
System of systems:
  about, 18–19
  and data, 188–189, 188*f*
  and hardware, 116–118, 117*f*
  importance of, 434–436
  portfolio and, 346–347
  services in, 116, 117*f*
  and software, 152–153, 153*f*
System test, 135
Systems, Constraint Management, 75–76
Systems of Engagement, 54, 185, 243
Systems of Innovations, 54, 185, 244
Systems of Insight, 54, 244
Systems of Record, 53, 185, 243
Systems software, 136
Tape drives, 176
Teaching institutions, 237
Teamwork, 253
Technical Architecture:
  about, 220–221
  importance of, 219–220
Technical debt principle, 465
  about, 64, 66
  crushing, 412
  data as source of, 177–178
  and drift, 232
  reducing, 490–491, 491*t*
  and software, 143–144
Technical obsolescence, 64
Technical questions, 240
Technical Services:
  about, 15–16, 54
  accounting for, 487–488, 487*t*
  methodologies and, 284–285
Technical strategy, defined, 441
Technical Support, 320
Technological Singularity, 203–204
Technology:
  constraint management on, 93–94
  partnerships for, 36
  rediscovering, 430–431
  shifts in, 43
  strategy affected by, 404–408
Technology principle, 96, 479–480
Technology-led organizations, 62–63
Terrestrial Broadcast, 209
Testing, software, 135–136
Theory of Constraints (*See* Constraint Management)
Thinking principle, 480
Thinking questions, 240
3D printers, 62–63
Throttling, 337–338
Throughput accounting, 94–95, 483
Time, 109
Time and Materials (T&M) contract, 360

Time buffer, 80
Time constraints, 385
Time principle, 96, 480
Time to value, 414
T&M (Time and Materials) contract, 360
Totally variable costs (TVC), 94
Traditional Computing, 59*f*, 60–61
Traditional strategy, 394–395
Transaction data, 224
Transaction record, 159
Transformation, data, 164–165
Tree, 159
Tribal knowledge, 199
T-shaped skills, 256
TVC (totally variable costs), 94
Typical buffers, 84

Unexpected use software, 125–126
Unit errors, 130
Unit test, 135
Unsigned integer, 159
Upgrades, software, 139
U.S. Health Insurance Portability and Accountability Act (HIPAA), 175
U.S. Internal Revenue Service, 106
Usability, software, 125
User stories, 277
User-developed software, 136
Users, 151–152
UTC (Coordinated Universal Time), 109
Utilization principle, 96, 481

Value, 380–381
Value chain squeeze, 43
Value migration, 57, 59*f*
Value price contract, 360
Variety principle, 67, 465
Velocity, 76
Video networks, 209
Virtual hardware, 108
Virtual machines, 108
Virtual memory, 108
Virtual Private Networks (VPNs), 208
Virtualization principle, 56–57, 67, 465
Voice networks, 208
Voice Over Internet Protocol (VOIP), 208
VPNs (Virtual Private Networks), 208

WANs (Wide Area Networks), 208
Waterfall Methodology, 268, 276, 465
Weakest link analogy, 70
Weather disruptions, 356
Whiteboarding, 241
Wholesalers, 114
Wide Area Networks (WANs), 208
Win-win principle, 481
Wireless LANs, 208
Work breakdown structure, 136, 265
Work rules, 290–292
"Working as designed," 125
"Working as desired," 125
Wrap-around days, 251
Wright, Frank Lloyd, 233

XP (Extreme Programming), 276

Year 2000 (Y2K), 145

ABOUT THE AUTHOR

Dr. John Arthur Ricketts is a distinguished engineer and Constraint Management practitioner. His career spans Manufacturing, Academia, and the Information field. John's job roles include professor of Information Systems, research manager in Information Technology, consulting partner in Business & Technical Services, chief technology officer in Computer Software, and venture capitalist in Corporate Strategy.

John is the author of *Reaching the Goal: How Managers Improve a Services Business Using Goldratt's Theory of Constraints*. His work has also appeared in *MIS Quarterly*, *Journal of Software Maintenance*, *Informatica*, *Computer Programming Management*, *Service Systems Implementation*, and *Theory of Constraints Handbook*.